Springer Series on Polymer and Composite Materials

Series Editor

The "Springer Series on Polymer and Composite Materials" publishes monographs and edited works in the areas of Polymer Science and Composite Materials. These compound classes form the basis for the development of many new materials for various applications. The series covers biomaterials, nanomaterials, polymeric nanofibers, and electrospun materials, polymer hybrids, composite materials from macro- to nano-scale, and many more; from fundamentals, over the synthesis and development of the new materials, to their applications. The authored or edited books in this series address researchers and professionals, academic and industrial chemists involved in the areas of Polymer Science and the development of new Materials. They cover aspects such as the chemistry, physics, characterization, and material science of Polymers, and Polymer and Composite Materials. The books in this series can serve a growing demand for concise and comprehensive treatments of specific topics in this rapidly growing field. The series will be interesting for researchers working in this field and cover the latest advances in polymers and composite materials. Potential topics include, but are not limited to:

Fibers and Polymers:

- Lignocellulosic biomass and natural fibers
- Polymer nanofibers
- Polysaccharides and their derivatives
- Conducting polymers
- Surface functionalization of polymers
- Bio-inspired and stimuli-responsive polymers
- Shape-memory and self-healing polymers
- Hydrogels
- Rubber
- Polymeric foams
- Biodegradation and recycling of polymers

Bio- and Nano- Composites:-

- Fiber-reinforced composites including both long and short fibers
- Wood-based composites
- Polymer blends
- Hybrid materials (organic-inorganic)
- Nanocomposite hydrogels
- Mechanical behavior of composites
- The Interface and Interphase in polymer composites
- Biodegradation and recycling of polymer composites
- Applications of composite materials

More information about this series at http://www.springer.com/series/13173

Ashok Kumar Nadda · Sajna K. V. ·
Swati Sharma
Editors

Microbial Exopolysaccharides as Novel and Significant Biomaterials

Springer

Editors
Ashok Kumar Nadda
Department of Biotechnology and Bioinformatics
Jaypee University of Information Technology
Waknaghat, Himachal Pradesh, India

Swati Sharma
University Institute of Biotechnology (UIBT)
Chandigarh University
Mohali, Punjab, India

Sajna K. V.
Department of Biochemistry
Indian Institute of Science
Bengaluru, Karnataka, India

ISSN 2364-1878 ISSN 2364-1886 (electronic)
Springer Series on Polymer and Composite Materials
ISBN 978-3-030-75288-0 ISBN 978-3-030-75289-7 (eBook)
https://doi.org/10.1007/978-3-030-75289-7

This Springer imprint is published by the registered company Springer Nature Switzerland AG
The registered company address is: Gewerbestrasse 11, 6330 Cham, Switzerland

Preface

In the last few decades, the demand of eco-friendly and bio-based products has been increased substantially. The issues of environmental sustainability and climate change can be resolved by replacing the majority of chemical and synthetic compounds with bio-based products. The present book is motivated by the current state of affairs of exopolysaccharides (EPSs) and their composites in wide range of applications. This book had been written to provide a framework of synthesis and production of EPSs using microbes and algae. This book mainly emphasizes on the range of applications of EPSs in various sectors. A variety of EPSs were reported to produce from microorganisms having remarkable properties to use for industrial purposes. These are heterogeneous polymeric substances which have immense applications in pharmaceuticals, medical, food and fabric industry. This polymeric nature also makes these as an alternative of synthetic plastic- and petroleum-based chemicals. Microbes present in marine or terrestrial ecosystem are efficient producers of EPSs. The biofilm forming bacteria are also a major source of the EPSs.

Their function in the aquatic microorganisms is attachment of cells to solid surface and also to defend the microbes from the predatory organisms. Researchers have explored the production of EPSs from microbes by media engineering, genetic engineering and recombinant DNA technologies. These are secreted by the cells in extracellular environment. So, their purification and large-scale production have some advantage over enzyme purification which is a tedious process. These are carbohydrate-rich compounds and produced in the excess of sugar-rich substrate. Now, microbes are quite efficient to utilize a variety of sugar-rich substrates available in the nature due to the presence of wide variety of enzymes encoding genes present in their genome. When these are secreted outside the cells, these may acquire the form of slimy layer or stable cohesive layer. These can be collected and produced at large scale from a number of algae, archea, thermophilic bacteria and microbes of extreme environment. Due to the superior performance and functional properties, microbial polysaccharides are the excellent choice over the plant and micro-algal-derived gums. Microbial polysaccharides are rheology-modifying agents, which can be thickening, stabilizing, emulsifying, flocculating, chelating and encapsulating agents as well. Microorganisms such as bacteria, yeast and fungi produce polysaccharide with various physiological roles. The immense functional properties of microbial polysaccharides undoubtedly rely on their structural conformation and physicochemical properties. Xanthan gum, gellan, dextran and pullulan are the most commercially used microbial polysaccharides. Xanthan gum is an omnipresent food ingredient, serving as thickener, leavener, stabilizer and texture enhancer. Gellan is a remarkable gelling agent that helps to rapidly set the food preparation at low concentration. Dextran is the most medically important polysaccharide used as an antithrombotic agent. Yeast-derived pullulan or its derivative have various biomedical applications such as drug delivery, plasma substitution, tissue engineering and so forth. β-glucan derived from baker's yeast *Saccharomyces cerevisiae* is a commercially available immunostimulatory agent.

The non-toxic nature and inherent biocompatibility have encouraged their applications in the tissue engineering, scaffolds or matrices, bone repair, drug delivery, wound healing and bio-plastic synthesis. These EPSs are also quite useful for in vivo applications as these have inherent capability to undergo auto-degradation in the body cells and tissue. The researchers from various countries have contributed their knowledge and recent progress in the synthesis, production and applications of these exopolysaccharides from various sources. We compiled the chapters written by various experienced researchers working in the microbiology and relevant areas. The rationale of this book is to provide a toolbox from which researchers, students, and industry professionals, can collect the information to utilize and EPSs in various fields. Another major reason for editing the book was the topic of the research area of our interest. Generally, we spend many hours to collect the information on a wide range of topic and were able to get little information or puzzling results. Thus, in the book, we complied the chapters on all the important issues which need to be solved urgently. Chapters "Microbial Exopolysaccharides: An Introduction" and "Techniques Used for Characterization of Microbial Exopolysaccharides" will introduce the various origin historical prospects of EPSs in nature and analytical techniques to study these bio-based compounds. Chapters "Molecular Basis and Genetic Regulation of EPS" and "Molecular Engineering of Bacterial Exopolysaccharide for Improved Properties" describe the molecular basis and modification of microbial EPSs. Chapters "Extremophiles: A Versatile Source of Exopolysaccharide" and "Pullulan: Biosynthesis, Production and Applications" focused on the sources and applications of microbial EPSs. Various pharmacological and industrial applications of the EPSs were described in the chapter "Exopolysaccharides in Drug Delivery Systems "–"Microbial EPS as Immunomodulatory Agents". Chapter "Novel Insights of Microbial Exopolysaccharides as Bio-adsorbents for the Removal of Heavy Metals from Soil and Wastewater" and "Applications of EPS in Environmental Bioremediations" emphasized on the environmental applications of microbial

EPSs. The last chapter summarizes the “Cost-Benefit Analysis and Industrial Potential of Exopolysaccharides”. We firmly hope that the present book will be beneficial for all the early stage researchers and industrialists.

Waknaghat, India — Ashok Kumar Nadda
Bengaluru, India — Sajna K. V.
Mohali, India — Swati Sharma

Contents

Editors and Contributors

About the Editors

Dr. Ashok Kumar Nadda is working as Assistant Professor in the Department of Biotechnology and Bioinformatics, Jaypee University of Information Technology, Waknaghat, Solan, Himachal Pradesh, India. He holds an extensive research and teaching experience of more than 8 years in the field of microbial biotechnology, with research expertise focusing on various issues pertaining to 'nano-biocatalysis, microbial enzymes, biomass, bioenergy' and 'climate change'. He is teaching enzymology and enzyme technology, microbiology, environmental biotechnology, bioresources and industrial products to the bachelor, master and Ph.D. students. He also trains the students for enzyme purification expression, gene cloning and immobilization onto nanomaterials experiments in his laboratory. He holds international work experiences in South Korea, India, Malaysia and People's Republic of China. He worked as a postdoctoral fellow in the State Key Laboratory of Agricultural Microbiology, Huazhong Agricultural University, Wuhan, China. He also worked as a Brain Pool Researcher/Assistant Professor at Konkuk University, Seoul, South Korea. He has a keen interest in microbial enzymes, biocatalysis, CO_2 conversion, biomass degradation, biofuel synthesis and bioremediation. His work has been published in various internationally reputed journals, namely *Chemical Engineering Journal, Bioresource Technology, Scientific Reports, Energy,*

International Journal of Biological Macromolecules, Science of Total Environment and Journal of Cleaner Production. He has published more than 100 scientific contributions in the form of research, review, books, chapters and others at several platforms in various journals of international repute. The research output includes 74 research articles, 25 chapters and 16 books. He is the main series editor of *Microbial Biotechnology for Environment, Energy and Health* that publishing the books under Taylor and Francis, CRC Press, USA. He is also a member of the editorial board and reviewer committee of the various journals of international repute. He has presented his research findings in more than 40 national/international conferences. He has attended more than 50 conferences/workshops/colloquia/seminars, etc., in India and abroad. He is also an active reviewer for many high-impact journals published by Elsevier, Springer Nature, ASC, RSC and Nature Publishers. His research works have gained broad interest through his highly cited research publications, book chapters, conference presentations and invited lectures.

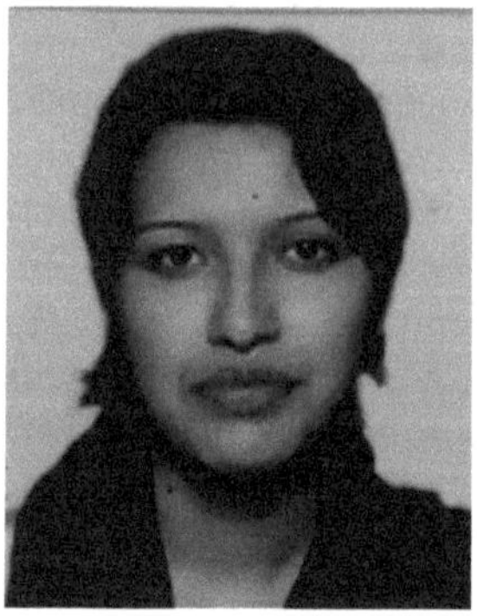

Dr. Sajna K. V. is currently working as a UGC-Kothari postdoctoral fellow at Department of Biochemistry, Indian Institute of Science, Bengaluru, India. She had completed her Ph.D. from the Department of Biotechnology, CSIR-National Institute for Interdisciplinary Science and Technology, Trivandrum, India, in 2016. Her areas of interest are biosurfactants, exopolysaccharides, bioremediation and sustainable technology. She has published her work in various internationally reputed journals such as Green Chemistry, Bioresource Technology, International Journal of Biological Macromolecules and Biochemical Engineering Journal. She has published 13 papers, three chapters and 15 conference communications. She was a university gold medallist and had received Business Plan Appreciation Award in CSIR Technology-led entrepreneurship program. She had presented papers at international conferences including the 5th IFIBiop Conference held at National Taiwan University, Taipei, and ESBES-IFIBiop 2014 Symposium held at Lille, France. She had also worked at the University of Naples, Italy, for three months as a part of BIOASSORT program under Marie Curie Actions—International Research Staff Exchange Scheme.

Dr. Swati Sharma is working as assistant professor in University Institute of Biotechnology, Chandigarh University, Mohali, Punjab, India. She is working extensively on the waste biomass, biopolymers and their applications in various fields. She has completed her Ph.D. from Universiti Malaysia Pahang, Malaysia. She worked as a visiting researcher in the college of life and environmental sciences at Konkuk University, Seoul, South Korea. She has completed her master's (M.Sc.) from Dr. Yashwant Singh Parmar University of Horticulture and Forestry, Nauni, Solan HP, India. She has also worked as a program co-coordinator at the Himalayan Action Research Center, Dehradun, and senior research fellow at India Agricultural Research Institute in 2013–2014. She has published her research papers in reputed international journals. Presently, her research is in the field of bioplastics, hydrogels, keratin nano-fibers and nano-particles, biodegradable polymers and polymers with antioxidant and anticancer activities and sponges. She has published 22 research papers in various internationally reputed journals, 5 books and a couple of book chapters.

Contributors

Thasneem Abdulla Department of Biotechnology, Sir Syed Institute for Technical Studies, Kannur, Taliparamba, Kerala, India

Dilhun Keriman Arserim Ucar Department of Nutrition and Dietetics, Faculty of Health Sciences, Bingöl University, Bingöl, Turkey

Manish Biyani Department of Bioscience and Biotechnology, Japan Advanced Institute of Science and Technology, Nomi City, Ishikawa, Japan

Dipti Deo Symbiosis School of Biological Sciences, Symbiosis International (Deemed University), Lavale, Pune, India

Joyleen Fernandes Symbiosis School of Biological Sciences, Symbiosis International (Deemed University), Lavale, Pune, India

Kenji Fukuda Research Center for Global Agromedicine, Obihiro University of Agriculture and Veterinary Medicine, Obihiro, Hokkaido, Japan

Mozafar Bagherzadeh Homaee Department of Biology, Farhangian University, Tehran, Iran

Ahmad Homaei Department of Marine Biology, Faculty of Marine Science and Technology, University of Hormozgan, Bandar Abbas, Iran

K. V. Jaseera ICAR-Central Marine Fisheries Research Institute, Kochi, Kerala, India

Siya Kamat Department of Biochemistry, Indian Institute of Science, Bengaluru, India

Nirat Kandwani Department of Biotechnology, Biyani Girls College, Jaipur, Rajasthan, India

Hiroichi Kono Department of Agro-environmental Science, Obihiro University of Agriculture and Veterinary Medicine, Obihiro, Japan

Dilara Konuk Takma Department of Food Engineering, Faculty of Engineering, Aydın Adnan Menderes University, Aydın, Turkey

Figen Korel Department of Food Engineering, Faculty of Engineering, İzmir Institute of Technology, İzmir, Turkey

Ram Kulkarni Symbiosis School of Biological Sciences, Symbiosis International (Deemed University), Lavale, Pune, India

Tarun Kumar Kumawat Department of Biotechnology, Biyani Girls College, Jaipur, Rajasthan, India

Varsha Kumawat Naturilk Organic & Dairy Foods Pvt. Ltd., Jaipur, Rajasthan, India

Naga Raju Maddela Instituto de investigación, Universidad Técnica de Manabí, Portoviejo, Ecuador;
Facultad la Ciencias de la Salud, Universidad Técnica de Manabí, Portoviejo, Ecuador

Ramchander Merugu Department of Biochemistry, University College of Science and Informatics, Mahatma Gandhi University, Nalgonda, India

Philippe Michaud Université Clermont Auvergne, CNRS, SIGMA Clermont, Clermont-Ferrand, France

Ashok Kumar Nadda Department of Biotechnology and Bioinformatics, Jaypee University of Information Technology, Waknaghat, Solan, Himachal Pradesh, India

Azita Navvabi Department of Marine Biology, Faculty of Marine Science and Technology, University of Hormozgan, Bandar Abbas, Iran

Monalisa Padhan Microbiology, School of Life Sciences, Sambalpur University, Burla, Sambalpur, India

Supriya Pandey Department of Bio-Engineering, Birla Institute of Technology, Mesra, Jharkhand, India

Matteo Radice Departamento de Ciencias de la Tierra, Universidad Estatal Amazónica, Puyo, Ecuador

Mozhgan Razzaghi Department of Marine Biology, Faculty of Marine Science and Technology, University of Hormozgan, Bandar Abbas, Iran

Shashwati Ghosh Sachan Department of Bio-Engineering, Birla Institute of Technology, Mesra, Jharkhand, India

Kuttuvan Valappil Sajna Department of Biochemistry, Indian Institute of Science, Bangalore, India

Rajesh Sani Department of Chemical and Biological Engineering, South Dakota School of Mines and Technology, Rapid City, SD, USA

Laura Scalvenzi Departamento de Ciencias de la Tierra, Universidad Estatal Amazónica, Puyo, Ecuador

Swati Sharma University Institute of Biotechnology (UIBT), Chandigarh University, Mohali, Punjab, India

Ishita Shreshtha Department of Bio-Engineering, Birla Institute of Technology, Mesra, Jharkhand, India

Rani Padmini Velamakanni Department of Biochemistry, University College of Science and Informatics, Mahatma Gandhi University, Nalgonda, India

Priyanka Vuppugalla Department of Biochemistry, University College of Science and Informatics, Mahatma Gandhi University, Nalgonda, India

Microbial Exopolysaccharides: An Introduction

Kuttuvan Valappil Sajna, Swati Sharma, and Ashok Kumar Nadda

Abstract Microbes secrete high molecular-weight polysaccharides of diverse structures into the surrounding environment termed exopolysaccharides (EPSs). EPSs serve multifarious roles which aid the microbes to thrive at different ecosystems. Many EPSs are industrially/clinically relevant polymers owing to their biocompatibility, biodegradability, non-toxic nature and distinct physicochemical properties. Considering their past success for various applications ranging from hydrocolloids to biomedical applications, microbial EPSs still hold considerable attention of biotechnologists. They are high-value products, and their market value will grow in the coming years due to their potential nutraceutical, therapeutic and industrial potential. The objective of the chapter is to update the readers with recent findings on microbial EPSs. This chapter also gives interesting insights into physiological roles and biosynthesis of microbial EPS. The chapter also discusses the recent advances in applications of microbial EPSs and their commercial prospects.

Keywords Microorganisms · Polysaccharides · Hydrocolloids · Polymers · Biomedical application

1 Introduction

Microbes are the source of many biotechnological products due to their metabolic diversity and ease of cultivation. One such product-exopolysaccharides (EPSs) are widely used as the polymers in various industries owing to distinct physicochemical

K. V. Sajna
Department of Biochemistry, Indian Institute of Science, Bangalore 560012, India

S. Sharma
University Institute of Biotechnology (UIBT), Chandigarh University, Mohali, Punjab, India

A. K. Nadda (✉)
Department of Biotechnology and Bioinformatics, Jaypee University of Information Technology, Waknaghat, Solan 173 234, Himachal Pradesh, India

A. K. Nadda et al. (eds.), *Microbial Exopolysaccharides as Novel and Significant Biomaterials*, Springer Series on Polymer and Composite Materials,
https://doi.org/10.1007/978-3-030-75289-7_1

properties, non-toxic nature, biocompatibility, biodegradability and the ease of production. Microbial polysaccharides are of two types—intracellular polysaccharides and extracellular polysaccharides. Extracellular polysaccharides are further classified into capsular polysaccharides that encapsulate the microbes (exocellular polysaccharide) and exopolysaccharide (EPS) which secreted into the surrounding environment [1]. Intracellular polysaccharides are the storage polysaccharides serving as a rapid carbon source under nutrient deprivation [2]. Capsular polysaccharides play a significant role in microbial pathogenesis. The immunogenic property of capsular polysaccharide makes them a good target for vaccine development [3]. EPSs play diverse roles from biofilm formation to pathogenesis.

The first EPS discovered was dextran by Louie Pasteur in the nineteenth century as a microbial product in the wine industry [4]. The contribution by Allene Jeanes in the mass level production of dextran and discovery of xanthan revolutionized the industrialization of microbial EPS. EPSs are high molecular weight compounds with the molecular weight ranging from 0.5×106 to 2×106 daltons. EPSs may be of homopolymeric or heteropolymeric in sugar composition and can be linear or branched, structurally [5]. Apart from the monosaccharide composition and structural complexity of EPSs, EPSs may contain functional groups such as acetyl, carboxyl, sulfate, phosphate, pyruvate and uronic acid groups, which all determine the physicochemical and biological properties of EPSs.

Microbial EPSs are inevitable for modern human lifestyle as the ingredient in food and personal care formulations. They have immense clinical applications including emergency medicine or an ingredient in pharmaceutical formulations. They are also used extensively in the petroleum industry, household product formulations and construction applications. Considering the current R&D scenario in microbial EPS, their clinical, lifestyle and other implications will be accentuated in the near future. Table 1 summarizes commercially available microbial EPS with potential industrial/clinical applications.

2 Novel Exopolysaccharides with Therapeutic/Industrial Significance

Considering the past success of EPSs for various applications ranging from hydrocolloids to biomedical applications, exopolysaccharide still holds considerable attention of biotechnologists. Many novel EPSs with significant clinical/industrial applications have been reported in the last decade (Table 2). Some of these microbial sources are already known for EPS production. Novel variation in EPS can be pinpointed by investigating the monosaccharide composition of EPS. Strain-specific EPS is encoded by unique EPS biosynthetic genes. Diversity of epsE gene in *Lactococcus lactis* strains result in strain-specific EPS production [25]. Some of the most common sources for the isolation of EPS producing microbes are dairy products, fermented products and plant parts. Identification of lactic acid bacteria

Table 1 Summarizes commercially available microbial EPS with potential industrial/clinical applications

EPS	Microbial strain	Structure	Industrial/ clinical uses	References
Dextran	*Leuconostocmesenteroids*	α-1,6-Glucan with branching of α-1,3-glycosidic linkage	Clinical applications—plasma volume extender, antithrombotic agent, blood substitute, vascular surgery, drug delivery agent, clinical management of iron deficiency anaemia, preservation solution for organs, and wound healing agent. Other uses—food packaging, photographic uses, separation technology, cell culture techniques and cryoprotectant agent	De Belder [6], Bhavani and Nisha [7], Abir et al. [8], Debele et al. [9], Rutherford et al. [10], Aman et al. [11], Candinas et al. [12], Zhu et al. [13]
Xanthan	*Xanthomonas campestris*	A polymer of D-glucose, D-mannose and D-glucuronic acid	Additive in food, medical and personal care formulations; used as drilling fluid in oil field drilling and building materials for construction applications	BeMiller [14], Akpan et al. [15, Plank [16]
Pullulan	*Aureobasidium pullulans*	Glucan of α-(1-6) and α-(1-4) glycosidic linkage	Food and pharmaceutical additive; oral care ingredient	Singh et al. [17]

(continued)

Table 1 (continued)

EPS	Microbial strain	Structure	Industrial/ clinical uses	References
Gellan	*Sphingomonas elodea*	A polymer of tetrasaccharide units comprised of D-glucose, D-glucuronic acid, D-glucose L-rhamnose	Food, pharmaceutical and personal care formulation; an additive in household products; also used in tissue culture media preparations	Iurciuc et al. [18]
Curdlan	*Agrobacterium* sp.	(1-3)-β-glucan	Food additive; used in pharmaceutical formulation and drug delivery system	Zhang and Edgar [19]
Scleroglucan	*Sclerotium rolfsii*	β-1,3-β-1,6-glucan	Petroleum recovery; used in nutraceutical and pharmaceutical industry; in food and personal care formulations; construction applications	Castillo et al. [20]
Schizophyllan	*Schizophyllum commune*	β-1,3-β-1,6-glucan	Therapeutic application, cosmetic application	Leathers et al. [21]
Bacterial cellulose	*Acetobacter xylinum*	β-1-4 glucan	Hydrocolloid dressing; cosmetic and textile industrial application	Wang et al. [22–24]

secreting a novel EPS composed of unusual monomer like *N*-acetylglucosamine from a fig leaf highlight the importance of bioprospecting of environmental source such as these for EPS producers [26]. Exploring the ecological hotspots and extreme environments can lead to the discovery of the microbes producing novel EPS with significant biotechnological implications. Delbarre-Ladrat et al. [27] that the majority of bacterial species inhabiting deep-sea hydrothermal vents has the potential of producing structurally diverse high-value EPS, which emphasized the bioprospecting of marine environment for EPS producing microbes.

Table 2 Novel EPS of therapeutic/industrial significance

EPS	Source organism	Monomeric composition	Potential application	Reference
EPS-NA3	*Lactobacillus coryniformis*	α-rhamnose, α-mannose, α-galactose, and α-glucose	Antioxidant and antibiofilm agents	Xu et al. [28]
α-mannan	*Pseudoalteromonas* SM20310	2-α- and 6-α-mannose	Cryoprotection	Liu et al. [29]
EPS-1 and EPS-2	*Bacillus amyliliquefaciens C-1*	Glucose, mannose, galactose and arabinose (EPS-1); Glucose and mannose (EPS-2)	EPS-1 as an antioxidant agent	Yang et al. [30]
Neutral EPS	*Lactobacillus paracasei* IJH-SONE68	*N*-acetylglucosamine	Anti-inflammatory agent	Noda et al. [26]
Acidic EPS	*Lactobacillus plantarum* SN35N	Glucose, galactose, and mannose	Anti-inflammatory agent	Noda et al. [31]
Pseudozyma EPS	*Pseudozyma* sp. NII 08165	Glucose, galactose, and mannose	Emulsifying and suspending agent	Sajna et al. [32, 33]
DM-1 EPS	*Bacillus licheniformis* strain DM-1	Proteoglycan	In situ microbial enhanced oil recovery	Fan et al. [34]
EPS	*Lactobacillus fermentum* R-49757	D-glucose and D-mannose	Not investigated	Do et al. [35]
EPS-S3	*Pantoea* sp. YU16-S3	Glucose, galactose, *N*-acetyl galactosamine and glucosamine	Wound healing applications	Sahana and Rekha [36]
EPS	*Lactobacillus paraplantarum*	Glucose, galactose and mannose	Emulsifying and texturing agent	Sharma et al. [37]
EPS-SN-1	*Bacillus velezensis SN-1*	Glucose, mannose and fructose	Antioxidant agent	Cao et al. [38]
EPS	*Bifidobacterium breve* lw01	Rhamnose, arabinose, galactose, glucose, and mannose	Anticancer activity	Wang et al. [22–24]
Nat-103	*Natronotaleasambharensis* AK103^T	Mannose, glucose and glucuronic acid	Antioxidant activity	Singh et al. [39]
EPS	*Lactobacillus mucosae* VG2	D galactan	Not investigated	Fagunwa et al. [40]

3 Physiological Roles and Ecological Aspects of EPS

EPS serve multifarious roles which aid the microbes to thrive in different ecosystems. EPS plays a varying role from biofilm formation, quorum sensing to pathogenesis and the functions depend on ecological niche of host organisms. Physiological roles of EPS are unravelled using the approach of knocking out EPS biosynthetic genes to create mutant deficiencies in EPS production. Pullulan produced by a desert isolate *Aureobasidium melanogenum* confers adaptation for living in the harsh desert environment by protecting from various abiotic stresses [41]. EPS produced by an arctic sea isolate *Pseudoalteromonas* strain SM20310, plays a significant role in environmental adaptation of strain in sea ice by providing high salinity tolerance and cryoprotection [29]. EPS has implication in the protection of plant growth-promoting *Rhodotorula* sp. from adverse environmental conditions [42]. Similarly, pH buffering property of cyanobacterial EPS matrix protects the dryland cyanobacteria from acid damage [43].

On solid surfaces, EPS facilitates the growth of bacterial communities as biofilm by leading bacterial cell adhesion and bacterial cell aggregation. Caro-Astorga et al. [44] revealed that each EPS produced by *Bacillus cereus* serve distinct roles. EPS1 contributes to bacterial motility, while EPS2 is involved in biofilm formation and gut colonization, thus playing a role in host-pathogen interaction. Being an integral part of biofilm, EPS makes the bacterial colonies recalcitrant to a wide range of antimicrobial agents. During the biofilm formation by *Pseudomonas aeruginosa*, production of matrix EPS 'psl' and the intracellular signalling molecule 'c-di-GMP' that stimulates the synthesis of biofilm matrix EPS is in the feedforward control loop. Hence, targeting the biofilm signalling mechanism can be an effective strategy to tackle chronic *P. aeruginosa* infections [45]. Another EPS, pel is cationic and hold the extracellular DNA in the biofilm matrix, apart from being the structural element of biofilm [46].

Studies on EPS produced by *Lactobacillus* species revealed the role of EPS in bacterial surface properties and host interaction. EPS affected the surface properties such as colony phenotypes and bacterial surface charge. Gene deletion studies revealed that EPS plays a significant role in bacterial cell aggregation. Concealing the surface structure with EPS might be one of the tactics to reduce the cell-cell interaction and the role of EPS in host cell interaction is strained specific [47–49]. EPS 1, a major virulence factor of a phytopathogenic bacteria *Ralstonia solanacearum* regulate the feedback loop of quorum sensing [50].

In the case of lactic acid bacteria, EPS protect the bacteria from bacteriophage, nisin and lysozyme [51]. EPS is the major arsenal for microbes to compete with each other for food and space. Toska et al. [52] suggested that EPS is involved in the antagonistic interaction between bacterial species and lead to the successful establishment of bacterial communities. In Gram negative bacteria such as *Vibrio cholerae*, EPS protect bacteria from other bacterial attacks by inhibiting the type 6-secretion system (T6SS). Type 6 secretion system by gram-negative bacteria is used to deliver the toxic protein into adjacent eukaryotic and bacterial cells. Deletion of EPS biosynthetic genes makes the *V. cholerae* more susceptible to T6SS attack by heterologous bacteria. On other hand, the same EPS of *V. cholerae* will not affect its T6SS attack on other bacteria [52].

EPS plays an important role in the establishment of plant microbial symbiosis. Plant root attachment of nitrogen-fixing bacteria *Paraburkholderia phymatum* is determined by the production of an EPS, cepacian [53]. Plant-growth promoting soil-borne *P. aeruginosa*, *P. syringae*, *P. putida*, and *P. fluorescens* produce EPS 'alginate'. Alginate play an important role in Zn^{2+} biosorption and phenazine biosynthesis, a biocontrol agent produced by fluorescent *Pseudomonas* strain. Increased alginate production affects the rhizosphere compatibility with improved biofilm formation and enhanced root colonization [54]. EPS helps to maintain the spore physiology and improve spore survival. *pzX* is an eps exclusively produced during sporulation of *Bacillus* species. Composition of amino sugar provides unique properties to *pzX* like lowering the surface tension and inhibiting cell-spore aggregates formation [55]. Metagenomic analysis of biological soil crust showed the presence of EPS and lipopolysaccharide (LPS) producing bacterial species. Here, EPS and LPS act as soil glue for soil aggregate formation that aid the formation of biological soil crust [56].

EPS plays a crucial role in etiology of dental caries. Demineralization of teeth by cariogenic biofilms leads to the formation of the oral cavity. In the presence of carbohydrates, cariogenic microbes produce organic acids that leach calcium from the teeth. A study showed that cariogenic microbes such as *Streptococcus mutans*, *Lactobacillus rhamnosus*, and *Candida albicans* produce EPS that have a high calcium-binding affinity, which attributes to the calcium tolerance of the microbes. Apart from structural anchorage to the biofilm, EPS also serve as a survival tool of cariogenic microbes to defuse high calcium concentration [57]. Targeting EPS can be an effective strategy to control cariogenic microbes [58]. However, in the case of catheter-associated urinary tack infection, EPS secreting *P. aeruginosa* adopt exopolysaccharide independent biofilm formation [59]. Hence, understanding the role of microbial EPS is crucial for developing therapeutic interventions against pathogenic microbes in which EPS production can be targeted. Furthermore, ecological functions of microbial EPSs promote their huge agronomical implications.

4 Biosynthesis and Metabolic Regulations of EPSs

Functional genomics analysis provides valuable information on EPS biosynthesis, export, and regulation. Identifying the gene targets can pave the ways to engineer high EPS producing strains or strains that produce tailor-made EPS [60]. Genomic analysis of microbes can reveal microbial potential to produce unknown exopolysaccharides. Borlee et al. [61] identified a novel EPS biosynthetic gene cluster involved in biofilm formation of *Burkholderia pseudomallei*. Genome annotation of EPS producing thermophilic bacteria *Geobacillus* may improve its prospects as a microbial cell factory for EPS production [22–24]. Padmanabhan et al. [62] studied differential gene expression during EPS biosynthesis by *Streptococcus thermophilus* ASCC 1275 in different sugar-containing media at stationery and log phases. They observed a correlation between high EPS production and upregulation of genes involved in sugar metabolism. A similar observation of increased UDP-glucose and UDP-galactose synthesis associated with a high yield of EPS, by *S. thermophilus* S-3 was reported by Xiong et al. [63]. Proteomic analysis revealed that upregulation of proteins involved in sugar transport, EPS assembly and amino acid metabolism was also associated with high EPS production [62, 64].

Availability of whole genome sequence of EPS producing microbes facilitates the metabolic engineering strategies for EPS production [65]. Evaluation of EPS production by gene knockout mutants, gene overexpression mutants and gene complementation mutants of EPS biosynthetic genes can shed light on the role of each EPS biosynthetic genes in EPS production [66]. CRISPR-Cas9 genome editing had enabled researchers to produce EPS variants with different monomeric composition from *Paenibacillus polymxa.* These EPS variants can give insights into the structure-function relationship of polysaccharides and aid to create customized EPS with desirable properties [67]. *Xanthomonas campestris* strains were engineered to produce xanthan gum variants with distinct secondary structure and rheological properties, which may be suitable for application in various industries. Structure-activity relationship of these tailor made-xanthan gums revealed that terminal mannose is one of the major determinants of rheological properties of xanthan gum, while the terminal mannose and internal acetyl group are integral to its double-helical conformation [68]. Genome editing and metabolic engineering could yield tailor-made EPS with improved stability and higher performance, which can have huge commercial potential when compared to native EPS.

5 Applications and Commercial Prospects of EPS

Due to the presence of a large number of hydroxyl groups, microbial EPS have been long used as hydrocolloids, which modify the rheology of the system by altering the flow behaviour and texture. In food and personal care industry, they serve as a thickening, gelling, stabilizing, emulsifying and water-binding agents [69, 70]. Xanthan gum is a widely used thickener in food formulation. In food and confectionary, xanthan gum has become more prominent in recent years due to its status as vegan-friendly. In gluten-free baking, xanthan gum provides structure and elasticity to dough or batter, and as an egg substitute, it emulsifies and thickens the food preparations. Xanthan gum based thickened fluid appears promising for treatment of patients with oropharyngeal dysphagia. Apart from safety and efficacy, it is resistant to α-amylase and preferred by patients, when compared to starch-based thickener [71, 72]. The concentration, type and setting time of xanthan gum-based food thickeners are the main factors in designing the infant food formulation used for paediatric dysphagia [73]. Gellan gum exhibit excellent gelling properties. To overcome the limitation of the gellan gum such as low mechanical strength and high gelation temperature, blending with natural or synthetic polymer has been employed [74]. Synergistic hydrogels of xanthan gum and gellan gum with other natural polymers are promising for the preparation of food packaging materials [75].

Antioxidant property and water-absorbing/retention properties are some of the features of EPS attractive for cosmetic applications [76]. 'Lubcan' an EPS with remarkable skin lubricating property produced by *Paenibacillus* sp. ZX1905 could be a low-priced replacement of hyaluronic acid in cosmetic formulations [77]. Extremophilic microbes may provide EPS with excellent keratinocyte protective ability from temperature or radiation-induced damage. An EPS of momomers-*N*-acetyl glucosamine, mannose and glucuronic acid produced by an artic marine bacterium *Polaribacter* sp. SM1127 could be an excellent cosmetic ingredient as it is dermatologically safe, possess better moisture retention properties than hyaluronic acid and good antioxidant activity, and protect human dermal fibroblast from low temperature-induced damage [78]. Radiation-resistant *Deinococcous radiodurans* derived EPS (deinopol) protect keratinocytes from radiation-induced ROS damage [79].

Potential bioactivities reported for EPS include antitumor, antioxidant, immunomodulatory, antiviral, antibacterial, anti-inflammatory, and cholesterol-lowering properties. Consumption of bioactive EPS can have potential health benefits [80]. Antitumor property of EPS stems from its ability to modulate oncogenic pathways. EPS produced by many lactic acid bacteria can induce apoptosis and cell cycle arrest in tumour cells, without any toxicity to normal cells [81]. EPS secreted by probiotic yeasts-*Kluyveromyces marxianus* and *Pichia kudriavzevii* were reported to induce apoptosis in colorectal cancer cells by inhibiting AKT-1, mTOR, and JAK-1 pathways [82]. Though many studies demonstrated the antitumor potential of EPS, the viability of EPS as a coadjuvant for cancer therapy needs to be addressed by in-depth in vivo studies. Some researchers observed that the sugar composition of EPS primarily determines its antitumor property. For instance, Tukunmez et al. [83] observed that the apoptotic induction by *Lactobacilli* EPS was related to the mannose content of EPS. The mode of action of *Lactobacilli* EPS is by upregulation of Bax, Caspase 2 and 9 and downregulation of Bcl-2 and Survivin leading to caspase-mediated apoptosis [83].

EPSs have been commonly used in pharmaceutical formulations for controlled and sustained release of drugs, coating of pills or as suspension stabilizers. Presence of hydroxyl groups and free carboxyl groups in EPS enables the structural modification of EPSs, improving the biostability and mechanical properties or impart novel functionality to EPSs, thus broadening their applications [84]. Adding hydrophobic moiety to xanthan gum reduces its solubility and porosity, and modified its rheology. The resulting amphiphilic xanthan gum reduced the surface tension/interfacial tension and stabilized the emulsion, which improves its prospects for pharmaceutical applications, in comparison to native xanthan gum [85]. Du et al. [86] reported an antibacterial hydrogel made of hydrophobically modified chitosan and oxidized dextran with improved wound healing properties than that of traditional gauze. Similarly, a thermoreversible hydrogel made with xanthan and konjac glucomannan appear promising for in situ would healing [87].

Non-immunogenicity, biocompatibility and biodegradability determine the applicability of EPS in biomedical application. Dextran is the most clinically used bioabsorbable EPS. Dextran has been used as a plasma extender and an antithrombotic agent. Dextran is neutral in charge, exhibit excellent pharmacokinetics and is easily degraded by dextranase enzyme in our body [88]. Acetalated dextran (Ac-Dex) is modified dextran with hydrophobic nature. It can be easily formulated to micro/nanoparticle, which can encapsulate a diverse payload. Its pH-sensitive nature makes it an effective drug delivery system for protein, miRNAs and chemotherapeutic drugs [89–91]. Studies with natural compound ganothalamine revealed promising application of Ac-Dex as an encapsulating agent for the sustained release of the anticancer drug [92]. Wannasarit et al., [93] synthesized a conjugated dextran-based polymeric nanoparticle which can mimic viral entry to the cell. Adapting a viral mode of delivery of therapeutics to the cytoplasm is a good approach to bypass the lysosomal degradation that happens after the internalization of the drug. The prepared poly(lauryl methacrylate-comethacrylic acid)-grafted acetalated dextran carrying the payload of asiatic acid showed improved therapeutic efficacy than treatment with asiatic acid alone. Pinho et al. [94] prepared a dextran-based photocrosslinked membrane which shows potential as implantable devices for biomedical application. In vivo studies using rat models indicated that the developed dextran-based membrane is biocompatible.

With the aim to restore or regenerate the damaged tissue, tissue engineering comprises of cells and growth factors in a biomaterial that acts as the scaffold for cell growth. Biocompatibility, gelation and mechanical properties are the attractive properties of EPSs for their use as biomaterials in tissue engineering [95]. Microbial EPS containing hexosamine and uronic acid as monomers and acetyl/sulfate groups as functional groups hold great therapeutic potential due to their structural resemblance with mammalian glycosaminoglycans (GAG). Using bacterial GAG-like polymers over mammalian GAG have the following advantages. Bacterial EPSs are produced by fermentation that is more feasible when compared to strenuous extraction of GAG from animal tissue. Bacterial EPSs are free of prions and viruses as in the case of mammalian GAGs. EPS produced by marine isolated *Vibrio diabolicus* and *Alteromonas infernus* are promising candidates for tissue repair and remodelling. Chemical modifications of these depolymerized polysaccharides using *N*-deacetylation and sulfation can yield heparin-like polymers [96–98]. Cross-linked dextran is an effective injectable hydrogel for cartilage regeneration [99]. Capsular alginate extracted from *Azotobacter agile* exhibit lower cytotoxicity on mesenchymal stem cells than algal alginate. Moreover, tailor-made alginate with attractive properties to serve as a biomaterial can be produced by metabolic engineering of host bacterium [100].

The petroleum industry has been using EPS as a viscosifier for the drilling purpose. In situ EPS production by *Pseudomonas stutzeri* XP1 isolated from oil reservoir could enhance the oil recovery that demonstrated the potential of EPS for enhanced oil recovery [101]. EPS can be a potential bioadsorbent for heavy metal removal. They are environmentally friendly, cost-effective, and require milder conditions to operate. Metal adsorption by EPS depends on ionic nature of metal, its size and charge density. Positively charged heavy metals can be sequestered using anionic charged EPS [102]. When arsenic degrading bacteria was cultivated in arsenic-containing media, they produced EPS that can effectively sequester arsenic. These EPS are rich in polyanionic functional groups, which result in electrostatic to covalent binding with arsenic [103]. Similarly, studies also demonstrated the excellent flocculating activity of microbial EPS [33, 104].

EPS can be effective and sustainable soil strength improver. Using EPS as the soil stabilizer can alleviate the negative environmental impact associated with traditional soil stabilizers such as lime and cement. Improvement in soil shear strength and soil fabric was noted on addition of xanthan gum to the organic peat matrix, due to the hydrogen and electrostatic binding between xanthan gum and clay particle [105, 106]. Xanthan gum and sodium alginate could alleviate soil erosion and reduce the collapsible potential of soil material [107, 108]. Water adsorption and moisture-retention abilities of soil can be greatly improved by the addition of xanthan gum [109].

Exopolysaccharide-derived oligosaccharides can be considered for sustainable agricultural practices. Plant growth-promoting biostimulants can greatly benefit agriculture by stimulating the nutrient uptake, enhancing the photosynthetic activity of plants and protect the plants by mitigating abiotic stress [110]. Low molecular weight oligo-gellan prepared by depolymerization of gellan gum is promising as a biostimulant, which improved the plant growth and survival of Red Perilla plants under normal and stress conditions. Biostimulatory activity may be due to elicitation of plant polyphenol content and other secondary metabolites, leading to high antioxidant activity [111]. Though gellan gum also confers some biostimulatory effects on plants, the oligo-gellan exhibited better performance [112].

6 Conclusions

Microbial EPSs are one of the industrially significant microbial products, which are used as the functional ingredient in the food, pharmaceutical, personal care and other industries. Functional application of EPS is correlated to their structural complexity, which determines their physicochemical properties and bioactivities. Besides, the structural modification of EPS and synergistic manipulation with natural or synthetic polymers to broaden the applications of EPS, researchers are actively searching for novel EPS with versatile physicochemical properties or unique bioactivities which can have industrial/therapeutic applications. For that, they pursue the bioprospecting of EPS producing microbes from different environmental samples, specifically extreme environment. The ability of microbes to produce unknown exopolysaccharide is also being studied by genomic analysis.

Some latest studies shed light on the role of EPS in host-microbial symbiosis and pathogenesis. Understanding the physiological roles of EPS secreted by pathogenic or opportunistic microbes is quite crucial for developing novel therapeutic strategies against these microbes. Employing multiple omics techniques and metabolic engineering strategies in the field of microbial EPS can greatly expand the knowledge in EPS biosynthetic pathways and also, leads to the generation of tailor-made EPS with superior properties.

Microbial EPS possess excellent rheological, emulsifying, and water-retention properties, which makes them highly sought-after industrial polymers in food, personal care, pharmaceutical and oil-drilling industry. In addition to this, they possess stability in a wide range of temperature and pH that heighten their commercial prospects. Furthermore, EPS may possess biological activities such as antioxidant, antitumor, immunomodulatory and antimicrobial properties and are promising for therapeutic and nutraceutical applications. Structural modified EPS with natural or synthetic polymers make an effective hydrogel with implications in clinical and biomedical field as wound dressing and tissue engineering applications. Growing researches demonstrate the potential use of EPS for bioremediation, soil conservation and sustainable agricultural practices.

Acknowledgements KVS would like to acknowledge Dr. D.S. Kothari Post-Doctoral Fellowship scheme of University Grant Commission, New Delhi, India for the fellowship. The financial support from the Jaypee University of Information Technology, Waknaghat to undertake this study is thankfully acknowledged. Further, the authors have no conflict of interest either among themselves or with the parent institution.

References

1. Castro-Bravo N, Wells JM, Margolles A, Ruas-Madiedo P (2018) Interactions of surface exopolysaccharides from *Bifidobacterium* and *Lactobacillus* within the intestinal environment. Front Microbiol 9:2426. https://doi.org/10.3389/fmicb.2018.02426
2. Sekar K, Linker SM, Nguyen J et al (2020) Bacterial glycogen provides short-term benefits in changing environments. Appl Environ Microbiol 86: e00049–20. https://doi.org/10.1128/aem.00049-20
3. Apicella MA, Post DM, Fowler AC et al (2010) Identification, characterization and immunogenicity of an O-antigen capsular polysaccharide of *Francisella tularensis*. PLoS ONE 5: https://doi.org/10.1371/journal.pone.0011060
4. Pasteur L (1861) On the viscous fermentation and the butyrous fermentation. Bull Soc Chim Fr 11:30–31
5. Flemming H, Wingender J (2010) The biofilm matrix. Nat Rev Microbiol 8:623–633. https://doi.org/10.1038/nrmicro2415
6. De Belder AA (1993) DEXTRAN. In: Whistler RL, Bemiller JB (eds) Industrial gums, 3rd edn. Academic Press, pp 399–425. ISBN 9780080926544, https://doi.org/10.1016/B978-0-08-092654-4.50018-8
7. Bhavani AL, Nisha J (2010) Dextran—the polysaccharide with versatile uses. Int J Pharm Med 1:569–573
8. Abir F, Barkhordarian S, Sumpio BE (2004) Efficacy of dextran solutions in vascular surgery. Vasc Endovasc Surg 38:483–491. https://doi.org/10.1177/153857440403800601
9. Debele TA, Mekuria SL, Tsai H (2016) Polysaccharide based nanogels in the drug delivery system: application as the carrier of pharmaceutical agent. Mat Sci Eng C-Mater 68:964–981. https://doi.org/10.1016/j.msec.2016.05.12
10. Rutherford RB, Jones DN, Bergentz SE, Bergqvist D, Karmody AM, Dardik H, Moore WS, Goldstone J, Flinn WR, Comerota AJ et al (1984) The efficacy of dextran 40 in preventing early postoperative thrombosis following difficult lower extremity bypass. J Vasc Surg 1:765–773

11. Aman A, Siddiqui NN, Qader SAU (2012) Characterization and potential applications of high molecular weight dextran produced by Leuconostoc mesenteroides AA1. Carbohydr Polym 87:910–915. https://doi.org/10.1016/j.carbpol.2011.08.094
12. Candinas D, Largiader F, Binswanger U et al (1996) A novel dextran 40-based preservation solution. Transpl Int 9:32–37. https://doi.org/10.1007/BF00336809
13. Zhu Q, Jiang M, Liu Q et al (2018) Enhanced healing activity of burn wound infection by a dextran-HA hydrogel enriched with sanguinarine. Biomater Sci 6:2472–2486. https://doi.org/10.1039/c8bm00478a
14. BeMiller JN (2019) In: BeMiller JN (ed) Carbohydrate chemistry for food scientists, 3rd edn. AACC International Press, pp 261–269. https://doi.org/10.1016/B978-0-12-812069-9.00011-X
15. Akpan EU, Enyi GC, Nasr GG (2020) Enhancing the performance of xanthan gum in water-based mud systems using an environmentally friendly biopolymer. J Petrol Explor Prod Technol 10:1933–1948. https://doi.org/10.1007/s13202-020-00837-0
16. Plank J (2005) Applications of biopolymers in construction engineering. Biopolym Online 10:29–39. https://doi.org/10.1002/3527600035.bpola002
17. Singh RS, Saini GK, Kennedy JF (2008) Pullulan: microbial sources, production and applications. Carbohydr Polym 73:515–531. https://doi.org/10.1016/j.carbpol.2008.01.003
18. Iurciuc CE, Lungu C, Martin P, Popa M (2017) Gellan-pharmaceutical, medical and cosmetic applications. Cellulose Chem Technol 51:187–202
19. Zhang R, Edgar KJ (2014) Properties, chemistry, and applications of the bioactive Polysaccharide Curdlan. Biomacromolecules 15:1079–1096. https://doi.org/10.1021/bm500038g
20. Castillo NA, Valdez AL, Fariña JI (2015) Microbial production of scleroglucan and downstream processing. Front Microbiol 6:1106. https://doi.org/10.3389/fmicb.2015.01106
21. Leathers TD, Nunnally MS, Stanley AM, Rich JO (2016) Utilization of corn fiber for production of schizophyllan. Biomass Bioenerg 95:132–136. https://doi.org/10.1016/j.biombioe.2016.10.001
22. Wang J, Goh KM, Salem DR et al (2019a) Genome analysis of a thermophilic exopolysaccharide-producing bacterium—*Geobacillus* sp. WSUCF1. Sci Rep 9:1608. https://doi.org/10.1038/s41598-018-36983-z
23. Wang J, Tavakoli J, Tang Y (2019b) Bacterial cellulose production, properties and applications with different culture methods—A review. Carbohydr Polym 219:63–76
24. Wang L, Wang Y, Li Q et al (2019c) Exopolysaccharide, isolated from a novel strain *Bifidobacterium breve* lw01 possess an anticancer effect on head and neck cancer—genetic and biochemical evidences. Front Microbiol 10:1044. https://doi.org/10.3389/fmicb.2019.01044
25. Suzuki C, Kobayashi M, Kimoto-Nira H (2013) Novel exopolysaccharides produced by *Lactococcus lactis* subsp. lactis, and the diversity of epsE genes in the exopolysaccharide biosynthesis gene clusters. Biosci Biotechnol Biochem 77:2013–2018. https://doi.org/10.1271/bbb.130322
26. Noda M, Sugimoti S, Hayashi I, Danshiitsoodol N, Fukamachi M et al (2018a) A novel structure of exopolysaccharide produced by a plant-derived lactic acid bacterium *Lactobacillus paracasei* IJH-SONE68. J Biochem 164:87–92. https://doi.org/10.1093/jb/mvy048
27. Delbarre-Ladrat C, Salas ML, Sinquin C, Zykwinska A, Colliec-Jouault S (2017) Bioprospecting for exopolysaccharides from deep-sea hydrothermal vent bacteria: relationship between bacterial diversity and chemical diversity. Microorganisms 5:63
28. Xu X, Peng Q, Zhang Y et al (2020) A novel exopolysaccharide produced by *Lactobacillus coryniformis* NA-3 exhibits antioxidant and biofilm-inhibiting properties in vitro. Food Nutr Res 64. https://doi.org/10.29219/fnr.v64.3744
29. Liu SB, Chen XL, He HL et al (2013) Structure and ecological roles of a novel exopolysaccharide from the arctic sea ice bacterium *Pseudoalteromonas* sp. Strain SM20310. Appl Environ Microbiol 79(1):224–230. https://doi.org/10.1128/aem.01801-12

30. Yang H, Deng J, Yuan Y et al (2015) Two novel exopolysaccharides from *Bacillus amyloliquefaciens* C-1: antioxidation and effect on oxidative stress. Curr Microbiol 70:298–306. https://doi.org/10.1007/s00284-014-0717-2
31. Noda M, Shiraga M, Kumagai T, Danshiitsoodol N, Sugiyama M (2018b) Characterization of the SN35N strain–specific exopolysaccharide encoded in the whole circular genome of a plant-derived *Lactobacillus plantarum*. Biol Pharm Bull 41:536–545
32. Sajna KV, Sukumaran RK, Gottumukkal LD, Jayamurthy H, Dhar KS, Pandey A (2013) Studies on structural and physical characteristics of a novel exopolysaccharide from *Pseudozyma* sp. NII 08165. Int J Biol Macromol 59:84–89. https://doi.org/10.1016/j.ijbiomac.2013.04.025
33. Sajna KV, Sukumaran RK, Gottumukkal LD, Sasidharan S, Pandey A (2020) Functional evaluation of exopolysaccharide from *Pseduozyma* sp. NII 086165 revealed the potential thickening and emulsifying applicability. Indian J Exp Biol 58:539–547
34. Fan Y, Wang J, Gao C et al (2020) A novel exopolysaccharide-producing and long-chain n-alkane degrading bacterium *Bacillus licheniformis* strain DM-1 with potential application for in-situ enhanced oil recovery. Sci Rep 10:8519. https://doi.org/10.1038/s41598-020-65432-z
35. Do TBT, Tran TAL, Tran TVT et al (2020) Novel exopolysaccharide produced from fermented bamboo shoot-isolated *Lactobacillus Fermentum*. Polymers 12:1531. https://doi.org/10.3390/polym12071531
36. Sahana TG, Rekha PD (2020) A novel exopolysaccharide from marine bacterium *Pantoea* sp. YU16-S3 accelerates cutaneous wound healing through Wnt/β-catenin pathway. Carbohydr Polym 238:116191. https://doi.org/10.1016/j.carbpol.2020.116191
37. Sharma K, Sharma N, Handa S et al (2020) Purification and characterization of novel exopolysaccharides produced from *Lactobacillus paraplantarum* KM1 isolated from human milk and its cytotoxicity. J Genet Eng Biotechnol 18:56. https://doi.org/10.1186/s43141-020-00063-5
38. Cao C, Liu Y, Li Y et al (2020) Structural characterization and antioxidant potential of a novel exopolysaccharide produced by *Bacillus velezensis* SN-1 from spontaneously fermented Da-Jiang. Glycoconj J 37:307–317. https://doi.org/10.1007/s10719-020-09923-1
39. Singh S, Sran KS, Pinnaka AK, Roy Choudhury A (2019) Purification, characterization and functional properties of exopolysaccharide from a novel halophilic *Natronotalea sambharensis* sp. nov. Int J Biol Macromol 136:547–558. https://doi.org/10.1016/j.ijbiomac.2019.06.080
40. Fagunwa O, Ahmed H, Sadiq S et al (2019) Isolation and characterization of a novel exopolysaccharide secreted by *Lactobacillus mucosae* VG1. Carbohyr Res 481: https://doi.org/10.1016/j.carres.2019.107781
41. Jiang P, Li J, Han F et al (2011) Antibiofilm activity of an exopolysaccharide from marine bacterium *Vibrio* sp. QY101. PLoS ONE 6:e18514. https://doi.org/10.1371/journal.pone.0018514
42. Silambarasan S, Logeswari P, Cornejo P, Kannan VR (2019) Evaluation of the production of exopolysaccharide by plant growth promoting yeast *Rhodotorula* sp. strain CAH2 under abiotic stress conditions. Int J Biol Macromol 121:55–62
43. Gao X, Liu LT, Liu B (2019) Dryland cyanobacterial exopolysaccharides show protection against acid deposition damage. Environ Sci Pollut Res 26:24300–24304. https://doi.org/10.1007/s11356-019-05798-4
44. Caro-Astorga J, Álvarez-Mena A, Hierrezuelo J et al (2020) Two genomic regions encoding exopolysaccharide production systems have complementary functions in *B. cereus* multicellularity and host interaction. Sci Rep 10:1000. https://doi.org/10.1038/s41598-020-57970-3
45. Irie Y, Borlee BR, O'Connor JR et al (2012) Self-produced exopolysaccharide is a signal that stimulates biofilm formation in *Pseudomonas aeruginosa*. Proc Natl Acad Sci 109:20632–20636

46. Jennings LK, Storek KM, Ledvina HE et al (2015) Pel is a cationic exopolysaccharide that cross-links extracellular DNA in the *Pseudomonas aeruginosa* biofilm matrix. Proc Natl Acad Sci 112:11353–11358. https://doi.org/10.1073/pnas.1503058112
47. Dertli E, Mayer MJ, Narbad A (2015) Impact of the exopolysaccharide layer on biofilms, adhesion and resistance to stress in *Lactobacillus johnsonii* FI9785. BMC Microbiol 15:8. https://doi.org/10.1186/s12866-015-0347-2
48. Horn N, Wegmann U, Dertli E et al (2013) Spontaneous mutation reveals influence of exopolysaccharide on *Lactobacillus johnsonii* surface characteristics. PLoS ONE 8: https://doi.org/10.1371/journal.pone.0059957
49. Lee IC, Caggianiello G, van Swam II et al (2016) Strain-specific features of extracellular polysaccharides and their impact on *Lactobacillus plantarum*-host interactions. Appl Environ Microbiol 82:3959–3970. https://doi.org/10.1128/aem.00306-16
50. Hayashi K, Senuma W, Kai K et al (2019) Major exopolysaccharide, EPS I, is associated with the feedback loop in the quorum sensing of *Ralstonia solanacearum* strain OE1-1. Mol Plant Pathol 20:1740–1747
51. Looijesteijn PJ, Trapet L, Vries E de, Abee T, Hugenholtz J (2001) Physiological function of exopolysaccharides produced by *Lactococcus lactis*. Int J Food Microbiol 64:71–80. https://doi.org/10.1016/s0168-1605(00)00437-2
52. Toska J, Ho BT, Mekalanos JJ (2018) Exopolysaccharide protects *Vibrio cholerae* from exogenous attacks by the type 6 secretion system. Proc Natl Acad Sci USA 115:7997–8002. https://doi.org/10.1073/pnas.1808469115
53. Liu Y, Bellich B, Hug S et al (2020) The exopolysaccharide cepacian plays a role in the establishment of the *Paraburkholderia phymatum–Phaseolus vulgaris* symbiosis. Front Microbiol 11:1600. https://doi.org/10.3389/fmicb.2020.01600
54. Upadhyay A, Kochar M, Rajam MV, Srivastava S (2017) Players over the surface: unraveling the role of exopolysaccharides in zinc biosorption by fluorescent *Pseudomonas* strain Psd. Front Microbiol 8:284. https://doi.org/10.3389/fmicb.2017.00284
55. Li Z, Hwang S, Bar-Peled M (2016) Discovery of a unique extracellular polysaccharide in members of the pathogenic bacillus that can co-form with spores. J Biol Chem 29:19051–19067. https://doi.org/10.1074/jbc.m116.724708
56. Cania B, Vestergaard G, Kublik S et al (2020) Biological soil crusts from different soil substrates harbor distinct bacterial groups with the potential to produce exopolysaccharides and lipopolysaccharides. Microb Ecol 79:326–341. https://doi.org/10.1007/s00248-019-01415-6
57. Astasov-Frauenhoffer M, Varenganayil MM, Decho AW et al (2017) Exopolysaccharides regulate calcium flow in cariogenic biofilms. PLoS ONE 12: https://doi.org/10.1371/journal.pone.0186256
58. Castillo PMC, de Oliveira Fratucelli ED et al (2020) Modulation of lipoteichoic acids and exopolysaccharides prevents *Streptococcus mutans* biofilm accumulation. Molecules 25:2232
59. Cole SJ, Records AR, Orr MW et al (2014) Catheter-associated urinary tract infection by *Pseudomonas aeruginosa* is mediated by exopolysaccharide-independent biofilms. Infect Immun 82:2048–2058. https://doi.org/10.1128/iai.01652-14
60. Patel A, Prajapat J (2013) Food and health applications of exopolysaccharides produced by lactic acid bacteria. Adv Dairy Res 1:2. https://doi.org/10.4172/2329-888X.1000107
61. Borlee GI, Plumley BA, Martin KH et al (2017) Genome-scale analysis of the genes that contribute to *Burkholderia pseudomallei* biofilm formation identifies a crucial exopolysaccharide biosynthesis gene cluster. PLoS Negl Trop Dis 11: https://doi.org/10.1371/journal.pntd.0005689
62. Padmanabhan A, Tong Y, Wu Q et al (2018) Transcriptomic insights into the growth phase- and sugar-associated changes in the exopolysaccharide production of a high EPS-producing *Streptococcus thermophilus* ASCC 1275. Front Microbiol 9:1919. https://doi.org/10.3389/fmicb.2018.01919

63. Xiong ZQ, Kong LH, Lai PF et al (2019) Genomic and phenotypic analyses of exopolysaccharide biosynthesis in *Streptococcus thermophilus* S-3. J Dairy Sci 102:4925–4934. https://doi.org/10.3168/jds.2018-15572
64. Padmanabhan A, Tong Y, Wu Q et al (2020) Proteomic analysis reveals potential factors associated with enhanced EPS production in *Streptococcus thermophilus* ASCC 1275. Sci Rep 10:807. https://doi.org/10.1038/s41598-020-57665-9
65. Ates O (2015) Systems biology of microbial exopolysaccharides production. Front Bioeng Biotechnol 3:200. https://doi.org/10.3389/fbioe.2015.00200
66. Song X, Xiong Z, Kong L et al (2018) Relationship between putative *eps* genes and production of exopolysaccharide in *Lactobacillus casei* LC2W. Front Microbiol 9:1882. https://doi.org/10.3389/fmicb.2018.01882
67. Rutering M, Cress BF, Schilling M et al (2017) Tailor-made exopolysaccharides—CRISPR-Cas9 mediated genome editing in *Paenibacillus polymyxa*. Synth Biol 2:ysx007. https://doi.org/10.1093/synbio/ysx007
68. Wu M, Qu J, Tian X et al (2019) Tailor-made polysaccharides containing uniformly distributed repeating units based on the xanthan gum skeleton. Int J Biol Macromol 15:646–653. https://doi.org/10.1016/j.ijbiomac.2019.03.130
69. Khan T, Park JK, Kwon JH (2007) Functional biopolymers produced by biochemical technology considering applications in food engineering. Korean J Chem Eng 24:816–826. https://doi.org/10.1007/s11814-007-0047-1
70. Saha D, Bhattacharya S (2010) Hydrocolloids as thickening and gelling agents in food: a critical review. Food Sci Technol 47:587–597. https://doi.org/10.1007/s13197-010-0162-6
71. Ortega O, Bolívar-Prados M, Arreola V et al (2020) Therapeutic effect, rheological properties and α-Amylase resistance of a new mixed starch and xanthan gum thickener on four different phenotypes of patients with oropharyngeal dysphagia. Nutrients 12:1873
72. Vilardell N, Rofes L, Arreola V et al (2015) A comparative study between modified starch and xanthan gum thickeners in post-stroke Oropharyngeal Dysphagia. Dysphagia 31:169–179
73. Yoon SN, Yoo B (2017) Rheological behaviors of thickened infant formula prepared with xanthan gum-based food thickeners for dysphagic infants. Dysphagia 32:454–462. https://doi.org/10.1007/s00455-017-9786-2
74. Zia KM, Tabasum S, Khan MF et al (2018) Recent trends on gellan gum blends with natural and synthetic polymers: a review. Int J Biol Macromol 109:1068–1087. https://doi.org/10.1016/j.ijbiomac.2017.11.099
75. Balasubramanian R, Kim SS, Lee J, Lee J (2019) Effect of TiO_2 on highly elastic, stretchable UV protective nanocomposite films formed by using a combination of k-Carrageenan, xanthan gum and gellan gum. Int J Biol Macromol 123:1020–1027. https://doi.org/10.1016/j.ijbiomac.2018.11.151
76. Chen T, Xu P, Zong S et al (2017) Purification, structural features, antioxidant and moisture-preserving activities of an exopolysaccharide from Lachnum YM262. Bioorg Med Chem Lett 27:1225–1232. https://doi.org/10.1016/j.bmcl.2017.01.063
77. Sun X, Wang l, Fu R et al (2020) The chemical properties and hygroscopic activity of the exopolysaccharide lubcan from *Paenibacillus* sp. ZX1905. Int J Biol Macromol PMID 32828891. https://doi.org/10.1016/j.ijbiomac.2020.08.129
78. Sun M, Zhao F, Shi M et al (2016) Characterization and biotechnological potential analysis of a new exopolysaccharide from the arctic marine bacterium *Polaribacter* sp. SM1127. Sci Rep 5:18435. https://doi.org/10.1038/srep18435
79. Lin SM, Baek CY, Jung JH et al (2020) Antioxidant activities of an exopolysaccharide (DeinoPol) produced by the extreme radiation-resistant bacterium *Deinococcus radiodurans*. Sci Rep 10:55. https://doi.org/10.1038/s41598-019-56141-3
80. Yildiz H, Karatas N (2018) Microbial exopolysaccharides: resources and bioactive properties, Process Biochem 72:41–46. https://doi.org/10.1016/j.procbio.2018.06.009

81. Wu J, Zhang Y, Ye L, Wang C (2021) The anti-cancer effects and mechanisms of lactic acid bacteria exopolysaccharides in vitro: a review. Carbohydr Polym 253:11708. https://doi.org/10.1016/j.carbpol.2020.117308
82. Sadaat YR, Khosroushahi AY, Movassaghpor AA, Talebi M, Gargari BP (2020) Modulatory role of exopolysaccharides of *Kluyveromyces marxianus* and *Pichia kudriavzevii* as probiotic yeasts from dairy products in human colon cancer cells. J Funct Foods 64: https://doi.org/10.1016/j.jff.2019.103675
83. Tukenmez U, Aktas B, Aslim B, Yavuz S (2019) The relationship between the structural characteristics of lactobacilli-EPS and its ability to induce apoptosis in colon cancer cells in vitro. Sci Rep:8268. https://doi.org/10.1038/s41598-019-44753-8
84. Kumar A, Rao KM, Han SR (2018) Application of xanthan gum as polysaccharide in tissue engineering: a review. Carbohydr Polym 180:128–144. https://doi.org/10.1016/j.carbpol.2017.10.009
85. Sara H, Yahoum MM, Lefnaouni S et al (2020) New alkylated xanthan gum as amphiphilic derivatives: synthesis, physicochemical and rheological studies, J Mol Struct 1207:127768. https://doi.org/10.1016/j.molstruc.2020.127768
86. Du X, Liu Y, Wang X et al (2019) Injectable hydrogel composed of hydrophobically modified chitosan/oxidized-dextran for wound healing. Mater Sci Eng C 104:109930. https://doi.org/10.1016/j.msec.2019.109930
87. Alves A, Miguel SP, Araujo ARTS et al (2020) Xanthan gum-konjac glucomannan blend hydrogel for wound healing. Polymers (Basel) 12:99. https://doi.org/10.3390/polym12010099
88. Fonseca AC, Serra AC, Coelho JFJ (2015) Bioabsorbable polymers in cancer therapy: latest developments. EPMA J 6:22. https://doi.org/10.1186/s13167-015-0045-z
89. Bachelder EM, Beaudette TT, Broaders KE et al (2008) Acetal-derivatized dextran: an acid-responsive biodegradable material for therapeutic applications. J Am Chem 130:10494–10495
90. Broaders KE, Cohen TT, Beaudette TT et al (2009) Acetalated dextran is a chemically and biologically tunable material for particulate immunotherapy. Proc Natl Acad Sci 106:5497–5502
91. Liu D, Cito S, Zhang Y, Wang CF et al (2015) A versatile and robust microfluidic platform toward high throughput synthesis of homogeneous nanoparticles with tunable properties. Adv Mater 27:2298. https://doi.org/10.1002/adma.201405408
92. Braga CB, Kido LA, Lima EN et al (2020) Enhancing the anticancer activity and selectivity of goniothalamin using pH-sensitive acetalated dextran (Ac-Dex) nanoparticles: a promising platform for delivery of natural compounds. ACS Biomater 6:2929–2942. https://doi.org/10.1021/acsbiomaterials.0c00057
93. Wannasarit S, Wang SQ, Figueiredo P et al (2019) A virus-mimicking pH-responsive acetalated dextran-based membrane-active polymeric nanoparticle for intracellular delivery of antitumor therapeutics. Adv Funct Mater 29:1905352. https://doi.org/10.1002/adfm.201905352
94. Pinho AC, Fonseca AC, Caseiro AR et al (2020) Innovative tailor made dextran based membranes with excellent non-inflammatory response: in vivo assessment. Mater Sci Eng C Mater Biol Appl 107: https://doi.org/10.1016/j.msec.2019.110243
95. Freeman FE, Kelly DJ (2017) Tuning alginate bioink stiffness and composition for controlled growth factor delivery and to spatially direct MSC fate within bioprinted tissues. Sci Rep 7:17042. https://doi.org/10.1038/s41598-017-17286-1
96. Rederstorff E, Weiss P, Sourice S et al (2011) An in vitro study of two GAG-like marine polysaccharides incorporated into injectable hydrogels for bone and cartilage tissue engineering. Acta Biomater 7:2119–2130
97. Senni K, Gueniche F, Changotade S et al (2013) Unusual glycosaminoglycans from a deep sea hydrothermal bacterium improve fibrillar collagen structuring and fibroblast activities in engineered connective tissues. Mar Drugs 11:1351–1369

98. Zanchetta P, Lagarde N, Guezennec J (2003) A new bone-healing material: a hyaluronic acid-like bacterial exopolysaccharide. Calcif Tissue Int 72:74–79
99. Wang X, Li Z, Shi T et al (2017) Injectable dextran hydrogels fabricated by metal-free click chemistry for cartilage tissue engineering. Mater Sci Eng C Mater Biol Appl 73:21–30. https://doi.org/10.1016/j.msec.2016.12.053
100. Akoulina E, Dudun A, Bonartsev A et al (2019) Effect of bacterial alginate on growth of mesenchymal stem cells. Int J Poly Mater Po 68:115–118. https://doi.org/10.1080/00914037.2018.152573
101. Zhao F, Guo C, Cui Q et al (2018) Exopolysaccharide production by an indigenous isolate *Pseudomonas stutzeri* XP1 and its application potential in enhanced oil recovery. Carbohydr Polym 199:375–381. https://doi.org/10.1016/j.carbpol.2018.07.038
102. Gupta P, Diwan B (2016) Bacterial exopolysaccharide mediated heavy metal removal: a review on biosynthesis, mechanism and remediation strategies. Biotechnol Rep (Amst). 13:58–71. https://doi.org/10.1016/j.btre.2016.12.006
103. Rehman SY, Ahmed M, Sabri AN (2019) Potential role of bacterial extracellular polymeric substances as biosorbent material for arsenic bioremediation,Bioremediat J 23:72–81. https://doi.org/10.1080/10889868.2019.1602107
104. Wang Y, Ahmed Z, Feng W (2008) Physicochemical properties of exopolysaccharide produced by *Lactobacillus kefiranofaciens* ZW3 isolated from Tibet kefir. Int J Biol Macromol 43:283–288. https://doi.org/10.1016/j.ijbiomac.2008.06.011
105. Latifi N, Horpibulsuk S, Meehan CL et al (2016) Xanthan gum biopolymer: an eco-friendly additive for stabilization of tropical organic peat. Environ Earth Sci 75:825. https://doi.org/10.1007/s12665-016-5643-0
106. Soldo A, Miletić M, Auad ML (2020) Biopolymers as a sustainable solution for the enhancement of soil mechanical properties. Sci Rep 10:267. https://doi.org/10.1038/s41598-019-57135-x
107. Ayeldeen M, Negm A, El-Sawwaf et al (2017) Enhancing mechanical behaviors of collapsible soil using two biopolymers. J Rock Mech Geotech Eng 9:329–339. https://doi.org/10.1016/j.jrmge.2016.11.007
108. Ivanov V, Chu J (2008) Applications of microorganisms to geotechnical engineering for bioclogging and biocementation of soil in situ. Rev Environ Sci Bio Technol 7:139–153
109. Im J, Tran ATP, Chang I, Cho GC (2017) Dynamic properties of gel-type biopolymer-treated sands evaluated by resonant column (RC) tests. Geo mech Eng 12:815–830
110. Van Oosten MJ, Pepe O, De Pascale S et al (2017) The role of biostimulants and bioeffectors as alleviators of abiotic stress in crop plants. Chem Biol Technol Agric 4:5. https://doi.org/10.1186/s40538-017-0089-5
111. Salachna P, Grzeszczuk M, Meller E, Mizielińska M (2019) Effects of gellan oligosaccharide and NaCl stress on growth, photosynthetic pigments, mineral composition, antioxidant capacity and antimicrobial activity in red perilla. Molecules 24:3925. https://doi.org/10.3390/molecules24213925
112. Salachna P, Mizielińska M, Soból M (2018) Exopolysaccharide gellan gum and derived oligo-gellan enhance growth and antimicrobial activity in eucomis plants. Polymers (Basel) 10:242. https://doi.org/10.3390/polym10030242

Techniques Used for Characterization of Microbial Exopolysaccharides

Rani Padmini Velamakanni, Priyanka Vuppugalla, and Ramchander Merugu

Abstract Exopolysaccharides (EPS) that are secreted by the bacteria are one of the main components of its structure. It is usually associated with the cell wall, or the bio film layer of the bacterial cell consisting of a monosaccharide as a framework or network. The study of its properties, structure, functions and interaction will enable us to gain knowledge about the potential benefits of Exopolysaccharides and the role played by them. Many techniques are available to study and analyze the structure. A multidisciplinary approach is needed that will cumulate data and give an insight and in-depth information about Exopolysaccharides. In this chapter, an effort is being made to introduce all the techniques with an emphasis on spectroscopic and microscopic techniques that can be used to characterize the Exopolysaccharides and analyze them. The techniques are described in brief providing subtle information about the principle, technique, advantages and limitations.

Keywords Exopolysaccharides · Spectroscopic techniques · Microscopy · Polysaccharides · Monosaccharide's · Sugar residues

1 Introduction

The bacterial cell wall surfaces, in addition to cell wall components abundantly contain carbohydrates molecules. These groups are unique and show diversity among certain groups of microorganisms. These are called Exopolysaccharides (EPS), or extracellular polysaccharides. They are produced by saprophytes, pathogenic microbes, microalgae and have good pharmacological and physical properties. They are found as a loose slime sheath, bio films, etc. Techniques like Scanning electron microscopy and Transmission electron microscopy will reveal to us details about the properties and organization of the sugar groups attached to the

R. P. Velamakanni · P. Vuppugalla · R. Merugu (✉)
Department of Biochemistry, University College of Science and Informatics, Mahatma Gandhi University, Nalgonda 508254, India

A. K. Nadda et al. (eds.), *Microbial Exopolysaccharides as Novel and Significant Biomaterials*, Springer Series on Polymer and Composite Materials,
https://doi.org/10.1007/978-3-030-75289-7_2

cell surface. The limitation of this technique is specimen preparation which is very difficult due to the very high water content of the polysaccharide and it owes to insufficient resolution of the image formed. The bacterial surface will be observed in uncollapsed state, with the radially extended fibres' from the bacteria. Sugars like D-glucose, D-galactose, D-mannose, fructose, rhamnose, ribose, xylose, etc. are frequently found as exopolysaccharides. These sugar groups are also associated with inorganic substituents like the phosphates, sulphates, polyanionic polymers, mono and divalent cations like sodium, calcium, barium, strontium, etc. are found associated. These groups correspond to the specific properties exhibited by the microbes. Some of the organic non-carbohydrate moiety includes acetate, pyruvate, glycerate, etc. and amino acids like glutamate and serine are common certain species. Thc study of the polysaccharides give us the valuable information about the potential therapeutic applications they harbour [1, 2]. Antioxidant and antibacterial activities of Exopolysaccharides from *Bifidobacterium bifidum* WBIN03 and *Lactobacillus plantarum* R315 were studied by [3].

Exopolysaccharides or the extracellular polysaccharide layers are molecules with potential industrial applications. Because of their wide number of as study on the production, distribution and the attachment and aggregation is very important. Consequently, a number of studies have been made in recent times focusing on specific components of the Exopolysaccharides and their applications. The present article aims to provide an overview understanding of the various techniques like spectroscopy and microscopy techniques, and various other techniques and a combination of the several techniques, sample preparation and the procedure that can be ideally adopted to study the Exopolysaccharides (Fig. 1).

2 Isolation Methods of Extracellular Polysaccharides

Some of the methods used for isolating the Exopolysaccharides includes Cationic ferritin treatment. The end products of the acid complex can be purified by reverse phase chromatography techniques or by Ion exchange chromatography. In another method, the crude preparation can be precipitated by alcohol like acetone or ethanol. 10% TCA (Trichloroacetic acid) is added and kept for 1 h at room temperature to precipitate the proteins. High concentration of TCA might result in loss of sample. The extract is further purified by dialysis and lyophilization of the extract gives the dry powder Exopolysaccharides [1, 2]. The Exopolysaccharides layer can be fragmented by acid hydrolysis and is the first step for the study of composition and to determine the structure. Trifluroacetic acid (TFA), hydrochloric acid, formic acid, or H_2SO_4 at a temperature of 100 °C is required to perform the acid hydrolysis. Important and crucial factors for acid hydrolysis are the concentration of acid, the temperature and time as acids at high concentrations may destroy the released monomeric sugar residues from polymer. Temperature is another defining factor for hydrolysis as the glycosidic bonds cannot be broken at low temperatures. A study by Grobben et al. [4] has observed that 1 M H_2SO_4 at 100 °C for 3 h gives

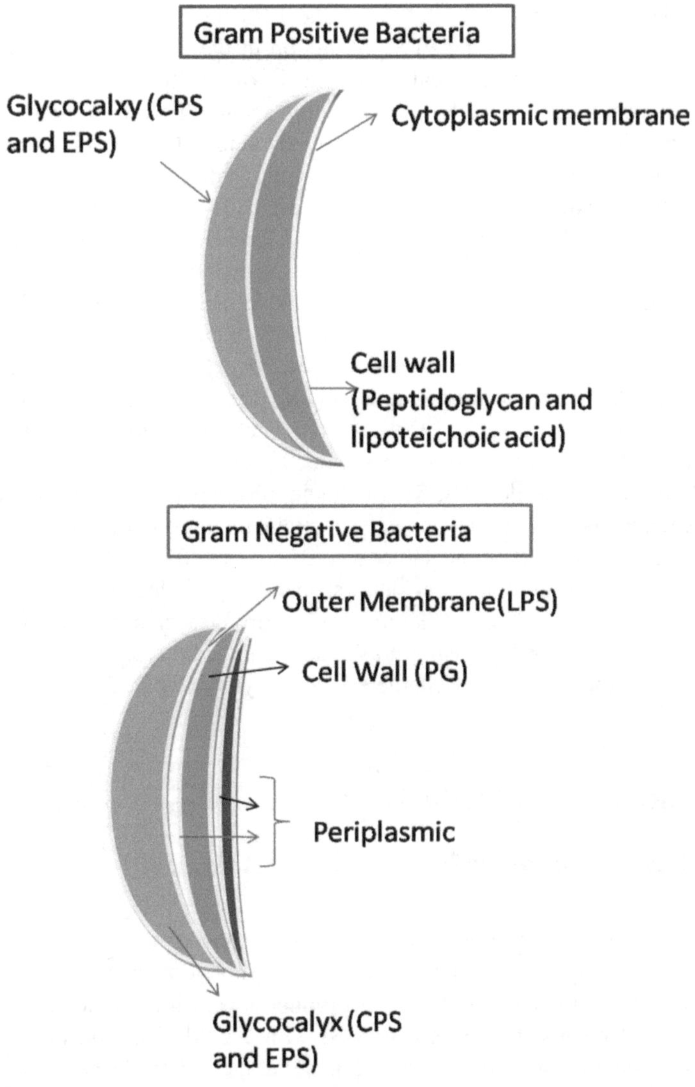

Fig. 1 Location of polysaccharides in the Gram-Negative and Gram-Positive Bacteria (CPS = cellular Polysaccharide, EPS = Exopolysaccharide)

a good yield of the monosaccharide residues. In another study by Levander et al. [5], 4 M HCl at 100 °C for 45 min and 85% formic acid at 100 °C for 6 h and 2 M TFA at 100 °C for 6 h was found to be better for acid hydrolysis of exopolysaccharides [6]. Ruas-Madiedo and De Los Reyes-Gavilán [7] reported different methods of isolation such as ultrafiltration, dialysis, precipitation, ion-exchange chromatography for isolation of exopolysaccharides. Gel permeation chromatography in an HPLC system allows the separation of the EPS polymers by size exclusion. The simultaneous detection of the EPS molecules by refractive index (RI) can be used to quantify the EPS production in the corresponding EPS elution peak [8]. The physicochemical properties of EPS were studied using Fourier-transform infrared (FTIR) spectroscopy, scanning electron microscopy, and atomic force microscopy (AFM). Trichloroacetic acid (TCA) was concentration of 14% and centrifugation, ethanol precipitation followed by centrifugation were used for isolation of exopolysaccharides produced by various *Streptococcus thermophilus* strains by Pachekrepapol et al. [9]. Mende et al. [10] reported that the molar mass of EPS from *S. thermophilus* strain St-143 was 4.3 × 106 g/mol when analyzed by gel filtration chromatography. Milk fermented by two EPS-producing *S. thermophilus* strains (Rs and Sts) yielded EPS of molar mass that ranged from 2.6 × 106 to 3.7 × 106 g/mol when analyzed by permeation chromatography combined with static light scattering and RI [11]. Rheo-chemical characterization of exopolysaccharides produced by plant growth promoting rhizobacteria was done by Shahzad et al. [12]. Liquid broths of secluded were treated with ethanol to precipitate Exopolysaccharides (EPS) for their physicochemical characterization. Anion-exchange and high-performance size-exclusion chromatographic showed the presence of Mannose (52%) and Glucose (29%). 0.5% EPS solution had low viscosity with pseudoplastic behaviour, least suspended particles producing less turbid solutions [12]. *Lactobacillus fermentum* Lf2 exopolysaccharide (EPS) functional aspects were characterized for using it as a dairy food additive. It was observed that it contributes to the rheological characteristics of yogurt (Fig. 2).

3 Characterization Methods

3.1 Pulsed Amperometric Detection (PAD)

Voltage is applied across an electrochemical cell causing the oxidation of carbohydrates, which helps in the identification of carbohydrates. Prevention of over adsorption of the oxidized carbohydrates product is very important, for satisfactory results. For best results, the system is used with a gold electrode [7]. Before the application of sample one cycle of oxidation and reduction of the electrode surface is required to clean the contaminants if any. The electrode has to return to oxidized state, which is the functional state before the actual experiment begins. This Technique detects the sample in picomoles and is one of the sensitive methods for quantification of carbohydrates [1, 2]. Products obtained can be subjected for the colorimetric detection by anthrone method. "The Method Requires the addition of sample to anthrone Reagent containing sulphuric acid. The reaction is completed by two very short incubation period at 90 °C for 3 min and at room temperature for 15 min, yielding greenish-blue colour" (Table 1) or phenol method. "To the sample sulphuric and Phenol is added is appropriate quantities and vortexed and the orange-yellow colour developed is measured at 490 nm. The method gives results in

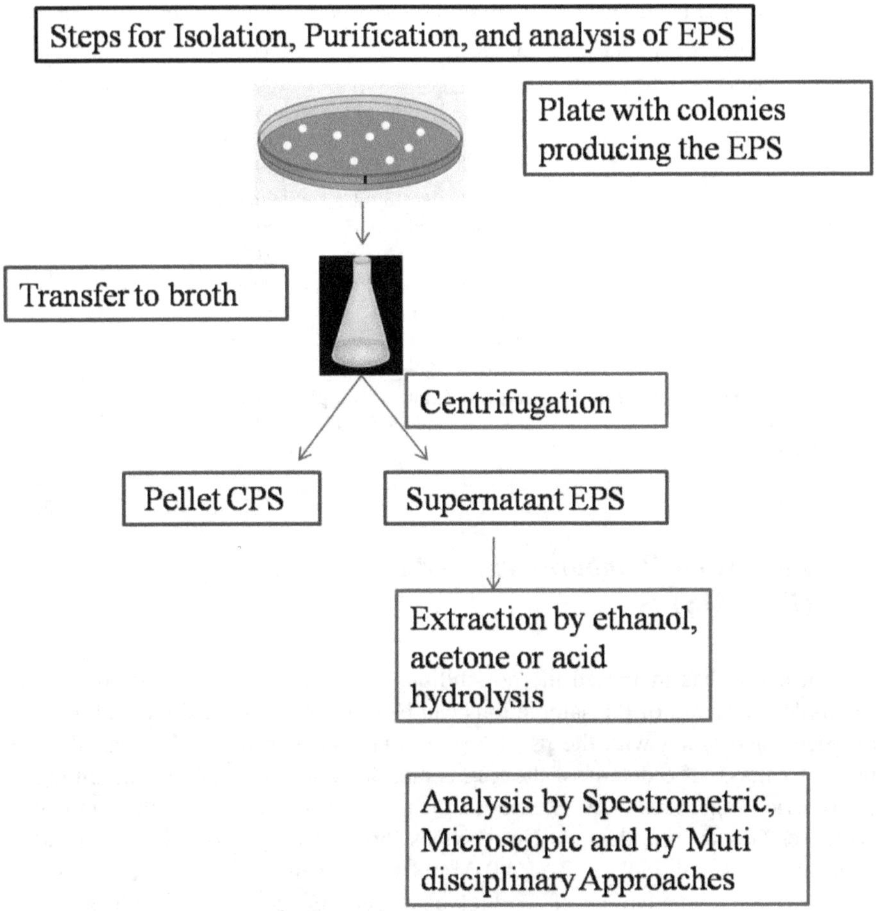

Fig. 2 Basic steps involved in the isolation, purification and characterization of EPS

less than 15 min. One heating step is required for the development of colour" [13], alkaline tetrazoleum blue method. "The methods require sodium hydroxide, ethanol, sodium potassium tartrate and tetrazolium blue as the reagents. It is one of the most common standard methods adopted for the determination of reducing sugars. Heating after addition of the reagents is required with in a range of 80–95 °C" etc. or the products can also be subjected to flourimetric [1, 2] detection to increase the sensitivity. The monomeric units of carbohydrates can be converted to alditols or acetylated alditols and can be analyzed by gas-liquid chromatography. Then the retention time is calculated which helps in the identification of the samples. The hyphenated technique of mass spectrometry gives information only about the class of sugars but does not reveal information about stereoisomers [1, 2].

Table 1 Tests used for qualitative analysis of polysaccharides calorimetrically

Name of the test	Concentration	Method	Absorbance	References
Phenol sulphuric acid	4.5–676 mg/l	50 μl sample + 150 μl sulphuric acid + 30 μl 5%phenol	Orange-yellow colour absorbance at 490 nm	[13]
Anthrone sulphuric acid	10–70 mg/l	1 ml sample + 3 ml sulphuric acid	Greenish-blue colour at 315 nm	[14]
Carbazole	20–2000 mg/l	50 μl sample + 200 μl of 25 mM sodium tetracarbonate in sulphuric acid	Brownish colour absorbance at 550 nm	[14, 15]
Phenol sulphuric acid coupled with glucose assay	50–5000 mg/l	20 μl of filtrate + 180 μl of phenol sulphuric acid	Absorbance at 480 nm	[14, 15]

3.2 *Fast Atom Bombardment—Mass Spectrometry (FAB-MS)*

This technique has increased the possibilities for analysis of the sugar molecules. The molecular mass of the sample ions can be obtained. The positive and negative ions generated along with the pseudo molecular ions can be used to calculate the molecular mass. The details of the fragmented ions help in establishing amount of monomeric sugar residues present and non-carbohydrate residues and their sequence. The specific degradative enzymes for the isolation of the sugar residues from microbial cell wall [1, 2]. FAB-Mass spectrometry can be more specifically used for the elucidation of endoglycosylated O-acetylated residues of an oligosaccharide. The technique also gives information on the polar molecules. The advantage of the method is that derivatization or volatilization of the sample is not mandatory. The oligosaccharide residues can be permethylated or per O-methylated with perdeuteroacetic acid under acidic conditions to remove any existing acyl groups and then the sample is run on FAB-MS to get the information about sequence of the oligosaccharide residues [1, 2]

3.3 *Turbidometric and Colorimetric Method*

Turbidometric and colorimetric are simple techniques which gives us information quantitatively and qualitatively. Exopolysaccharide layer is expressed as dextran per millimetre in turbidometric method. It is defined as the amount of polymer that produces the same turbidity at 720 nm as one milligram of dextran (molar mass: 2×10^3 kDa) under the same measurement conditions. The limitation to the technique is that only pure samples free from contaminants can be used, while the presence of other molecules like proteins interferes in the turbidometric measurement [6]. Like turbidometry, colorimetry is also used for both qualitative and quantitative identification. Most common method used is the anthrone-sulphuric acid method, one of the sensitive, simple methods. The blue-green characteristic colour developed can be calorimetrically measured due to the reaction of the anthrone reagent under acidic conditions with the sugar molecule. The anthrone reaction will not give satisfactory results with methylated sugars. Phenol-sulphuric acid is another widely used test for the identification of total carbohydrates [13]. The sugars give an orange-yellow characteristic colour in this method. Using both anthrone and phenol sulphuric acid method gives better results and help us in knowing the reducing sugars present in the Exopolysaccharides [6]. The reducing sugars are estimated quantitatively by DNSA (Dinitrosalisylic acid) method. This method also gives us information about the free carbonyl groups if any are present. The reaction takes place in alkaline conditions and the absorbance is measured at 540 nm. The colour observed is reddish-brown due to the oxidation of the aldehyde and keto groups of the sugar residues. Therefore, this is another potential test used for the identification of depolymerized sugar residues of the EPS.

3.4 *Carbazole Method*

The method is used to detect the uronic acids in glycosaminoglycans. The sugar residues on reaction with the tetraborate and sulphuric acid results in the development of colour which can be read calorimetrically. The test is modified to quantify the presence of neutral sugars where in the hydrolysis with acid like sulphuric acid remain the same, the brownish colour developed is a result of the reaction [16, 17, 18, 19] with *m*-hydroxydiphenyl and uronic acids which can be read colorimetric ally at 540 nm. The assays that are described are capable of measuring for uronic acid-containing polymers [20].

3.5 *Uronic Acid Screening*

The presence of specific carbohydrate residues can be identified by looking for the presence of uronic acids. The method is an extension of Blumenkrantz and Asboe-Hansen method [21]. Different sugar residues like Mannuronic, glucuronic and galacturonic acids show different calibration curves in the presence of uronic acids hence the method is a reliable one. On hydrolysis of the polysaccharide into constituent monosaccharide followed by specific colour reaction is which can be read at 520 nm is the characteristic of the method [22].

3.6 96-Well Assay

It is an extension of the phenol sulphuric acid method. For the rapid identification of various carbohydrates presents this method was first described by Dubois et al. [23] and later was adapted to a 96-well application. The method was optimized for rapid identification method (<15 min) with only one heating step and without any mixing step. The major disadvantages of this method are the need for a carcinogenic phenol reagent and the incubation in a water bath [13].

3.7 High-Performance Liquid Chromatography (HPLC)

High-performance liquid chromatography is an optimized conventional column chromatography technique. The greatest advantage of HPLC is that it can employ the principles of other techniques like adsorption, partition, ion exchange, affinity chromatography, etc. High pH anion exchange chromatography coalesced with the pulsed amperometric detector (PAD) is most commonly used for the detection of monomeric sugar residues in food, beverages, dairy, canned foods, and biotechnological products. The technique cannot be used for exopolysaccharides isolated by acid hydrolysis. The detector used is Ultraviolet (UV) or refractive index (RI) detector. In case of gradient elution RI detector cannot be applied, while UV detectors have this advantage over the RI detectors. Therefore, HPLC with RI detectors is less commonly used. Sugars do not absorb light, hence to increase the sensitivity of the sample near the detector they should be subjected to pre-column derivatization with a fluorescently labelled tags like aromatic amines to enable them to UV absorption [6]. The disadvantage of the technique is that derivatization reduces the yield and there is loss of reproducibility of the method. The advantage of the HPLC with derivatization is that the sample can be measure in femtomoles, while when it is combined with PAD picomoles of sample is required for analysis [6]

3.8 Ultra-High Performance Liquid Chromatography-Electrospray Ionization/Mass Spectrometry(UHPLC-UV-ESI-MS)

The exopolysaccharides are isolated from the cell wall by centrifugation and filtration and further purified by gel filtration method. The low molecular weight sugar residues like glucose are separated. The sample is then derivatized with the HT-PMP method (HT-1-phenyl-3-methyl-5-pyrazolone). The method allows the detection of various sugar residues hexoses, pentoses, deoxy and amino sugars, uronic acids, etc. in single run. The technique is equipped with two detectors like UV and ESI-MS which makes the identification and quantification reliable. The method is very fast and can be considered as a carbohydrate finger print enabling the detection of 96 different strains in a single day [14, 15]. Purification and analysis of monosaccharide composition of exopolysaccharide (EPS) produced by *Halorubrum* sp. TBZ112 was done to evaluate antiproliferative activity against human gastric cancer (MKN-45) cell line [24]. An additional ultraviolet detector can provide information on the presence of protein in the sample [25]. High-pressure size exclusion chromatography (HPSEC) with multi-angle laser light scattering (MALLS) and high-pressure anion exchange chromatography (HPAEC) were used to determine the molecular weight and monosaccharide composition. It was observed that the EPS had a molecular weight of 5.052 kDa and were a heteropolysaccharide containing mannose (19.95%), glucosamine (15.55%), galacturonic acid (15.43%), arabinose (12.24%), and glucuronic acid (12.05%)[24]. EPS was analyzed by a UV-visible spectrophotometer in the wavelength range of 200–600 nm from a probiotic strain of *L. plantarum* WLPL04 by Liu et al [26].

3.9 Multi Laser Light Scattering Spectrometer (Malls)

The samples can also be further analyzed by Multi laser Light Scattering Spectrometer (MALLS). The mobile phase of the sample was 0.2 M $NaNO_2$ with 200 ppm NaN_3 and the flow velocity was maintained at 0.5 mL/min. A 250 μL sample at 5 mg/mL concentration was injected for each run [27]. For optimum results, SEC coupled with MALLS is commonly used. The isolated exopolysaccharide layer is dissolved in the buffers like phosphate/chloride buffer whose pH is to be maintained at 6.8 with an ionic strength of 0.1. To the buffer, diamino-tetraacetic acid–disodium salt (Na_2EDTA) and 0.01% sodium azide are added for chelation. The filter clear buffer solution is then used to treat the sample. The chromatographic system is equipped with degasser and a high-pressure pump, that maintains a constant flow rate of 0.5 ml/min. The sugar residues are separated and analyzed based on the (Dn/Dc) Specific Refractive index, which is specific for each sugar residue. The method can be used to study the molecular weight, Root Mean Square gyration (RMS) of the monosaccharide residues along with the identification with the help of the computational tool associated with the technique. From the double logarithmic plot of RMS radius vs molecular mass of exopolysaccharides can be calculated. The conformation of exopolysaccharides in aqueous solution could be identified according to the following equations:

$$\log ri = k + a \log Mi$$

where ri is the RMS radius of an EPS molecule, Mi is the molar mass of EPS, k is the intercept at the Y-axis (RMS radius), and a is a critical slope values to determine the molecular conformation of each Exopolysaccharides. For polymers of sphere-like structure, random coils, or rigid rods, their corresponding values of slope are known as 1/3, 1/2 and 1, respectively, according to the following dimensional relationships: for globular shape, $r_3\ i \propto Mi$; for flexible coil, $r_2\ i \propto Mi$; rod-like chain, $ri \propto M$. Exopolysaccharide sugars with low value of polydispersity ratio will usually form large aggregates in aqueous solution [28]. Partial characterization of exopolysaccharides from *Bifidobacterium longum* NB667 and the cholate-resistant strain *B. longum* IPLA B667dCo was done by Salazar et al [29]. Analysis by size exclusion chromatography-multi angle laser light scattering indicated that the EPS crude fractions of both strains contained two polymer peaks indicating them to be heteropolysaccharides composed of glucose, galactose and rhamnose. Size-exclusion chromatography coupled with multi-angle laser light scattering (SEC-MALLS) has been used to determine the molar mass of the isolated EPS [30].

3.10 Fourier-Transform Infrared Spectroscopy

Fourier-Transform Infrared (FT-IR) spectrum was used to determine the functional group and glycosidic bonds of the monosaccharide's residues. The technique is often described as the blue print of the molecule as the spectrum depends on the vibrations that an atom experiences in a field and thus is highly specific for every molecule. At the regions where the molecule absorbs radiation and is excited to higher energy state, there the fingerprints are obtained. The exopolysaccharides must be dry before FTIR analysis because the water absorbs in the mid Infrared region [31]. The purified exopolysaccharides layer was grinded with KBr powder (EPS:KBr approximately 1:100) and pressed into flats or discs and these discs are then subjected to infrared spectroscopy between 400 and 4000 cm^{-1}. The existence of C–O–C glycosidic bonds and C–O–H link bonds of the pyranose unit in each polysaccharide can be identified by the characteristic specific absorption in the region of 1000–1200 cm^{-1} indicates the existence of C–O–C glycosidic bonds and C–O–H link bonds of the pyranose unit in each polysaccharide. The presence of α-type configuration of the mannose units can be identified by characteristic peak at 810 cm^{-1}. Furthermore, the signal at 890 cm^{-1} suggests the existence of a β-type glycosidic linkage. The characteristic band at a region of 1645–1655 cm^{-1} could be correlated to the stretching vibration of the carbonyl group (C=O) of the polysaccharide [27]. A method for the simultaneous quantification of lactic acid, lactose, and EPS yield directly in culture media has been developed recently using near infrared spectroscopy (NIRS) [32].

3.11 Attenuated Reflection Microscopy (ATR)

FT-IR spectroscopy has been combined with the attenuated reflection microscopy to extend the internal reflection spectroscopy to the microscopy. It is a non-destructive technique. The sample can be studied in hydrated conditions [33]. Various functional groups associated with the exopolysaccharides and the structural modifications of the exopolysaccharides can be monitored by this method. The ATR crystal has high refractive index. ATR-FTIR uses multiple detectors which result in the spatial localization and give numerical aperture resulting in better spatial resolution [34]. ATR-FTIR cannot give satisfactory results when the specimen is thick. The acceptable thickness is below 1.0 μm [35]. ATR-FTIR is a zero-dimensional technique that can collect the data only from the surface of the molecule [36]. The recent developments in the IR Microspectrometry were highlighted by Bhargava [34] who opined that these advancements have got the potential to change the imaging spectroscopy. They also mentioned that microscope integrated with the Planar Infra Red Spectrometers (PAIR) can magnify even the smallest depth with only a small signal. e.g.; FTIR cannot easily do the monolayer mapping, while PAIR can analyze and is useful in detecting Exopolysaccharides.

3.12 Nuclear Magnetic Resonance spectroscopy

Nuclear Magnetic Resonance is technique that measures the absorption frequencies in the presence of magnetic field. NMR is a non-invasive and non-destructive technique [37]. Slight variations that are observed as the proton or neutron jumps from the lower energy state to higher energy and shielding of electrons in the presence of magnetic fields are the factors contributing to the signal and is the characteristic of the given nucleus. ^{1}H is high in its natural abundance and sensitivity [38]. The limitations of the technique can be that the energies that arise due to transition results are low when compared to the thermal fluctuations. Therefore, when compared to optical methods the technique is insensitive and also the proton NMR requires the labelled substrates which might affect the physiology of the Exopolysaccharides [31]. ^{1}H and ^{13}C Nuclear Magnetic Resonance (NMR) analysis can be useful in the identification of the sugar residues. The dry sample was redissolved in deuterium oxide (D_2O, 99.9%) at about 5 mg/ml to 10 mg/mL, and the chemical shifts were represented as "parts per million" (ppm) [27]. Proton spectra were run at a probe temperature of 25 °C, while carbon spectra were obtained at 25 °C. The 1H NMR spectra was complex due to the convergence of most sets of the resonance signals. Resonances signals were identified in the region for anomeric protons in the lower field in most cases for the sugar residues. The numbers of the signals generated in the anomeric regions correspond to the number of monosaccharide residues. The strong signals at 5.10 and 4.53 ppm were attributed to -D-glucose and -D-mannose residues. 4.32 ppm was assigned to reduced end-chain of -d-glucose. The high-resolution 13C NMR spectra in the anomeric region at 102.40, 95.25 and 91.31 ppm, which were assigned to the mannose (C-1), glucose (C1) and reduced end-chain of glucose (C-1) residues, respectively [28]. NMR helps us to know the chemical functionalities in the sample and the spectrum gives us information regarding the intact matrix, the amount of

Exopolysaccharides layer, the mass of each component of the EPS, a detailed data about the functional groups presents and their bonds with the neighbouring molecules, the mechanism of metal binding and their affinities, mobility, etc. [39]. Two methods of NMR technique, which are solid and liquid state, can be applied to study the exopolysaccharide of microbes. Therefore, it is an important tool to study the molecules in the biochemical analysis [40]. Solid-state NMR gives us promising insights about chemical composition, structure, linkage analysis, but cannot distinguish the β-O-4 and β-O-3 linkages of glycosidic carbon atoms.

4 Microscopic Techniques

The microscopic techniques that have been developed utilizes the labelled and unlabeled methods for imaging. Some of the techniques that can be used to study the Exopolysaccharides can be Scanning Electron Microscopy and Confocal Laser Scanning Microscopy, Atomic Force Microscopy.

4.1 Scanning and Electron Microscopy

The EPS can be studied by SEM as a two-dimensional image. The technique involves the study of sample in vacuum conditions; the specimen preparation is an important step in the SEM. The specimen since it is a biological sample and lacks conductance should be very small, 50 nm in thickness and includes rigorous fixation steps onto a copper, or gold metals which are having electrical conductivity [41]. The sample is subjected to dehydration with solvents like ethanol and acetone, which results in the lowering the surface tension due to loss of water molecules and less hydrogen bonding ability [42]. This intense specimen preparation might result in the loss of morphology to avoid that the samples can be freeze-dried hexamethyldisilazine and then coated with the conductive metal which is followed by glutaraldehyde fixation results in the preservation of Exopolysaccharides [43] The SEM images give good comparative information under various altered experimental conditions [44]. The limitation to the technique is that it does not provide any information about the chemistry of the molecules only the structural elucidation is possible [45, 42]. To circumvent this limitation, SEM can be operated in wet mode, ESEM. The conditions in ESEM are moderate vacuum without prior fixation, dehydration of the sample and the coating of the sample with the conductive metal. The limitation of ESEM is that less electron-dense specimens like exopolysaccharides do no resolve well [42]. The samples are stained in the ESEM method the staining techniques increases the image contrast improving the image analysis [46].

4.2 *Multiple Fluorescence Staining and Confocal Laser Scanning Microscopy (CLSM)*

CLSM can be performed in real-time and is a non-destructive method and is a sensitive method. The images produced by this microscopy technique can be visualized as 3D structures and can also be quantified. The 3D images are obtained due to a combination of multiple optical regions which helps in magnifications at different depth and analyzed by software to produce a final image. The images given by CLSM can be analyzed under the three criteria, visualization, quantification and deconvolution [47, 48, 49]. The technique is equipped with multichannel mode, which enables us to map individual components of the Extracellular polysaccharide layer. The technique is also equipped with multiple staining techniques and has been increasing used to study various components of Exopolysaccharides; hence the distribution of the components of the Exopolysaccharides can be studied. The EPS data can be collected both qualitatively and quantitatively which is a result of the fluorescence intensities. The fluorescence labelling is species from which the Exopolysaccharides is studied [50]. The fluorochromes used for the labelling the EPS components are selected based on pH of the sample, Excitation and emission parameters. The staining method should be appropriate, which results in mobilization and fixation of the sample. The selection of proper stain, selection of buffer, incubation time, washing time, are all important parameters for the specimen which ultimately demonstrate the image formed. The criterion to be followed is that there should be minimum spectral interference. When performing these procedures, the excitation spectra should not overlap because of the fluorochromes and if the regions of the emission spectra do not overlap of all the fluorochromes used, then the spectra emitted can be observed using a limited observation wavelength band one after one. As the emission and excitation wavelength overlap, the Nile red and Tetramethylrhodamine isothiocyanatae cannot be used a fluorochromes at once [51].

In order to identify the emission signals of multiple fluorochromes, CLSM uses green, red and blue regions of the spectrum, that results in good resolution of the components observed [52]. The serious limitation with the multiple fluorochromes on the same sample is that it might cause interference. The factors for good image formation are the thickness and the density of the specimen as thickened specimen might allow less dye penetration and results in is responsible for unsatisfactory image formation. A good specimen is only with a thickness of up to hundreds of μm [53, 54]. The specimen must be embedded in the paraffin, nanoplast, tissue freezing medium and then using a microtome or cryotome the sectioning must be done [55, 56]. Samples analyzed by two-photon laser scanning microscopy are magnified with high resolution when compared to the conventional single-photon laser microscopy [52]. The two-photon laser scanning microscopy has an advantage over the normal confocal microscopy in increased resolution, reduced photobleaching and reduced phototoxicity of the fluorescent probes [57] which will finally improve the 3D image formed. Some components of exopolysaccharides

Table 2 Example of the fluorochromes used for the identification of polysaccharides

Name of the stain	Sugar residue identified/labelled	Colour observed	References
TMR-Con A	Mannopyranosyl and α Glucopyranosyl	Green	[51, 60]
ConcavalinA	α-D Glucose	Green	[55]
TexasRed	αD-Mannose	Green	[55]
SYTOX Blue	Dead cells	Green	[51]
SYTO 63	Total cells	Red	[60]
Calcoflour white	β D Glucopyranose	Green	[59]

cannot be stained, in such a case CLSM provides information only about the stainable components. In particular, the person using the technique should be able to identify such defects of the technique and be able to gather the data in the most suitable mode to minimize these effects [58]. Some of the Examples used for the Identification of various Sugar Residues and the Colour observed are described in (Table 2) are mostly selected based on the emission spectra [59].

4.3 Energy Dispersive X-Ray Analysis

Using the scanning electron micrographs and elemental composition studies by the Energy Dispersive X-Ray analyzed the Exopolysaccharides from *L. plantarum* was observed [61]. The following results were obtained. The EPS is having a 3D structure and many irregular lumps were observed with varied sizes. The surface is very coarse. On magnifying to a higher end the spherical structure of EPS is clearly observed. The layer is porous and web-like [62]. SEM micrographs of Exopolysaccharides from *L. fermentum* CFR2195 showed flake-like structural units and were highly compact [63]. It was analyzed that the different morphology can be because of the different polysaccharides and the sample extraction, preparation also might cause some structural modifications.

4.4 Atomic Force Microscopy

The electron microscopy cannot provide the information about the samples that are hydrated. The sample must be completely dry for us to get the image. Specimen preparation stands as a major problem in electron microscopy. This problem can be circumvented with the AFM, Atomic force microscopy. It uses liquid phase, gas phase, ultra-high vacuum, and electrochemical environments. The technique can be used in three modes; they are contact, non-contact or tapping modes. For many biological samples, the scanning probe measurements are performed successfully in air using AFM. The tapping mode is used for the identification of the sample only [64]. The surface morphology is better understood in the tapping mode which can be operated at the nanoscale level both in air and fluid environments. AFM can give the 3D images with a nanoscale resolution (less than 1.0 nm). Molecular organization of EPS attached to solid substratum is characterized by AFM. The topographic images and force-distance curves enables the characterization and to understanding the morphology and molecular interactions of the Exopolysaccharides. The studies suggest that proteinaceous EPS accumulated as a thin, continuous layer at the surface of solid substratum. Van der Aa and Dufrêne [65] using AFM studied the interactions within the exopolysaccharides and cell. Outer exopolysaccharide layer is loosely associated with the cell surface. The inner layer of exopolysaccharide is tightly associated with the cell surface and contains less calcium. The studies also suggested that the calcium binding affinities are with inner and outer exopolysaccharide layers [66]. AFM gives information about the morphological details and also chemical composition limitation of the technique is the time period for the image formation. The technique is expensive and low light efficiency. The sample gets easily damaged due while specimen preparation [65, 67]. Through atomic force microscopy it was observed that Intra- and intermolecular aggregation takes place in EPS units [68].

4.5 Raman Microscopy

Spectral Microscopy is an extension of the standard spectroscopy. The tools available with the technique enable us to fetch the information about the spatially resolved spectra, thus the information which is not available with the normal Microscopy techniques is collected. Each analyte will have its own unique absorption spectrum which will be used to identify and analyze the atomic interaction and their distribution in the space.

Raman spectroscopy is a non-destructive technique which is based on the Raman scattering of light. In the technique, the polychromatic light from the source is chopped to monochromatic light which will fall on the sample and scattering of monochromatic laser light by the sample provides the data which is the fingerprint of the molecule with the resolution of optical microscope. Integrating the Raman microscopy with the common Raman spectroscopy enables the spectral analysis enables the analysis of spectrum at micrometer spatial resolution. Raman Microscopy helps in understanding the chemical composition and the structure of Exopolysaccharides, and also its formation stages can be studied. [69]. The advantage of the technique is the molecules with the symmetrical modes which cannot be studied by infrared spectroscopy can be analyzed [70]. Experiments by Ivleva et al. [71] using Raman microscopy to study the chemical composition of

Exopolysaccharide during the formation suggested that Raman Microscopy can sufficiently replace the CLSM. The technique does not require tedious sample preparation. It is a label free technique, and the EPS can be analyzed in the hydrated forms. Since the whole spectrum can be collected with a fixed laser wavelength by excitation, a separate tunable excitation source is not required. Cations induce several changes on binding to polysaccharides, during such state also the effect of photobleaching is observed [54, 71]. The limitation to Raman microscopy is that it is time-consuming. The time can be reduced by combining Raman Microscopy to confocal microscopy, hence called as Confocal Raman Microscopy (CRM). The disadvantage of CRM is that it does not remove the autofluorescence of the sample. The advantage of confocal Raman microscopy is that the samples can be studied in different stages of growth, without affecting the metabolic activity [45, 72].

4.6 Thermo Gravimetric Analysis

Thermogravimetric analysis of the polysaccharide was conducted in a TA Q5000IR TGA apparatus using 15 mg EPS fraction of the test material. TGA curve plot TGA signal (converted to percent weight change on the *Y*-axis) against the reference material temperature (on the *X*-axis). The polysaccharides are having potential role in the drug delivery and they are being used in the solid dosage forms. So when the polysaccharides are being used to the treatment they is a need for pre-treatment like usage of high temperatures. Hence it is important for us to understand the changes in the chemistry like decomposition behaviour of the polysaccharides. The data regarding the thermo stability and decomposition will help us to a great extent in the drug discovery process. This technique helps us to understand and study about the changes in the mass of the molecules with respect to changes in the temperatures like gain or loss of mass with respect to temperature. The experiments on 3 fragments of Exopolysaccharides from *Boletus aereus* by Zheng et al. [28] suggested that there is a significant mass loss at 260 °C, and the degradation temperatures are identified to be 170, 156, and 155 °C, respectively and it is concluded that the exopolysaccharide fragments are thermally stable up to a temperature of 150 °C. In another experiment on *Leuconostoc pseudomesenteroides* similar kind of results were obtained, when the exopolysaccharides were extracted and subjected to Thermogravimetric analysis. From the experiment and results obtained on analysis, it was understood that the degradation behaviour is related to its structure. EPS produced by a hot spring origin *Pseudomonas* species was structurally elucidated. It was found to be thermostable (T_d 289.98 °C), has low molecular weight (1510 Da) and is rhamnoglucose type consisting α-(1 $\rightarrow$ 3) linked to L-Rha*p* and α-(1 $\rightarrow$ 6) linked to D-Glc substituted with *O*-acetyl groups.

4.7 *Gas Chromatography (GC)*

Gas chromatography an important analytical method that is used for the analysis of free monosaccharide constituents that are dissociated from the polysaccharide. The technique is highly selective, quick and has high accuracy. Monomeric residues of the sugars need to derivatized which renders them to become stable volatile compounds. The monosaccharides are trimethylsilylated usually with Hexamethyldisilane (HMDS), Trimethylchlorosilane (TMCS), Trimethylsilyl or by using solvents like pyridine, Dimethylsulfoxide (DMSO) they are converted into alditol acetates. On treatment with sodium borohydride in anhydrous DMSO and acetylated with 1-methylimidazole followed by reaction with acetic anhydride will result in the reduction of monosaccharides. The trimethylsilyl derivatized product of monosaccharide when isolated using gas chromatography having a moderate polar stationary liquid gives satisfactory separation. While when a strong polar stationary liquid is used they give tailing peaks for the saccharides that are acetate derivatized, a strong polar stationary liquids provides maximum yield. For derivatization packed column and capillary columns are used. Some of the commonly used capillary columns are as OV-1, OV-225, HP-5, SE-30, Superleco SP 2380, BPB-5 Silar 10C Support-coated Open Tubular (SCOT) glass capillary column. Gas chromatography together with mass spectrometry is very efficient method and most widely used for methylated polysaccharides (Wang et al. 2003). Methylation can be done by dimsyl sodium/methyl iodide or sodium hydroxide/methyl iodide followed by hydrolysis to cleave the furanosidic or pyranosidic linkages helps in the analyzing the linkage pattern of the sugar residues of the exopolysaccharide. The hydrolysis can be done in two steps. In step one at 100 °C for 6 h using 90% formic acid and in step two with 2 M TFA at 100 °C for 3 h. The products obtained are subjected to treatment with alditol acetates to get partially methylated alditol acetates which can be analyzed with the gas-liquid chromatography-Mass spectrometry [73]. The polysaccharide producing bacteria was obtained from the marine sediment region and was characterized using FT - IR, XRD and GC—MS [74].

4.8 *Capillary Zone Electrophoresis (CE)*

The depolymerized monosaccharide's which are derivatized with Fluorescent or UV can be separated with capillary electrophoresis. The advantage of the technique is it requires small amounts of sample and gives good resolution. 8-Amino-1,3,6-Naphthalenetrisulfonate (ANTS), 7-nitro-2,1,3-Benzoxadiazole-Tagged Methylglycamine are some of the compounds used for the fluorescence detection [75]. For capillary zone electrophoresis, sugar molecules can be even treated with enzymes for derivatization. Two enzymes that have been selectively used are glucose oxidase which catalyzes the oxidation of glucose and horse radish peroxidase which catalyzes the oxidation of the Homovanillic acid (HVA) which is

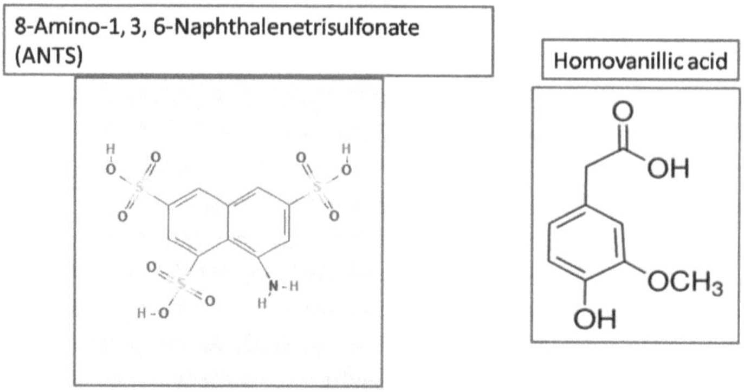

Fig. 3 Structure of ANTS (8-amino 1,3,6 naphthalene trisulfonic acid), Homovanillic acids

non-fluorescent to the 2,29-dihydroxy-3,39-dimethoxybiphenyl-5,59-diacetic acid (HVA) which is fluorescent and can be analyzed further by capillary electrophoresis. Capillary electrophoresis is improvised to Chip Capillary Electrophoresis (CCE), a new micro-total analysis system (μ-TAS). In this technique, on a chip made by micro-fabrication technique are placed the separation channel, reaction container and detector (Fig. 3).

4.9 Bioinformatics Methods

The study of carbohydrates is called as glycomics which includes identification of sugar residues, structural elucidation, sequence determination, study of isomers, anomeric configuration, and analysis of linkage. The interaction between carbohydrate and protein that is termed as glycoprotein and is very crucial to the function. The bioinformatics method gives us information regarding glycan-binding proteins. Some of the bioinformatics resources are CarbBank that gives us information of Oligosaccharide residues. This database furnishes the data regarding the text-oriented IUPAC description of glycans for information retrieval. Other examples include LINUCS, GLYDE, KEGG (Kyoto Encyclopaedia of Genes and Genomes) a database that is equipped with the metabolic pathway and the enzymes involved in the reaction along with the cofactor molecules. KegDraw is a tool that gives us topological information about the glycan molecules. The structures resolved by the NMR are uploaded into the database. By using the computational tools their dynamic behaviour and the spatial configuration is analyzed. Systematic searches are one of the most widely used computational tools that elucidate the torsion angles between the molecules and the energies associated with them. Metropolis Monte Carlo provides the information regarding the molecular dynamic simulations, the results often depicted in 3D conformations with their lowest possible energies. About 700 Φ, Ψ free energy maps can be retrieved from

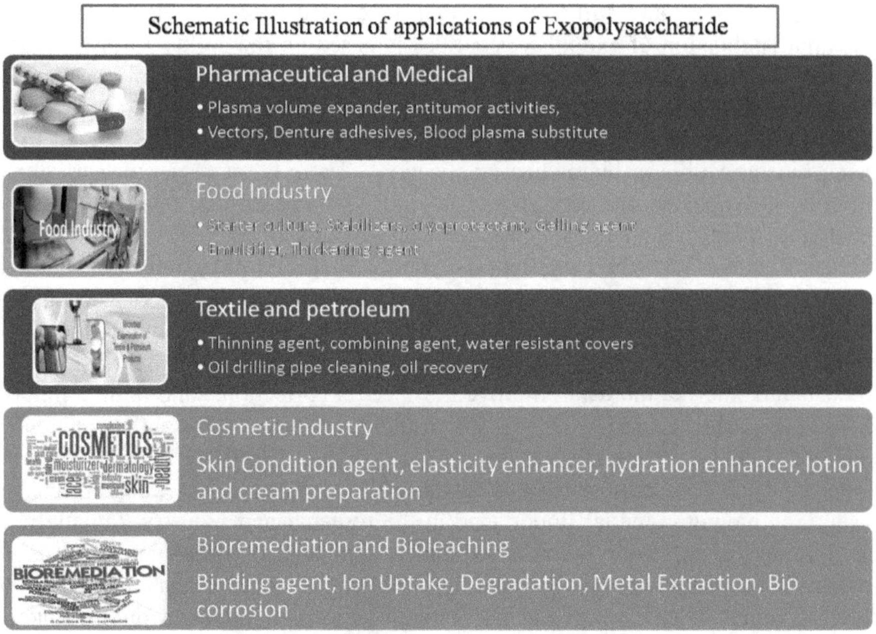

Schematic Illustration of applications of Exopolysaccharide
Pharmaceutical and Medical
• Plasma volume expander, antitumor activities,
• Vectors, Denture adhesives, Blood plasma substitute
Food Industry
Food Industry
• Starter culture, Stabilizers, cryoprotectant, Gelling agent
• Emulsifier, Thickening agent
Textile and petroleum
• Thinning agent, combining agent, water resistant covers
• Oil drilling pipe cleaning, oil recovery
Cosmetic Industry
COSMETICS
moisturizer
dermatology
skin
face
beauty
Skin Condition agent, elasticity enhancer, hydration enhancer, lotion and cream preparation
Bioremediation and Bioleaching
BIOREMEDIATION
Binding agent, Ion Uptake, Degradation, Metal Extraction, Bio corrosion

Fig. 4 Schematic Illustration of applications of Exopolysaccharides

GlycoMaps DB which is temperature stable disaccharides simulations. In order to generate one glycan structure that is validated and energetically minimized can be done by a database called SWEET-II web interface that uses a collection of unrefined conformational maps. The generated 3D structures are further refined by computational tools like the 'Dynamic Molecules' which is comprehensive of other methods, taking the 3D structure from SWEET II provides access to manipulate the structure and analyze and study it. It was mainly thought to be used as starting points for further refinement using more comprehensive computational techniques. The interactive analysis of time dependencies of any interesting degrees of freedom, free energy conformational maps and support for the interpretation of experimental findings derived from NMR spectroscopy are generated by the web interface 'Dynamic Molecules' [76] (Fig. 4).

5 Future Perspectives and Conclusions

The wide use and potential applications of exopolysaccharides are dependent on the techniques used for the characterization of these polysaccharides. Identification of these exopolysaccharides requires very economical methods and rapidly characterization techniques. Methylation analysis for linkages [77] and NMR technique for 2 or 3-dimensional structural analysis have been proposed earlier [78]. The potential application of exopolysaccharides is based on the ability to make modifications to different forms [79] and biodegradable nature of the exopolysaccharides [1, 2, 80–88]. Techniques should be developed which will be able to modify the obtained Exopolysaccharides to withstand physical and chemical stress apart from being biocompatible in biomedical applications. There is much yet to learn about the role of exopolysaccharides and its functions, characteristics, applications. The techniques help us to elucidate the growth, structure of exopolysaccharides and their interactions with the bacterial cells. Furthermore, the integration of multidisciplinary technologies is enabling us to study the structure of exopolysaccharides during the various growth phases which help us to understand the signal processing events and use of the algorithms and bioinformatics methods will help us to store the data and analyze the data, improve the spectral corrections, and high definition 3D images. The techniques those are available till date helps us to gain new knowledge and to correlate the new data with the already available data and understand the new changes that have taken place. Since it is a known fact that bacteria is having potential importance in the research and industry, the study of exopolysaccharides with multi disciplinary approaches furnish more in-depth insights about structure, function and applications. Thus our understanding of the role and functions of exopolysaccharides with various techniques would enable us to use them for different important applications in various fields.

References

1. Sutherland IW (1990) Biotechnology of microbial exopolysaccharides. Cambridge University Press, pp 1–17. ISBN: 0-521-36350-0
2. Sutherland IW (1990) Biotechnology of microbial exopolysaccharides. 9, Cambridge University Press, pp 379–403
3. Li S, Huang R, Shah NP, Tao X, Xiong Y, Wei H (2014) Antioxidant and antibacterial activities of exopolysaccharides from *Bifidobacterium bifidum* WBIN03 and *Lactobacillus plantarum* R315. J Dairy Sci 97:7334–7343
4. Grobben GJ, van Casteren WHM, Schols HA, Oosterveld A, Sala G, Smith MR, Sikkema J, de Bont JAM (1997) Analysis of the exopolysaccharides produced by *Lactobacillus delbrueckii* subsp. *bulgaricus* NCFB 2772 grown in continuous culture on glucose and fructose. Appl Microbiol Biotechnol 48:516–521
5. Levander F, Svensson M, Rådström P (2001) Small-scale analysis of exopolysaccharides from *Streptococcus thermophilus* grown in a semidefined medium. BMC Microbiol 1:23
6. Madhuri KV, Prabhakar KV (2014) Recent trends in the characterization of microbial exopolysaccharides. Int J Pure Appl Chem 30(2). http://dx.doi.org/10.13005/ojc/300271
7. Ruas-Madiedo P, De Los Reyes-Gavilán CG (2005) Invited review: methods for the screening, isolation, and characterization of exopolysaccharides produced by lactic acid. J Dairy Sci 88(3):843–856
8. Ruas-Madiedo P, Tuinier R, Kanning M, Zoon P (2002) Role of exopolysaccharides produced by *Lactococcus lactis* subsp. *cremoris* on the viscosity of fermented milks. Int Dairy J 12:689–695

9. Pachekrepapol U, Lucey JA, Gong Y, Naran R, Azadi P (2017) Characterization of the chemical structures and physical properties of exopolysaccharides produced by various *Streptococcus thermophilus* strains. J Dairy Sci 100:3424–3435
10. Mende S, Mentner C, Thomas S, Rohm H, Jaros D (2012) Exopolysaccharide production by three different strains of *Streptococcus thermophilus* and its effect on physical properties of acidified milk. Eng Life Sci 12:466–474
11. Faber EJ, Zoon P, Kamerling JP, Vliegenthart JFG (1998) The exopolysaccharides produced by *Streptococcus thermophiles* RS and STS have the same repeating unit but differ in viscosity of their milk cultures. Carbohydr Res 310:269–276
12. Shahzad H, Iqbal M, Khan QU (2018) Rheo-chemical characterization of exopolysaccharides produced by plant growth promoting rhizobacteria. Turk J Biochem 43(6)
13. Masuko T, Minami A, Iwasaki N, Majima T, Nishimura S-I, Lee YC (2005) Carbohydrate analysis by a phenol-sulfuric acid method in microplate format. Anal Biochem 339:69–72. https://doi.org/10.1016/j.ab.2004.12.001
14. Rühmann B, Schmid J, Sieber V (2015) Methods to identify the unexplored diversity of microbial exopolysaccharides.Front Microbiol. https://doi.org/10.3389/fmicb.2015.00565
15. Rühmann B, Schmid J, Sieber V (2015) Automated modular high throughput exopolysaccharide screening platform coupled with highly sensitive carbohydrate fingerprint analysis. J Visualized Exp
16. Felz S, Vermeulen P, van Loosdrecht MC, Lin YM (2019) Chemical characterization methods for the analysis of structural extracellular polymeric substances (EPS). Water Res 157:201–208
17. Dische Z (1946) A new specific color reaction of hexuronic acids. J Biol Chem 167:189–198
18. Filisetti-Cozzi TM, Carpita NC (1991) Measurement of uronic acids without interference from neutral sugars. Anal Biochem 197:157–162. https://doi.org/10.1016/0003-2697(91)90372-Z
19. Galambos JT (1967) The reaction of carbazole with carbohydrates. Anal Biochem 19:133–143. https://doi.org/10.1016/0003-2697(67)90142-X
20. Van Den Hoogen BM, Van Weeren PR, Lopes-Cardozo M, Van Golde LMG, Barneveld A, Van De Lest CHA (1998) A microtiter plate assay for the determination of uronic acids. Anal Biochem 257(2):107–111. https://doi.org/10.1006/abio.1997.2538
21. Blumenkrantz N, Asboe-Hansen G (1973) New method for quantitative determination of uronic acids. Anal Biochem 54(2):484–489
22. Mojica K, Elsey D, Cooney MJ (2007) Quantitative analysis of biofilm EPS uronic acid content. J Microbiol Methods 71:61–65. https://doi.org/10.1016/j.mimet.2007.07.010
23. Dubois M, Gilles KA, Hamilton JK, Rebers PT, Smith F (1956) Colorimetric method for determination of sugars and related substances. Anal Chem 28(3):350–356
24. Hamidi M, Mirzaei R, Delattre C, Khanaki K, Pierre G, Gardarin C, Petit E, Karimitabar F, Faezi S (2019) Characterization of a new exopolysaccharide produced by *Halorubrum* sp. TBZ112 and evaluation of its anti-proliferative effect on gastric cancer cells. 3 Biotech 9 (1)
25. Tuinier R, Zoon P, Olieman C, Cohen-Stuart MA, Fleer GJ, de Kruif CG (1999) Isolation and physical characterization of an exocellular polysaccharide. Biopolymers 49:1–9
26. Liu Zhengqi, Zhang Zhihong, Qiu Liang, Zhang Fen, Xiongpeng Xu, Wei Hua, Tao Xueying (2017) Characterization and bioactivities of the exopolysaccharide from a probiotic strain of *Lactobacillus plantarum* WLPL04. J Dairy Sci 100:6895–6905
27. Hu X, Pang X, Wang PG, Chen M (2018).Isolation and characterization of an antioxidant exopolysaccharide produced by Bacillus sp. S-1 from Sichuan Pickles. J Carbohydr Polym
28. Zheng J-Q, Wang J-Z, Shi C-W, Mao D-B, He P-X, Xu C-P (2014) Characterization and antioxidant activity for exopolysaccharide from submerged culture of Boletus aereus. Process Biochem 49(6):1047–1053. https://doi.org/10.1016/j.procbio.2014.03.009
29. Salazar N, Ruas-Madiedo P, Prieto A, Calle LP, de Los Reyes-Gavilán CG (2012) Characterization of exopolysaccharides produced by *Bifidobacterium longum* NB667 and its cholate-resistant derivative strain IPLA B667dCo. J Agric Food Chem 60(4):1028–1035

30. Wang T, Lucey JA (2003) Use of multi-angle laser light scattering and size-exclusion chromatography to characterize the molecular weight and types of aggregates present in commercial whey protein products. J Dairy Sci 86:3090–3101
31. Reuben S, Banas K, Banas A, Swarup S (2014) Combination of synchrotron radiation-based Fourier transforms infrared microspectroscopy and confocal laser scanning microscopy to understand spatial heterogeneity in aquatic multispecies biofilms. Water Res 64:123–133. https://doi.org/10.1016/j.watres.2014.06.039
32. Macedo MG, Laporte MF, Lacroix Ch (2002) Quantification of exopolysaccharide, lactic acid, and lactose concentrations in culture broth by near-infrared spectroscopy. J Agric Food Chem 50:1774–1779
33. Ojeda JJ, Romero-Gonzalez ME, Pouran HM, Banwart SA (2008) In situmonitoring of the biofilm formation of Pseudomonas putida on hematite using flow-cell ATR-FTIR spectroscopy to investigate the formation of inner-sphere bonds between the bacteria and the mineral. Mineral Mag 72(1):101–106
34. Bhargava R (2012) Infrared spectroscopic imaging: the next generation. Appl Spectrosc 66 (10):1091–1120. https://doi.org/10.1366/12-06801
35. Kavita K, Mishra A, Jha B (2013) Extracellular polymeric substances from two biofilm forming Vibrio species: characterization and applications. Carbohydr Polym 94(2):882–888. https://doi.org/10.1016/j.carbpol.2013.02.010
36. Paquet-Mercier F, Safdar M, Parvinzadeh M, Greener J (2014) Emerging spectral microscopy techniques and applications to biofilm detection. In: Méndez-Vilas A (ed) Microscopy: advances in scientific research and education. Formatex Research Center, Badajoz, Spain, pp 638–649
37. Wolf G, Crespo JG, Reis MAM (2002) Optical and spectroscopic methods for biofilm examination and monitoring. Rev Environ Sci Biotechnol 1(3):227–251. https://doi.org/10.1023/a:1021238630092
38. Neu TR, Manz B, Volke F, Dynes JJ, Hitchcock AP, Lawrence JR (2010) Advanced imaging techniques for assessment of structure, composition and function in biofilm systems. FEMS Microb Ecol 72(1):1–21. https://doi.org/10.1111/j.1574-6941.2010.00837.x
39. Jiao Y, Cody GD, Harding AK, Wilmes P, Schrenk M, Wheeler KE, Banfield JF, Thelen MP (2010) Characterization of extracellular polymeric substances from acidophilic microbial biofilms. Appl Environ Microbiol 76(9):2916–2922. https://doi.org/10.1128/aem.02289-09
40. McCrate OA, Zhou X, Reichhardt C, Cegelski L (2013) Sum of the parts: composition and architecture of the bacterial extracellular matrix. J Mol Biol 425(22):4286–4294. https://doi.org/10.1016/j.jmb.2013.06.022
41. Weber K, Delben J, Bromage TG, Duarte S (2014) Comparison of SEM and VPSEM imaging techniques with respect to Streptococcus mutans biofilm topography. FEMS Microbiol Let 350(2):175–179. https://doi.org/10.1111/1574-6968.12334
42. Hannig C, Follo M, Hellwig E, Al-Ahmad A (2010) Visualization of adherent micro-organisms using different techniques. J Med Microbiol 59(1):1–7. https://doi.org/10.1099/jmm.0.015420-0
43. Ratnayake K, Joyce DC, Webb RI (2012) A convenient sample preparation protocol for scanning electron microscope examination of xylem-occluding bacterial biofilm on cut flowers and foliage. Sci Hortic-Amsterdam 140:12–18
44. Simões M, Pereira MO, Sillankorva S, Azeredo J, Vieira MJ (2007) The effect of hydrodynamic conditions on the phenotype of Pseudomonas fluorescens biofilms. Biofouling 23(4):249–258
45. Sandt C, Smith-Palmer T, Pink J, Brennan L, Pink D (2007) Confocal Raman microspectroscopy as a tool for studying the chemical heterogeneities of biofilms insitu. J Appl Microbiol 103(5):1808–1820
46. Priester JH, Horst AM, Van De Werfhorst LC, Saleta JL, Mertes LAK, Holden PA (2007) Enhanced visualization of microbial biofilms by staining and environmental scanning electron microscopy. J Microbiol Methods 68(3):577–587

47. Lawrence JR, Neu TR, Swerhone GDW (1998) Application of multiple parameter imaging for the quantification of algal, bacterial and exopolymer components of microbial biofilms. J Microbiol Methods 32(3):253–261. https://doi.org/10.1016/s0167-7012(98)00027-x
48. Beyenal H, Donovan C, Lewandowski Z, Harkin G (2004) Three-dimensional biofilm structure quantification. J Microbiol Methods 59:395–413. https://doi.org/10.1016/j.mimet.2004.08.003
49. Neu TR, Lawrence JR (2015) Innovative techniques, sensors, and approaches for imaging biofilms at different scales. Trends Microbiol 23(4):233–242. https://doi.org/10.1016/j.tim.2014.12.010
50. Schlafer S, Meyer RL (2016) Confocal microscopy imaging of the biofilm matrix. J Microbiol Meth (in press). https://doi.org/10.1016/j.mimet.2016.03.002
51. Adav SS, Lin JCT, Yang Z, Whiteley CG, Lee DJ, Peng XF, Zhang ZP (2010) Stereological assessment of extracellular polymeric substances, exoenzymes, and specific bacterial strains in bioaggregates using fluorescence experiments. Biotechnol Adv 28:255–280. https://doi.org/10.1016/j.biotechadv.2009.08.006
52. Neu TR, Woelfl S, Lawrence JR (2004) Three-dimensional differentiation of photo-autotrophic biofilm constituents by multi-channel laser scanning microscopy (single-photon and two-photon excitation). J Microbiol Methods 56(2):161–172. https://doi.org/10.1016/j.mimet.2003.10.012
53. Barranguet C, Beusekom SAMV, Veuger B, Neu TR, Manders EMM, Sinke JJ, Admiraal W (2004) Studying undisturbed autotrophic biofilms: still a technical challenge. Aquat Microb Ecol 34(1):1–9. https://doi.org/10.3354/ame034001
54. Wagner M, Ivleva NP, Haisch C, Niessner R, Horn H (2009) Combined use of confocal laser scanning microscopy (CLSM) and Raman microscopy (RM): investigations on EPS—matrix. Water Res 43(1):63–76. https://doi.org/10.1016/j.watres.2008.10.034
55. Battin TJ, Kaplan LA, Newbold JD, Cheng X, Hansen C (2003) Effects of current velocity on the nascent architecture of stream microbial biofilms. Appl Environ Microbiol 69(9):5443–5452. https://doi.org/10.1128/aem.69.9.5443-5452.2003
56. Savidge T, Pothoulakis C (2004) Microbial imaging. In: Bergan T, Norris J (eds) Methods in microbiology. Academic Press, New York, USA, pp 89–137
57. Choi O, Yu CP, Fernández GE, Hu Z (2010) Interactions of nanosilver with Escherichia coli cells in planktonic and biofilm cultures. Water Res 44(20):6095–6103. https://doi.org/10.1016/j.watres.2010.06.069
58. Yu GH, Tang Z, Xu YC, Shen QR (2011) Multiple fluorescence labeling and two dimensional FTIR–^{13}C NMR heterospectral correlation spectroscopy to characterize extracellular polymeric substances in biofilms produced during composting. Environ Sci Technol 45(21):9224–9231. https://doi.org/10.1021/es201483f
59. Baird FJ, Wadsworth MP, Hill JE (2012) Evaluation and optimization of multiple fluorophore analysis of a *Pseudomonas aeruginosa* biofilm. J Microbiol Methods 90:192–196
60. Chen MY, Lee DJ, Yang Z, Peng XF (2006) Fluorescent staining for study of extracellular polymeric substances in membrane biofouling layers. Environ Sci Technol 40(21):6642–6646. https://doi.org/10.1021/es0612955
61. Krishna R, Muddada S (2017) Characterization of exopolysaccharide produced by *Streptococcus thermophilus* CC30. BioMed Res Int (5):1–11. https://doi.org/10.1155/2017/4201809
62. Wang J, Zhao X, Tian Z, Yang Y, Yang Z (2015) Characterization of an exopolysaccharide produced by Lactobacillus plantarum YW11 isolated from Tibet Kefir. Carbohydr Polym 125:16–25
63. Yadav V, Prappulla SG, Jha A, Poonia A (2011) A novel exopolysaccharide from probiotic Lactobacillus fermentum cfr 2195: production, purification and characterization. Biotechnol Bioinf Bioeng 1:415–421, 1
64. Jalili N, Laxminarayana K (2004) A review of atomic force microscopy imaging systems: application to molecular metrology and biological sciences. Mechatronics 30:907–945. https://doi.org/10.1016/j.mechatronics.2004.04.005

65. van der Aa BC, Dufrêne YF (2002) In situ characterization of bacterial extracellular polymeric substances by AFM. Colloids Surf B 23(2):173–182. https://doi.org/10.1016/s0927-7765(01)00229-6
66. Ahimou F, Semmens MJ, Novak PJ, Haugstad G (2007) Biofilm cohesiveness measurement using a novel atomic force microscopy methodology. Appl Environ Microbiol 73(9):2897–2904. https://doi.org/10.1128/aem.02388-06
67. Halan B, Buehler K, Schmid A (2012) Biofilms as living catalysts in continuous chemical syntheses. Trends Biotechnol 30(9):453–465. https://doi.org/10.1016/j.tibtech.2012.05.003
68. Banerjee A, Das D, Rudra SG, Mazumder K, Andler R, Bandopadhyay R (2019) Characterization of exopolysaccharide produced by *Pseudomonas* sp. PFAB4 for synthesis of EPS-coated AgNPs with antimicrobial properties. J Polym Environ. https://doi.org/10.1007/s10924-019-01602-z
69. Janissen R, Murillo DM, Niza B, Sahoo PK, Nobrega MM, Cesar CL, Temperini MLA, Carvalho HF, de Souza AA, Cotta MA (2015) Spatiotemporal distribution of different extracellular polymeric substances and filamentation mediate Xylella fastidiosa adhesion and biofilm formation. Sci Rep 5:1–10. https://doi.org/10.1038/srep09856
70. Neugebauer J, Reiher M, Kind C, Hess BA (2002) Quantum chemical calculation of vibrational spectra of large molecules—Raman and IR spectra for buckminsterfullerene. J Comput Chem 23(9):895–910. https://doi.org/10.1002/jcc.10089
71. Ivleva NP, Wagner M, Horn H, Niessner R, Haisch C (2008) Towards a nondestructive chemical characterization of biofilm matrix by Raman microscopy. Anal Bioanal Chem 393 (1):197–206. https://doi.org/10.1007/s00216-008-2470-5
72. Virdis B, Harnisch F, Batstone DJ, Rabaey K, Donose BC (2012) Non-invasive characterization of electrochemically active microbial biofilms using confocal Raman microscopy. Energ Environ Sci 5(5):7017–7024. https://doi.org/10.1039/c2ee03374g
73. Ciucanu I, Kerek F (1984) A simple and rapid method for the permethylation of carbohydrates. Carbohydr Res 131:209–217
74. Maheswari P, Arjun Kumar K, Sankaralingam S, Sivakumar N (2020) Optimization and characterization of exopolysaccharide from marine soil bacteria. J Pharm Technol 13(6)
75. Honda S, Okeda J, Iwanaga H, Kawakami S, Taga A, Suzuki S, Imai K (2000) Ultramicroanalysis of reducing carbohydrates by capillary electrophoresis with laser-induced fluorescence detection as 7-nitro-2,1,3-benzoxadiazole-tagged N-methylglycamine derivatives. Anal Biochem 286:99–111
76. Sudhamani SR, Tharanathan RN, Prasad MS (2004) Isolation and characterization of an extracellular polysaccharide from Pseudomonas caryophylli CFR 1705. Carbohydr Polym 56:423–427
77. Rodrıguez-Carmona E, Villaverde A (2010) Nanostructured bacterial materials for innovative medicines. Trends Microbiol 18:423–430
78. Pratt CW, Cornely K (2014) Essential biochemistry (3rd edn). John Wiley and Sons Inc
79. Dhillon GS, Kaur S, Brar SK, Verma M (2013) Green synthesis approach: extraction from fungus mycelia. Crit Rev Biotechnol 33(4):379–403
80. Casillo A, Lanzetta R, Parrilli M, Corsaro MM (2018) Exopolysaccharides from marine extremophilic bacteria: structure, properties, ecological roles and applications. Marine Drugs 16(2):69
81. Reichhardt C, Cegelski L (2013) Solid-state NMR for bacterial biofilms. Mol Phys 112 (7):887–894. https://doi.org/10.1080/00268976.2013.837983
82. https://pubchem.ncbi.nlm.nih.gov/compound/1_3_6-Naphthalenetrisulfonic-acid_-8-amino
83. Pan M, Zhu L, Chen L, Qiu Y, Wang J (2016) Detection techniques for extracellular polymeric substances in biofilms: a review. BioResources 11(3):8092–8115
84. Wang L, Zhang ZJ (2009) Dalian Nat Univ 5:54
85. Shukla A, Mehta K, Parmar J, Pandya J, Saraf M (2019) Depicting the exemplary knowledge of microbial exopolysaccharides in a nutshell. Eur Polym J 119:298–310

86. Khanal SN, Lucey JA (2017) Evaluation of the yield, molar mass of exopolysaccharides, and rheological properties of gels formed during fermentation of milk by *Streptococcus thermophilus* strains St-143 and ST-10255y. J Dairy Sci 100:6906–6917
87. Enikeev R (2012) Development of a new method for determination of exopolysaccharide quantity in fermented milk products and its application in technology of kefir production. Food Chem 134:2437–2441
88. Ale EC, Perezlindo MJ, Burns P, Tabacman E (2016) Exopolysaccharide from *Lactobacillus fermentum* Lf2 and its functional characterization as a yogurt additive. J Dairy Res 83(4):487–492

Molecular Basis and Genetic Regulation of EPS

Siya Kamat

Genes are like the story and DNA is the language that the story is written in.
Sam Kean.

Abstract The diverse world of exopolysaccharides (EPS) is contributed by bacteria, archaea, fungi, micro, and macroalgae. EPS are biosynthesized via Wzx/Wzy-dependent pathway, ABC transporter-dependent pathway, or synthase-dependent pathway, that connect to sugar and amino acid metabolism pathways. Relatively conserved gene clusters and regulons govern the EPS biosynthesis. Branching and functional group decorations introduce structural and functional diversities. Exceptional EPS synthesized by the extremophiles indicate the modifications in these operons. Since EPS production is a strategy for protection in hostile environments, several transcriptional and translational regulatory networks control its synthesis. Even though whole-genome analysis of some EPS-producing bacteria has annotated the essential genes and regulators of EPS biosynthesis, there is a lot to uncover concerning cyanobacteria and eukaryotic producers. These steps will work toward the manufacture of tailored EPS variants with the potential to be used as valuable renewable and high-performance products for industrial applications.

Keywords Exopolysaccharide · Gene clusters · Operons · Exopolysaccharide biosynthesis · Genetic regulation · Glycosyltransferase · Biofilm

S. Kamat (✉)
Department of Biochemistry, Indian Institute of Science, Bengaluru, India
e-mail: siyakamat@iisc.ac.in

A. K. Nadda et al. (eds.), *Microbial Exopolysaccharides as Novel and Significant Biomaterials*, Springer Series on Polymer and Composite Materials,
https://doi.org/10.1007/978-3-030-75289-7_3

1 Introduction

Polysaccharides can be categorized into three groups based on their biological roles, such as storage polysaccharides located inside the cell, capsular polysaccharides linked to the cell surface, and extracellular polysaccharides (EPS). Several microorganisms secrete structurally and chemically diverse EPS, which contribute to surface attachment, protection against abiotic or biotic stress factors, and nutrient assembling [8]. Xanthan, sphingan, alginate, and cellulose are standard EPS that define the bacterial biofilm's complex biophysical world [102]. EPS is also produced by cyanobacteria, diatoms, seaweeds, fungi, and extremophilic bacteria. Due to their diverse, sustainable applications, there has been a marked upsurge of interest in these biopolymers from several industrial sectors, including pharmaceutical, surfactant, food, textiles, cosmetology, wastewater, agriculture, etc. [95]. They possess unique properties as viscosifiers, stabilizers, emulsifiers, flocculating agents, and gelling agents [115].

Prokaryotic microbial groups produce the most extraordinary diversity of EPS. The biotope of microorganism governs the physiological and ecological significance of the respective EPS [29]. It is a direct response to selective environmental burdens such as salinity, light, temperature, and pressure. This adaptation feature contributes to biofilm formation and helps in protection, colonization, host-microbe interactions, and virulence [65, 117]. There has been a lot of advancement in deciphering the molecular basis and genetic regulation of EPS production. EPS genes are arranged into clusters to facilitate the stages responsible for their biosynthesis, including regulation, nucleotide sugar precursors formation, glucosyl transfer, polymerization of sugar monomers, and finally, secretion. General strategies of mutation or overexpression have been utilized to study biopolymers' modifications, including rheology adjustments, alteration of pathway, acetylation adjustments, molecular weight adjustments, and composition ratio modification [86].

The study of EPS biosynthetic pathways and genes is too disparate and usually refers to a particular product or a microorganism. Typically, microorganisms produce a mixture of polymers regulated by several gene clusters. There has been a lot of literature on specific EPS like xanthan, alginates, succinoglycan, lactic acid bacteria EPS. Such microbial strains have a well-documented genome sequence and deciphered biosynthetic pathways [29].

The physical and rheological properties of EPS are influenced by its molecular mass, stiffness, presence of sidechains, and non-polysaccharide substituents such as the organic groups (succinyl, pyruvyl, or acetyl) or inorganic groups (phosphate, sulfate) [3].

2 EPS Biosynthesis Pathways

EPS biosynthesis occurs in different growth stages, depending on the organism and the environmental conditions [85]. The organism utilizes sugars as its carbon or energy source and amino acids and ammonium salts as its nitrogen source. A high EPS production corresponds to low nitrogen concentration. The most favorable C/N ratio for higher EPS yield in *Tricholoma Mongolicum* M-1 was 10:1 [127]. In a study by Zhang et al. [136] on the EPS-producing *Alteromonas* sp JL2910, the elemental analysis of its EPS accounted for 33.84% carbon while only 0.88% nitrogen of the total EPS mass.

The canonical EPS biosynthesis mechanisms in bacteria include the

(a) the Wzx/Wzy-dependent pathway;
(b) the ATP-binding cassette (ABC) transporter-dependent pathway
(c) the synthase-dependent pathway
(d) the extracellular synthesis pathway by use of a single sucrase protein.

Several enzymatic reactions inside the cell generate precursor molecules required for the stepwise elongation of the EPS polymer strands. The first three pathways follow the common strategy of making activated sugars/sugar acids. However, the fourth pathway involved in the extracellular synthesis, the polymer chain gets extended by direct addition of monosaccharides produced by cleavage of di- or trisaccharides (Fig. 1) [102]. However, these pathways are regulated at standard points on a broad scale, including sugar transport into the cytoplasm, synthesis of activated sugars, synthesis of sugar nucleotides, and polymerization into EPS repeat units. Finally, the first three pathways end with EPS polymerization and export into the surrounding medium [64], thus connecting the EPS biosynthesis pathways with central carbon metabolism. To date, limited information is existing for EPS biosynthesis in fungi, and minimum is known for archaeal and cyanobacterial polysaccharide biosynthesis pathways. However, some evidence states that the microbial EPS routes are preserved across other EPS-producing species [93, 101]. Table 1 summarizes the different EPSs, genes and clusters involved in the biosynthesis, and corresponding pathways.

2.1 Wzx/Wzy-Dependent Pathway

This pathway is functioned by several glycosyltransferases (GTs) and Wzx protein, a flippase. GTs assemble the individual repeating sugars linked to an undecaprenol diphosphate anchor (C55) at the inner membrane of the cell. They are further translocated across the inner membrane by the Wzx protein, a polymerase. In the periplasmic space, the sugars are polymerized by Wzy proteins and then exported to the cell surface. The polymer transport across the outer membrane is assisted by additional protein(s) of the OPX outer membrane polysaccharide export (formerly

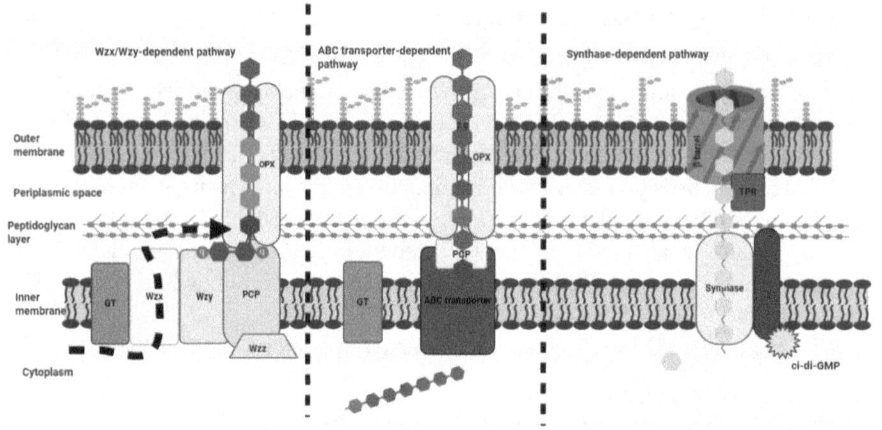

Fig. 1 General representation of the three EPS biosynthesis pathways. Created using BioRender.com

OMA) and polysaccharide copolymerase (PCP) families [41]. This pathway produces heteropolymers of highly diverse sugar patterns, as seen in xanthan and succinoglycan. The genes for Wzx and Wzy are commonly observed in strains using this pathway within their EPS operons [8]. A recent study on *Myxococcus xanthus*, a Gram-negative deltaproteobacterium, observed that homologs of Wzx flippase and Wzy polymerase were found and renamed as ExoM and ExoJ, respectively. The study also reported three GTs necessary to produce spore coat polysaccharide, synthesized b y Exo-1 proteins [79]. A putative EPS biosynthesis mechanism in cyanobacteria was reported after studying the *Escherichia coli* Wzx dependent model, together with information gained from *Anabaena* sp. PCC 7120 [132].

2.2 *ATP-Binding Cassette (ABC) Transporter-Dependent Pathway*

This pathway contributes to capsular polysaccharides (CPS) biosynthesis, in which case the polymer produced is attached to the cell surface. As seen in the previous pathway, the GTs assemble the repeating sugars at the cytoplasmic side of the inner membrane. Homopolymers are produced when only a single GT-containing operon is involved, and heteropolymers when multiple GTs are involved in the assembly step. The export is different since it involves the tripartite efflux pump-like complex built of ABC transporters spanning the inner membrane and the OPX and PCP and family's periplasmatic proteins. The polymers produced by this pathway bear a conserved glycolipid at the reducing end comprised of phosphatidylglycerol and a

Table 1 Examples of exopolysaccharides concerning the type, biosynthesis pathway, and genes involved

EPS	Type	Biosynthesis pathway	Genes	References
Alginate	Heteropolysaccharide	Synthase-dependent	*alg* cluster	[63, 89]
Cellulose	Homopolysaccharide	Synthase-dependent	*bcs* cluster	[124]
Colanic acid	Heteropolysaccharide	Wzx/ Wzy-dependent Curdlan	CA gene cluster (19 genes) *manC, manD, WcaA, WcaC, WcaE, WcaI, WcaL*	[67, 108]
Curdlan	Homopolysaccharide	Synthase-dependent	*crdA, crdS, scrdR, crdB*	[81, 135]
Gellan	Heteropolysaccharide	Wzx/ Wzy-dependent Curdlan	*sps* cluster	[96]
Hyaluronan	Heteropolysaccharide	Synthase-dependent	*has* cluster	[40]
Levan	Homopolysaccharide	Levansucrase	*levansucrase M1FT*	[31, 105]
Succinoglycan	Heteropolysaccharide	Wzx/ Wzy-dependent Curdlan	*exo* cluster	[101, 125]
Welan	Heteropolysaccharide	Wzx/ Wzy-dependent Curdlan	*wel, atr, rml, urf*	[60]
Xanthan	Heteropolysaccharide	Wzx/ Wzy-dependent Curdlan	*gum* cluster	[50, 101]
Cordyceps polysaccharide	Homo and Heteropolysaccharide	Unknown	Unknown	[112, 130]

poly-2-keto-3-deoxyoctulosonic acid (Kdo) linker. This feature is one of the key differences between the Wzx/Wzy- and ABC-dependent pathways [102, 123].

2.3 Synthase-Dependent Pathway

This EPS-producing route is independent of the flippase protein involved in translocation, as seen in the previous two pathways. In this pathway, the polymerization and translocation tasks are achieved by a single synthase protein, which is a subunit of a transmembrane multiprotein complex. This pathway is generally observed in the biosynthesis of homopolymers, which require only a single type of sugar precursor. Several modifications are encountered in this pathway to produce a variation of the polymer [62]. For instance, in curdlan biosynthesis, only β-(1-3)-linked glucose unite is found in the polymer. This feature is also observed in bacterial cellulose, a homopolymer comprising only of β-(1-4)-linked glucose residues. However, alginates are produced by polymannuronic acid, a preliminary polymer, which further gets modified by several epimerases and modifying enzymes. These modifications are necessary to produce glucuronic/mannuronic (G/M) acid block-polymers, that can vary in the sequence and ratio of G/M building blocks. A distinct variation is observed in the production of hyaluronic acid (HA) since a sole enzyme, hyaluronan synthase, accomplishes dual tasks of polymerization and secretion. The assembly of the polymeric disaccharide to form HA is carried out by two precursors, GlcNAc and glucuronic acid [22, 102].

2.4 Extracellular Synthesis Pathway by Use of a Single Sucrase Protein

In this pathway, the precursors of EPS are synthesized inside the cell, followed by polymerization and secretion in the periplasmic space, with certain exceptions like devan or levan. These are extracellular synthesized polysaccharides, whose biosynthesis takes place via the GTs, that are secreted and covalently bonded to the cellular surface [106].

2.5 Unidentified Pathways

Cordyceps is an EPS-producing parasitic edible fungus. Earlier studies have described it as a medical mushroom and a caterpillar fungus. However, according to the latest taxonomical classification, these species belong to Ascomycota, Hypocreales, and Cordycipitaceae [112]. Cordyceps polysaccharide (CP), composed of one or more glycosyl residue compositions of mannose, galactose, glucose, arabinose, rhamnose, xylose, and sorbose, hold a large proportion in the organisms' fruiting bodies, in the range of 3–8% of the total weight. Since the last two decades, CP has received increasing attention due to the several medicinal properties such as anti-tumor, immunomodulatory, antioxidation, anti-inflammatory, anti-aging, and anti-fatigue. CP obtained from Cordyceps such as *O. sinensis* and *C. militaris* possess abundant medicinal activities. However, due to the strict growing habitat and over-collection, the use of high yielding wild *Cordyceps* is rapidly reducing [130]. *Cordyceps* can synthesize both intracellular polysaccharides (IPS) and EPS. There is a gap between the CP and its biosynthesis due to its trace levels and the unclear relation between structure and function [129], CP synthesized by different species have different compositions (hetero or homo polysaccharide). Even though fungal EPS transmembrane transport data is limited, it is anticipated that CP could follow a Wzy-dependent or ABC-dependent pathway. The available information also reflects that the search for GT genes and EPS clusters in the complex and large Cordyceps genome data is tough [130].

3 Genes Directing EPS Biosynthesis

In microorganisms, functionally connected genes involved in metabolic pathways and large protein complexes are often encoded in operons (Fig. 2). These are commonly conserved across phylogenetically diverse species. Diverse phenomena such as horizontal transfer, gene duplication, rearrangement, and loss frequently increase the evolution frequency of novel biological pathways. Sometimes, it also ensures pre-existing pathways to adjust to the necessities of particular environments [16]. Bundalovic-Torma et al. [16] developed an innovative approach to study the evolution of operons grounded on phylogenetic clustering of operon-encoded protein families and genomic-proximity network visualizations of operon constructions. The study demonstrated that cellulose operons follow phyla-specific operon lineages due to gene rearrangement, loss, acquisition of accessory loci, and the existence of whole-operon duplications brought through horizontal gene transfer.

In this section, we will discuss gene clusters/genes responsible for the production of its respective polymers.

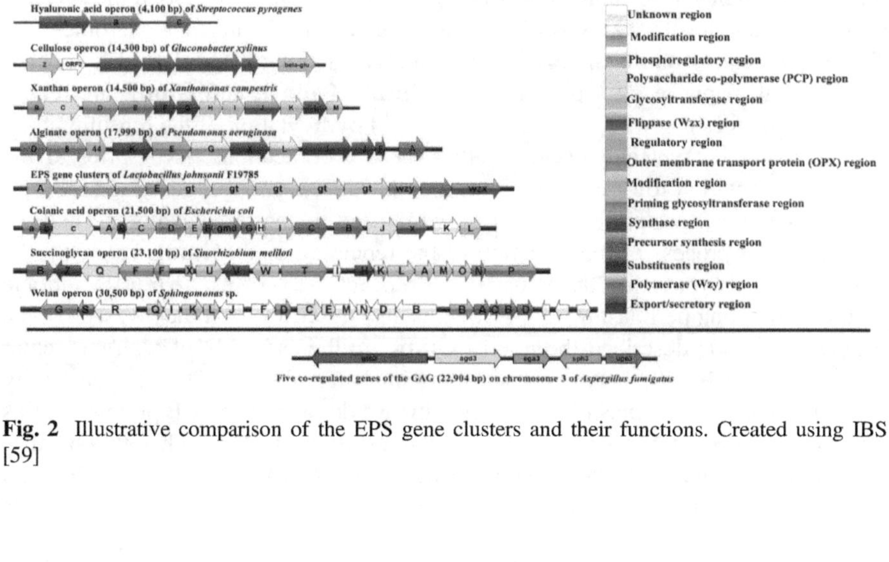

Fig. 2 Illustrative comparison of the EPS gene clusters and their functions. Created using IBS [59]

3.1 Xanthan—The Most Diverse Heteropolysaccharide Bearing a Long Side Chain

Commonly produced by *X. campestris*, xanthan gum is composed of cellulose-like backbone, [β-(1-4)-linked glucose], a side chain composed of one glucuronic acid and two mannose residues. Its two precursors, glucose- and fructose 6-phosphate, are the important intermediates of the central metabolism cycle. Vorholter et al. [116] compared the genomes of three strains of *X. campestris* pv. *campestris* (Xcc) and identified a common core of 3800 genes. These included ~500 unique genes, a highly conserved xanthan operon, and *xanAB* genes. In due course of time, draft genomes of several strains of *X. campestris* have been published, which will further add to the knowledge of the xanthan biosynthesis pathway. Schatschneider et al. [98] applied systems biology tools to *Xcc* B100 and constructed a large-scale metabolic network. This reconstructed network included 352 genes, 437 biochemical reactions, ten transport reactions, and 338 internal metabolites. This comprehensive model was utilized to predict the biomass generation, xanthan production, and essential genes.

Xanthan gum biosynthesis is controlled by the *gum* gene cluster comprising of thirteen genes. These are involved in the assembly of the repeat units, polymerization, translocation, and modification with substituent groups. The nucleotide precursors genes (*xanAB*) are positioned outside the *gum* cluster. Most of the GTs following the Wzx/Wzy–pathway seem to be monofunctional, and the same has been observed in the xanthan biosynthesis mechanism [14]. The pathway starts with the **GT GumD** priming the pentasaccharide assembly by transferring the first glucose unit to the phosphorylated lipid linker attached to the inner membrane. Further, the next glucose unit is attached by a β-(1-4)-bond to the first glucose unit by the action of the cytosolic **GT GumM.** Next, the catalytic activity of **GT GumH** links the first mannose unit via a α-(1-3)-glycosidic link, which is trailed by the action of the cytosolic **GT GumK** that attaches a β-(1-2)-linked glucuronic acid. The EPS repeating unit is finalized by the catalytic action of **GumI**, which attaches the terminal mannose by a β-(1-4)-bond. **GumF, GumG,** and **GumL** are recognized as enzymes contributing in acetylation and pyruvylation of the repeat units of xanthan. It is not yet elucidated whether these modifications occur prior to the spatial reorientation toward the periplasm or inside the periplasmic space [88].

GumJ is the Wzx protein (flippase), composed of ten transmembrane helices, completes the process of translocation of the repeating units. **GumE** catalyzes the next task of polymerizing the translocated repeats. It contains 12 transmembrane helices and a long periplasmic loop, which is also described for other Wzy proteins. Bianco et al. [11] described the role of **GumC** in the assembly of high molecular weight EPS. It is an inner membrane protein that, unlike its PCP-2a sub-family proteins, lacks the cytoplasmic C-terminal kinase domain. The PCP proteins in this pathway are accountable for the control of the resulting EPS' chain length. **GumB's structure and** role were described by Jacobs et al. [42]. It is essential for the final secretion step and hence contains the characteristic OPX-C family polysaccharide export sequence (PES). **GumB and GumC** are comprised of a molecular scaffold protein that spans the cell envelope. Earlier studies have revealed the indispensable role of genes *gumB* and *gumC* in the biosynthesis of xanthan. When both the genes were co-overexpressed, the resulting xanthan polymer produced was of higher molecular weight and higher viscosity, indicating the interplay of both these proteins. Alkhateeb et al. [1] studied the transcriptomic characteristics of *Xcc*-B100. The study revealed a genome-wide identification of 3067 transcription start sites (TSSs). Out of these, 1545 were mapped upstream of an actively transcribed CDS while 1363 were categorized as new TSSs demonstrating internal, antisense, and TSSs, thus unleashing previously unknown genomic characteristics. The authors reported the possible presence of an intricate transcription design of the *gum* genes with several TSSs and a pertinent role of antisense transcription. The study also presented gene *gumB* as evidence for genes that controlled a strong antisense TSS.

3.2 Sphingan—A Family of Diverse Heteropolysaccharides

The sphingans family, commonly produced by microorganism *Sphingomonas* and *Pseudomonas* strains by the Wzx/Wzy-dependent pathway, bear closely similar chemical structures but differ in material properties. This family includes EPSs such as welan, gellan, rhamsan, diutan, S-7, and S-88 where the basic differences in the chemical structures are contributed by the variations in the configuration of gene operons [82]. These bear a fairly common backbone of Rha-Glc-GlcA-Glc with minor variations in Rha or Man and sidechain decorations. The precursor genes (*rmlABCD*) of Rha are positioned apart of the vastly conserved operon, while the genes corresponding to other sugar precursor units are distributed randomly [118]. The various sphingan biosynthesis genes (*spn*) include *wel* for welan, *gel* for gellan, *dsp* for diutan. A priming GT SpnB helps in the assembly of the repeat units in welan, gellan, and diutan by transferring glucose toward the C55-anchor. Next, SpnK catalyzes the linking of glucuronic acid to the priming glucose via a β-(1-4)-bond. SpnL, the third GT, transfers the next glucose unit to the newly synthesized repeat, followed by the transfer of the rhamnose residue by linking it through α-(1-3)-bond by SpnQ. The subsequent stages differ for welan, gellan, and diutan, depending on the branching and location of the glycerol and acetyl decorations [102]. A study by Coleman et al. [23] described the possible roles of *urf31.4, urf31,* and *urf34I* genes to incorporate the side chains for diutan and welan. The subsequent stages of sphingan biosynthesis follow the same order for all derivatives and including roles of the Wzx protein flippase (SpnS), the Wzy polymerase (SpnG), and PCPs, that are assumed to be SpnE and SpnC, involved in chain length control, having the distinctive kinase domains. Wu et al. [126] investigated the evolution of polysaccharide sanxan biosynthesis gene cluster in *Sphingomonas sanxanigenens* NX02. The authors also compared the distribution of genes among *Sphingomonadaceae* strains and other sphingan-producing strains. Their findings suggested that the microbial EPS biosynthesis gene clusters' evolution is an extensive and dispersive cyclic development comprising cluster 1 → scatter → cluster 2. It was also observed that SpnA, a positive regulator of sphingans biosynthesis, could not regulate the sanxan. All the evidence proved that the sanxan pathway's evolution was independent, maybe through long-term gene acquisitions and adaptive mutations.

3.3 Succinoglycan (SG)—A Heteropolysaccharide Composed of a Large Repeating Unit

SG, an acidic EPS, commonly synthesized by several strains of *Agrobacterium, Rhizobium, Pseudomonas,* and *Alcaligenes*. *Rhizobium meliloti* RM 1201, the model organism for SG production, revealed the branched heteropolysaccharide nature of SG, comprised of an oligosaccharide repeating unit bearing numerous decorations, including acetate, succinate, and pyruvate. SG biosynthesis is governed by 19 genes called *exo* genes and 2 *exs* genes [8]. Its precursors UDP-galactose and UDP-glucose are encoded by *exoC*, enzymes phosphoglucomutase and UDP-glucose-4-epimerase are encoded by *exoB*, and UDP-pyrophosphorylase is encoded by *exoN*, respectively. ExoY, the priming GT, unlike GumD, requires the product of gene *exoF* to transport galactosyl residues toward the inner membrane. The subsequent steps include the elongation of the octasaccharide repeating units by adding a glucose residue, carried out by *exoA, exoL, exoM, exoO, exoU,* and *exoW* which encode the respective GTs. The proteins involved in translocation and polymerization, including regulation of chain length of SG, are represented by ExoP (PCP-2a), ExoQ (Wzy), and ExoT (OPX) [72]. The acetyl, succinyl, pyruvyl decorations are added by the transmembrane proteins ExoZ, ExoH, and a ketalase ExoV. Jones et al. [45] observed that SG synthesizing bacteria contain genes for ExoK and ExsH, the extracellular endogycanases, to lower the molecular weight of the polymer.

Various studies have reported that while 21 genes of the SG synthesis pathway are grouped on a megaplasmid, *exoC* is the only gene located on the chromosome. The *exs* and *exo* genes are found adjacent to each other upstream of *exoB*. This gene in *R. melioti* was reported to negatively regulated SG biosynthesis. In *Rhizobium* sp the gene cluster for the EPS galactoglucan is localized ~200 kb away on the same megaplasmid. This is a classic example of genetic machinery to produce more than one EPS, a common phenomenon in microorganisms, especially *Lactobacilli* strains [9].

In a recent study [86] a large EPS biosynthesis gene cluster of a salt-tolerant bacterium *Rhizobium radiobacter* SZ4S7S14 was identified. For the first time, transposon insertions were recognized near or within the coding regions of genes *exoK* and *exoM*. Under different salt-stress levels, varying expression of *exoK* and *exoM* was detected, resulting in structural modification of EPS. The study proved that salinity stress not only causes downregulation of genes *exoK* and *exoM* or a large EPS biosynthesis gene cluster, but also impacts their relative expression levels, re-groups the monomers within the EPS matrix, and finally dictates the molar ratio of monomers in the final product.

3.4 Colanic Acid (CA)—An Antigenic EPS

CA or the M antigen is an EPS commonly found in *Enterobacteria*, composed of glucose, fucose, galactose, and glucuronic acid repeat unit with non-stoichiometric decorations of acetyl and pyruvate groups. A cluster of 19 genes governs the CA biosynthetic pathway. While the genes (*manB* and *manC*) governing the production of the fucose nucleoside sugar precursors are located in the cluster, those encoding other precursors are dispersed in the genome. The WcaJ protein is a GT, which initiates the repeat unit's assembly and transfers the glucose residue toward the C55 lipid linker. The subsequent transfer of sugars is brought about by WcaA, WcaC, WcaE, WcaI, and WcaL [78]. The acetylation is assumed to be catalyzed by WcaB and WcaF. The Wzx protein in the cluster is WcaD that is predicted to be responsible for the polymerization of the repeat units. Temel et al. [113] mapped the regulatory interactions between the two proteins, Wzb (protein tyrosine phosphatase) and Wzc (related tyrosine kinase).

3.5 Curdlan—A Water-Insoluble Homopolysaccharide

Commonly produced by *Agrobacterium*, this EPS is water-insoluble composed of β-(1-3)-glucan glucose homopolymer lacks any substituent groups synthesized via the synthase-dependent pathways. The synthesis is governed by four essential genes *crdA, crdB, crdS, scrdR*. The key enzyme of this pathway is curdlan synthase (CrdS), which shows high similarity to cellulose synthases. CrdS is cytoplasmic and belongs to the GT2 family of glycosyltransferases. Much of the genomic and transcriptomic information from the curdlan producing *Agrobacterium* strains indicates that the major challenge for improving curdlan EPS yield is substrate availability, rather than energy efficiency [44]. Periasamy et al. [81] used nanodiscs and X-ray scattering experiments to obtain structural data of CrdS using a cell-free protein synthesis.

3.6 Cellulose-Bacterial EPS of Biomedical Value

A significant component of various bacterial biofilms, cellulose (β-(1-4)-Glucan), has sprung major interest as a biomaterial for medical use. Wong et al. [124] discovered the cellulose synthesizing genes (*celS*) in *Acetobacter xylinum*. The bacterial cytoplasmic membrane cellulose synthase proteins (Bcs) belonging to the GT2 family, are built of BcsA, BcsB, and BcsC subunits. Certain species have been found where BcsA and BcsB are fused as a single polypeptide. The *BcsA* encodes the catalytic subunit while the regulatory subunit is encoded by the *BcsB*. The *BcsC* encodes a pore forming protein that enables cellulose secretion. *BcsD* encoded protein controls the crystallization stages of cellulose nanofibrils. Upstream of the cellulose synthase operon are two essential genes of the biosynthesis pathway: *bcsZ* that encodes an endo-β-(1-4)-glucanase and *orf2* encodes a protein for cellulose completion whose function is not yet fully decoded. Downstream from the operon is the *β-glucosidase* gene (*bglxA*), which hydrolyzes the glucose units comprising of more than three monomers. Even though this enzyme is not essential for cellulose production, its disruption was observed to reduce the cellulose production. In *A. xylinum*, BglxA regulates the levels of glucose and oligosaccharides, which are precursors of cellulose production [48, 56]. Kubiak et al. [53] identified the cellulose synthesis operon in *Gluconacetobacter xylinus* E25.

3.7 Alginates-Heteropolysaccharide Synthesized Via an Envelope-Embedded Multiprotein Complex

Alginates are a gel-forming EPS built of different ratios of comonomers, (1-4)-linked β-D-mannuronic acid, and its C5-epimer α-L-guluronic acid. The monomers are organized as a continuous chain of either mannuronic acid units (M-blocks), guluronic acid units (G-blocks), or as alternating units (MG-blocks). Alginates are typically produced by bacteria of the genus *Azotobacter* and *Pseudomonas* and brown seaweeds [87].

Based on G and M block arrangement, alginates adopt diverse material properties that corroborate their biological significance, such as the viscous biofilm in *Pseudomonas* sp or the rigid cyst wall in the dormancy of *A. vinelandii* [102].

While the biosynthetic pathway beginning with the GDP-mannuronic acid, the activated precursor, is well understood, its underlying molecular mechanisms of polymerization and modification are less illuminated. Various in silico studies including stability and protein-protein interaction studies have demonstrated the link between polymerization and secretion. This connection is through an envelope spanning multiprotein complex of around six subunits, Alg8 and Alg44 (cytoplasmic membrane), AlgG, AlgL, AlgK, AlgX (periplasm), and AlgE (outer membrane). The membrane spanning pore protein responsible for secretion, observed in the outer membrane of *Psedomonas aeruginosa* was contributed by AlgE [38, 89].

The GDP-mannuronic acid synthesized from fructose-6-phosphate of the gluconeogenesis pathway via alginate-specific enzymes is the precursor for alginate synthesis. The membrane-anchored GT Alg8, helps in polymerizing GDP-mannuronic acid to alginate. The periplasmic proteins of the multiprotein complex have been projected to lead the nascent alginate chain through the periplasmic space and further for secretion by the specific channel protein AlgE. The periplasmic scaffold is linked to Alg8 via Alg44, which has been identified as a copolymerase. For initiation of alginate polymerization, it binds with c-di-GMP [102]. In studies by Hay et al. [38, 39], the authors defined the role for MucR, a membrane-anchored sensor protein, required specifically to initiate alginate synthesis by producing a local pool of c-di-GMP around Alg44. The study also mentioned that the synthesis and regulation of alginate are intricate and complex, involving several cascades of interlinked regulatory processes. The study reviewed the involvement of several transcriptional regulatory proteins, FimS/AlgR and KinB/AlgB two-component signal transduction systems, several σ/anti-σ factors, the Gac/Rsm sRNA system for posttranscriptional regulation, the antisense transcript (MucD-AS) that helps alginate synthesis by obstructing the translation of mucD mRNA along with c-di-GMP mediated activation.

The overexpression of alginate synthesis genes could increase the alginate yield in bacteria. By modulating the expression of various modifying enzymes including lyases, epimerases, acetyltransferases, alginates of desired M/G organization, molecular weight, and acetylation level can be obtained [89].

3.8 Hyaluronan—A Hydrophilic Heteropolysaccharide

Hyaluronic acid or hyaluronan is a hydrophilic linear EPS produced via the synthase-dependent pathway. It consists of alternating residues of β-D-*N*-acetylglucosamine and β-D-glucuronic acid linked via β-(1-3) and β-(1-4)-glycosidic bonds. Vertebrates and prokaryotes can synthesize hyaluronan. It finds several applications in the cosmetic industry and medical biomaterial industry owing to its water retention capacity, biocompatibility and viscosity [22].

The hyaluronan operon comprises of three genes *hasA, hasB,* and *hasC*. The HasA protein, an inner membrane linked hyaluronan synthase governing the assembly of the polymer, is a vital player in the biosynthesis pathway. The gene *hasC* encodes for UDP-glucose pyrophosphorylase, which catalyzes the production of UDP-glucose from glucose-1-phosphate and UTP. The gene *hasB* encodes for UDP-glucose dehydrogenase, that catalyzes the oxidation near UDP-glucuronic acid [102]. An in vitro study by Hubbard et al. [40] demonstrated through proteoliposomes the spatial alignment of synthesis and translocation processes. The detailed mechanism is still being solved.

3.9 *Amylovoran-Acidic Exopolysaccharide of* Erwinia amylovora

Erwinia amylovora, a Gram-negative plant pathogen, is known to produce three EPS types: levan, amylovoran, and a low molecular weight glucan. This solid coating of EPS on the bacterial cell surface is known to disguise elicitors of the plant defense reactions and contribute to its pathogenicity. Amylovoran-deficient mutants of *E. amylovora* are reported to be not pathogenic [51]. In an early study [15], a 16 kb transcript of the *ams* operon regulating the amylovoran biosynthesis was detected. RcsA, a transcriptional regulator, enhanced its transcription. The operon consists of several genes including *amsA* to *amsI*. AmsG is a GT, AmsI a phosphatase, AmsA membrane linked-ATP-binding protein. The existence of a secretory signal sequence in the N-terminus of AmsF indicated its involvement in the periplasmic processing the repeats. Kharadi et al. [51] reported the regulatory role of phosphodiesterases in *E. amylovora* using deletion studies. The study reported redundancy among the phosphodiesterase genes, necessary for intracellular degradation of c-di-GMP.

3.10 Dextran and Other Glucan Homopolymers

Dextran is an extracellularly synthesized sucrase activity-based EPS, composed primarily of α-(1-6) linked glucose units. Dextransucrase (aka glucansucrases), is a family of enzymes that biosynthesize dextrans of different structures from sucrose by initiating extracellular polymerization [91]. Dextran derivatives comprising a minor percentage of α-(1-3) or α-(1-2) linkages, mutan containing frequent α-(1-3) linkages, alternan built of strictly alternating α-(1-6) and α-(1-3) linkages, and reuteran containing primarily α-(1-4) linkages that are scattered with α-(1-6) linkages have also been found. Diversity in dextrans is contributed by different degrees of branching at several hydroxyl groups [18].

The sucrases are initially secreted and eventually attached to the cell wall. These enzymes have multiple domains comprising the catalytic core composed of 3 domains, two domains attached to the core and the C-domain, consisting of a single contiguous polypeptide contributing in glucan binding or glucan production or both. Several mechanisms have been introduced to understand the stages of chain initiation and elongation [102]. A study by Robyt et al [91] explained the key properties of the enzyme such as high processivity and the inverse relationship between number of enzymes and polymer length. A more straightforward mechanism proposed by Janecek et al. [43] explained that polymer branching is a statistical phenomenon triggered by the factors such as the movement of the EPS or by its intermediate release and rebinding.

3.11 Pullulan—A Glucose-Based Homopolysaccharide

Pullulan is also a linear homopolysaccharide composed of glucose with different linkages such as α-(1-4)-Glu- α-(1-4)-Glu- α-(1-6). These linkages connect the repeating units of maltotriose. The fungus *Aureobasidium pullulans*, a saprophytic yeast-like fungus, is described as the highest producer of pullulan at concentrations of 52.5 g/L. Although biosynthetic pathways of pullulan are not well established, it is observed that the polymer is synthesized in the cytosol [83]. *Aureobasidium melanogenum* P16 is a high-pullulan producer. The glucosyltransferase gene (*UGT1*) from *A. melanogenum* was characterized and encoded a protein with 381,199.9 Da. The deletion of this gene renders the organism pullulan negative. It was observed that pullulan synthesis rises under nitrogen starvation, and a high yield is attained only if a high carbon/nitrogen ratio is maintained [20]. Hamidi et al. [36] isolated *A. pullulans* from Iran's Khyroodkenar forest and optimized it for pullulan production at a higher yield. They observed a direct association between pullulan EPS yield and the expression of glucosyltransferase (*fks*), α-phosphoglucose mutase (*pgm*), and UDP-glucose pyrophosphorylase (*ugp*) as the critical genes in contributing in pullulan biosynthesis.

3.12 Levan and Inulin-Polyfructan Polymers

Levan, synthesized by levansucrases, primarily consists of α-(2-6) chain and infrequent α-(2-1) brancheing. Another EPS, inulin is synthesized by inulinosucrase, displays the reverse α-(2-1) chain and α-(2-6) branches. Such sucrases behave as fructose transferases producing polyfructan polymers [106]. While levansucrases are commonly found in the Gram-positive bacteria and numerous plant pathogens, inulinosucrases are found in lactic acid bacteria only [102]. In a study by Okonkwo et al. [74], the authors inactivated the levansucrase gene using double-crossover homologous recombination to prevent EPS production in *Paenibacillus polymyxa* DSM 365. This diminished the considerable economic and technical challenges that occurred during the 2,3-butanediol fermentation. These results also provide a new context toward the connection between EPS production and 2,3-butanediol.

Kozakia (K.) baliensis a recent addition in the acetic acid bacteria (AAB) group, can produce ultra-high molecular weight levan using sucrose. Brandt et al. [13] reported novel EPS upon the cultivation of two *K. baliensis* strains (DSM 14400, NBRC 16680) on sucrose-deficient media. Whole-genome sequencing of the two strains revealed the unusual presence of *gum*-like clusters in addition to the anticipated presence of levansucrases. The *gum*-like clusters exhibited the highest similarity with *gum*-like heteropolysaccharide clusters from other AAB, including *Komagataeibacter xylinus* and *Gluconacetobacter diazotrophicus.* Mutation studies indicated the essential role of the critical gene *gumD*. The novel

EPS was built of galactose, glucose, and mannose. This study established the potential role of *K. baliensis* strains as unique microbial cell factories for the production of biotechnologically relevant and novel polysaccharides.

The enzymatic reaction is similar to dextransucrases. Fructansucrases can also produce maltosylfructosides when solely fed with maltose as the sole priming substrate. A constitutive expression of these genes is observed in some *Leuconostoc mesenteroides* strains, while in some, it is made in the presence of sucrose [106].

3.13 Lactic Acid Bacteria (LAB)—A Versatile Group of EPS Producers

Lactococcus spp has been reported to synthesize two distinct phenotypic forms of EPS, ropy (excreted in the growth medium) or capsular or both [27]. The diversity in EPS structure and composition indicates the presence of a pool of GTs. The *eps* gene cluster has been investigated by several researchers to identify strains producing novel EPS. A mesophilic capsular EPS-producing strain was isolated from raw milk, called *Lactococcus lactis* subsp. *cremoris* SMQ-461. Its *eps* genes consisting of 15 ORFs were sited on the chromosome in a 13.2 kb region. Three novel GTs were reported from the *epsGIK* gene, while *epsD* encoded the priming GT [24]. In other LAB including lactobacilli and streptococci, the *eps* gene cluster is located on the chromosome, with exceptions of *Lactobacillus casei* subsp. *casei*, wherein the cluster is located within a plasmid. In either case, a disruption of the *epsD* generated a reversible EPS-negative mutant. The organization of the *eps* genes is conserved and similar in *L. lactis* (a 12 kb gene cluster on a 40 kb plasmid) and *Streptococcus thermophilus* (15.25 kb region of 16 ORFs on a chromosome) [2, 104]. In order to understand how the EPS production is associated with the regulatory eps genes in a probiotic strain *Lactobacillus casei* LC2W, Ai and co-workers deleted several genes in the cluster, overexpressed and complemented. Their study proved that the glucose-1-phosphate thymidyl transeferase gene (*LC2W_2179*), two uncharacterized proteins (*LC2W_2188* and *LC2W_2189*) were the key players in the EPS biosynthesis pathway [104].

Narbad and colleagues suggested a novel homopolysaccharide EPS biosynthesis pathway for in the Gram-positive *Lactobacillus johnsonii* F19785. *L. johnsonii* synthesizes two capsular EPS including a heteropolysaccharide (EPS2) and a branched glucan homopolysaccharide (EPS1). It was observed that homopolysaccharide EPS1 is produced in the without the presence of sucrose, along with the absence of specific glucansucrase genes in the genome. The authors employed quantitative proteomics, deletion, and complementation studies and found the presence of a flippase and a putative bactoprenol glycosyltransferase. In *L. johnsonii,* these were indispensable for homopolysaccharide biosynthesis indicating the possibility of an alternate route of glucan production to the glucansucrase pathway. A slower growth phenotype was observed if these genes were disturbed [65].

Streptococcus thermophilus, a LAB member, is a common bacterium as a dairy starter that is also acknowledged for its EPS. The presence of this polysaccharide can improve the viscosity, texture and properties of fermented dairy products. The whole-genome sequencing of *S. thermophilus* 1275 showed the presence of unique genes determining the EPS chain length in the EPS gene cluster, not observed in other strains of *S. thermophilus* [128]. This strain possesses *epsC* and *epsD*-genes in its *eps* cluster and synthesizes twice the EPS yield in sucrose medium than glucose and lactose. An uncommon extracellular proteinase, PrtS that acts on casein and cleaves it into oligo-peptides, was one of the several effective proteolytic systems discovered in *S. thermophilus* 1275 along with proteins of the Wzy/Wzx pathway accountable for the transport of EPS [71]. To reveal the global proteomic deviations happening under the influence of sugars at log phase strain, Padmanabhan et al. [75] used isobaric tags for relative and absolute quantitation (iTRAQ)-based proteomic analysis. Most of the 98 differentially expressed proteins in sucrose medium belonged to the EPS biosynthesis (*epsG1D*), and sugar transport (phosphoenolpyruvate) and amino acid metabolism (cysteine/arginine, methionine) pathways. This study provided valuable insights into the well-co-ordinated EPS assembly and regulation via amino acid metabolism and also into understanding high yielding EPS strains.

A plant-derived *Lactobacillus brevis* KB290, is a natural producer of EPS, tolerates aggregation and vigorous mixing in broth medium. Due to these traits, this strain has been used in Japanese fermented foods since 1993. The strain harbors nine plasmids out of which, one encodes two putative glycosyltransferase genes. Fukao et al. [32] demonstrated the function of these genes and other *eps* gene in pKB290-1. The study proved that *gtf27 and gtf28* and *orf29* (ABC transporter) are enough for EPS production with *N*-acetylglucosamine and glucose.

3.14 Biofilms—A Matrix-Encase

Biofilms are a common feature of opportunistic bacterial pathogens. EPSs contribute significantly to biofilm formation. Wozniak and colleagues carried out the first-ever experiment to investigate the structural determinants of the Psl EPS. *P. aeruginosa*, an opportunistic pathogen, synthesizes the Psl polysaccharide from the polysaccharide synthesis locus (psl). The authors performed several complementation and deletion experiments on specific *psl* genes and discovered that there are eleven genes *pslACDEFGHIJKL* are necessary for the production and attachment of Psl. The study also reported that the Psl polysaccharide is built of pentasaccharide repeat units containing D-glucose, D-mannose and L-rhamnose. They also identified that UDP-D-glucose, GDP-D-mannose and dTDP-L-rhamnose were involved in the biosynthesis [17]. *P. aeruginosa* possess the ability to produce another polymer called Pel along with Psl and alginate, which also plays a key role in cell surface interactions and biofilm development. In the Gram-negative bacteria, this cationic EPS is produced by the synthase-dependent polysaccharide secretory pathway. The three key proteins involved are an inner membrane-embedded synthase protein, a periplasmic tetratrico-peptide repeat-containing scaffold protein, and an outer membrane β-barrel porin. Whitfield et al. [120] defined the Pel polymerization by examining over 500 *pel* gene clusters across several *Proteobacteria.* This study revealed the presence of a set of four syntenic genes. *PelD*, *PelE*, and *PelG* are the three genes that encode proteins that reside inside the inner membrane while PelF, the fourth gene encodes a GT-containing domain protein. The authors also reported the presence of a complex for the proteins PelD, PelE, and PelG and also disclosed that this inner membrane complex interacts with the PelF protein. Diaz et al. [28] reported Pel biosynthesis pathway in sulfur-oxidizing species *Acidithiobacillus thiooxidans*. The study demonstrated the presence of 93 ORFs that encoded c-di-GMP metabolism proteins and signal transduction proteins. These included PelD, the c-di-GMP effector protein, of Pel EPS production.

The *pel* gene clusters have been identified even in Gram-positive bacteria. Whitfield et al. [121] also discovered the presence of one of the several *pel* gene clusters $pelDEA_{DA}FG$, in *Bacillus cereus* ATCC 10987. Out of the several genes identified were *CdgF*, a diguanylate cyclase, and *CdgE,* c-di-GMP phosphodiesterase that post-translationally control biofilm development like *PelD* in *P. aeruginosa*. There have been several reports on the presence of *Pel* gene clusters in halophilic and acidophilic bacteria such as *Halomonas smyrnensis* AAD6^{T} and *Leptospirillium ferriphilium,* in addition to levan and cellulose genes respectively.

Caro-Astorga et al. [19], reported the structural organization of two genomic regions for two different EPSs (*eps1* and *eps2*) in *B. cereus* ATCC14579 (CECT148) that supposedly encoded EPS synthesizing proteins. While *eps2* is devoted to the synthesis of capsular polysaccharide and observed to be upregulated in a biofilm cell population, *eps1* is also likely to be contributing in the EPS synthesis; however, no change was observed in its expression. To further understand the function of these genes in *B. cereus* multicellularity, preliminary biofilm assays and bacterial-host cell interaction experiments were performed. The results indicated an insignificant part for the proteins of the *eps2* gene in biofilm development while no role for the consequences of the *eps1* gene. This was because the *eps2* region contributed to the pathogenicity of the bacterium by functioning in cell-to-cell interaction and aggregation, adhesion to surfaces, and biofilm development. While the *eps1* gene appeared to be contributing to bacterial motility evolved to enhance its immunomodulatory properties while trading off its structural roles. The study demonstrated the multiple roles played by two EPS-producing genes.

A ubiquitous fungal pathogen *Aspergillus fumigatus,* transitions into biofilm-forming mode during its infective phase. The basic EPS of this biofilm is an α-1,4-linked linear EPS of galactose (Gal) and *N*-acetylgalactosamine (GalNAc) called Galactosaminoga-lactan (GAG), and plays a key role in virulence. The synthesis of GAG is coregulated by a five gene cluster located on chromosome 3. It starts with the production of activated monosaccharide repeating units by the action of an epimerase Uge3. The second step includes the synthesis and export by the

Gtb3, an integral membrane GT. The GAG EPS gene cluster also encodes the production of two glycoside hydrolases: an endo-α-1,4-galactosaminidase, Ega3 and an endo- α-1,4-*N*-acetylgalactosaminidase, Sph3;. Finally, GAG is moderately deacetylated by the activity of the secreted protein Agd3. Although the Δagd3 mutants lacking acetylation produce normal amounts of GAG, this strain has distinctly decreased biofilm formation and also lacks the cell wall decorations associated with GAG production as well as lower virulence. Agd3 homologs of GAG synthesis gene clusters have been reported in several animal and plant fungal pathogens [4].

3.15 Cyanobacterial and Diatoms' EPS

The marine and limnic environments harbor a rich ecosystem, which is a reservoir of unique chemical diversity. These organisms have been explored for several properties [46, 47, 97], including secondary metabolites and EPS. It is already established that cyanobacteria, microalgae, and diatoms are excellent sources of a diverse EPS range. Its synthesis follows a complex regulatory network with EPS genes either positioned on large plasmids or clustered within the genome. Microalgal EPS is either Capsular polysaccharides (CPS) or Released polysaccharides (RPS). CPS can be either in the form of sheath, capsule or slime. This is also reported in Bacillariophyta (Diatoms) polysaccharides, depending on their solubility in water. Even though there is limited information on EPS biosynthesis and regulation in cyanobacteria, the general schemes of biosynthetic pathways are the same as bacterial EPS, including synthase-dependent, ABC transporter, Wzx/Wzy-dependent pathways. Keidan et al. [49] reported that polysaccharide synthesis occurred in the Golgi bodies, whereas the same happens in the cytoplasm for cyanobacteria. The synthesis is initiated by intracellular hexokinases, which convert glucose into glucose-6-phosphate (Glc-6-P). In the next step, the enzyme (UDP)-glucose pyrophosphorylase helps in catalysis of Glc-1-P to UDP-Glc, which is the crucial stage of the biosynthesis pathway. As is observed in bacteria, there are GT genes reported in cyanobacterial genome, which are likely to be responsible for assembling of repeating units during the biosynthesis, processing and finally export of EPS to the extracellular compartment. The presence of *rfb* genes is also observed in cyanobacteria, which could implicate rhamnose in most cyanobacterial EPS [10].

Reports on cyanobacterial genome sequences suggested that the EPS biosynthesis genes do not always occur in small clusters or operons as detected in the EPS-producing bacteria but are dispersed and isolated throughout the genome. Yoshimura et al. [133] reported 18 ORFs involved in EPS biosynthesis in *Anabaena* sp. PCC 7120. The same study established a putative EPS biosynthesis mechanism after several experiments with the *E. coli* Wzx dependent model. The mechanism involves glycosyltransferases, which catalyze the synthesis and eventually the transfer of repeating units onto a lipid carrier, which then gets "flipped" across the inner membrane by the function of Wzx, the membrane protein. Further,

Wzy catalyzes the process of polymerization of other repeating units in the cyanobacterial periplasmic compartment. The multiprotein complex Wzc, Wzb, and Wza then translocates and exports the growing polysaccharide across the outer membrane. The cellular differences between the mucilaginous structures in microalgae and its cell envelopes contribute to the involvement of several enzymes in the export process. They also proved that a group-3-sigma-factor gene, *sigJ*, a crucial controller of desiccation tolerance in *Anabaena* sp. PCC 7120, also plays a significant role in EPS synthesis. The expressional changes in *sigJ* have also been observed in dehydration in terrestrial *Nostoc* HK-01. The new investigation of microalgal genomes has resulted in the identification and annotation of numerous glycosyl transferases genes and other proteins possibly contributing in EPS processing and central metabolism pathways [26].

4 Genes and Regulators that Influence EPS Biosynthesis

EPS regulation is a highly intricate network of processes that involves several biotic and abiotic factors. EPS biosynthesis is also controlled in response to conditions such as nutrient levels, including nitrogen, phosphate, and sulfate. Apart from the abiotic factors, every EPS-producing organism possesses an intricate network of regulatory proteins that control EPS synthesis (Table 2). The significance of natural EPS is echoed in the ongoing discovery of new regulators in various organisms.

4.1 RSI Circuit-Complex Network of Multiple Transcriptional Regulators

Various studies have suggested the involvement of the robust RSI circuit (ExoR periplasmic regulator; membrane localized ExoS histidine kinase; cytoplasmic ChvI response regulator) regulating the *exo* genes. This network modulates the production of acidic EPS: the EPS-I (succinoglycan) and the EPS-II (galactoglucan). The ExoS protein triggers a two-component system ChvI through phosphorylation, negatively regulated by interaction with ExoR. SyrA and SyrM are other regulators that interact with components of the RSI circuit. SyrM is a transcriptional regulator that modulates the expression of its another protein SyrA and thus the EPS-I level. It has also been revealed that EPS-I production is co-ordinately regulated with the expression of *nod* gene through SyrM, indicating the complex nature of EPS regulation [5]. The authors reported the isolation of several *Sinorhizobium meliloti* mucoid mutants with higher EPS-I production while co-ordinately downregulated motility. Thus, re-examining the bias of a strong relation between EPS production and low motility. The mutant screen is likely to provide evidence on new regulatory mechanisms [5].

Table 2 Examples of regulatory elements, their associated EPS and respective organisms

Sr. No.	Genes and Regulators	Organism	Associated EPS	References
1	RSI circuit	*Sinorhizobium meliloti*	EPS-I and EPS-II	[5]
2	RcsAB Box	Enteric bacteria	Colanic acid	[66, 77, 119, 134]
3	CysB	*X. campestris* pv. campestris B100	Xanthan	[103, 116]
4	KinB/AlgB	*P. aeruginosa*	Alginates	[25]
5	PilB	*M. xanthus*, *P. aeruginosa*	Biofilm	[12, 21]
6	IrrE	*Sphingomonas* sp	Wellan gum	[60]
7	HpaR1	*Xcc*	Xanthan	[111]
8	MetR	*Serratia marcescens*	Biofilm	[76]
9	EmmABC	*Sinorhizobium meliloti*	EPS-I	[6, 70]
10	GacS/A system	Enteric bacteria, *A. vinelandii*	Alginates	[55, 61]
11	PcoR and RfiA	*Pseudomonas corrugata*	Alginates	[58]
12	GlnA	*Agrobacterium* spp	Curdlan	[135]
13	AmrZ and FleQ	*Pseudomonas syringae* pv. Tomato DC3000	Cellulose	[80]
14	Multicomponent regulation	*Ralstonia* (*Pseudomonas*) *solanacearum*	EPS-I	[34]

4.2 RcsAB Box (Regulator of Capsule Synthesis AB)

A commonly observed phosphorelay system in the initiation of capsule biosynthesis pathway in enteric bacteria is the interaction of two proteins RcsA and RcsB, transcriptional regulators with a specific promoter of the colanic acid operon. The authors demonstrated this interaction in *Salmonella typhi, Escherichia coli K-12*, and *Klebsiella pneumoniae*, which revealed a conserved core sequence TaAGaatatTCctA, termed as RcsAB box, a regulon. Site-specific mutations of certain conserved bases in the RcsAB boxes of the *E. coli* and its corresponding rcsA promoters resulted in an EPS-negative phenotype. This strain also demonstrated a reduction in the reporter gene expression levels [119]. The Rcs system comprises of three core proteins including, RcsC (transmembrane sensor kinase), RcsD (transmembrane protein), and RcsB (response regulator) and several auxiliary regulators such as RcsA [35]. A recent study by Meng et al. [66] further defined the regulon through transcriptomic analysis in *Yersinia enterocolitica.* The study indicated the function of RcsB protein in overturning bacterial chemotaxis, motility, c-di-GMP levels, biofilm formation, and eventually the pathogenicity. The RcsAB regulon is also observed to interact with Ferric-uptake regulator (Fur) in enteric bacterium *Klebsiella pneumoniae* NTUH-K2044 in response to iron accessibility to coregulate the production of enterobactin [134]. In *Salmonella enterica* serovar Typhimurium, RcsAB activated 19 genes of the colonic acid biosynthesis operon, thus promoting the biofilm development [77].

4.3 3 CysB—A Transcriptional Regulator in Xanthan Biosynthesis

Schulte et al. [103] studied the association between methionine and biosynthesis of xanthan in *Xanthomonas campestris* pv. campestris B100 (*Xcc*). Methionine is known to reduce the EPS productivity of *Xcc* indicating its vital role beyond the metabolic constraints. The authors shed light on the methionine effect at a molecular level using microarray measurements, growth experiments, pull-down assays, qRT-PCR, and revealed the role of *XCCB100_0996* gene encoding CysB, the transcriptional regulator. CysB could trigger the transcription of the genes *gumB, gumC* and *gumD* indicating that CysB may influence xanthan biosynthesis and export [116]. Schulte et al. [103] introduced a model which demonstrates that CysB controls the expression of around 45 genes, including the xanthan biosynthesis genes, and sulfur metabolism-related genes including *cysD, cysI, cysJ, cysK, cysNC, metY, tauA, tauB* and *tauD.* Once methionine binds to the effector-binding domain of CysB, it disengages promoter region resulting in a reduction in the transcription of the sulfur and xanthan genes. This eventually leads to reduction in sulfur assimilation and xanthan synthesis.

4.4 KinB and AlgB—Regulators of Alginate Biosynthesis

AlgB, an activator of alginate biosynthesis, belongs to the superfamily of the response regulators of the two-component regulatory system, strictly essential for high-level alginate production in *P. aeruginosa. KinB*, located immediately downstream of *algB*, encodes for a 66 kDa protein, homologous to the members of the histidine protein kinase of the two-component regulatory systems. It was observed that *kinB* encodes the AlgB cognate histidine protein sensor kinase in response to environmental signals [63, 73]. Mucoidy phenotype of *P. aeruginosa* which over produces alginate was associated with the deletion of KinB [25].

4.5 PilB-Type IV Pilus Assembly ATPase

The motility of bacterium, *Myxococcus xanthus*, is motorized by the retractive action of the type IV pilus (T4P) triggered by the exopolysaccharides of its biofilm. PilB is a T4P assembly ATPase which is an intermediate protein in the EPS regulatory pathway. Various mutational studies indicated that the function of PilB in EPS regulation is autonomous of its contribution as the T4P assembly ATPase. The apo form of PilB (low c-di-GMP) positively regulates EPS production, thus indicating its dual roles in the regulation of biofilm formation and motility [12, 79].

4.6 IrrE—A Global Transcriptional Regulator

Wellan gum is an industrially relevant EPS commonly synthesized by *Sphingomonas* sp. However, *Sphingomonas* sp bears low tolerance to fermentation conditions such as microbial metabolism, pigment formation from carotenoid synthesis, high viscosity and weak acidity from welan accumulation, and heat from mechanical agitation. IrrE is a global transcriptional regular factor, whose expression is known to improve tolerance to radiation, osmosis, heat, oxidation in *E.coli.* Therefore, Liu et al. [60], expressed heat shock proteins and IrrE from an extremophile in the welan gum producing-*Sphingomonas.* sp. NX-3. The genetic elements were introduced into the strain's genome at key enzyme sites of the carotenoid pathway, thereby constructing a pigment free-stress resistant strain. The use of such global regulators could be help in strain improvement for other industrially relevant EPSs.

4.7 HpaR1—A GntR Family Transcription Regulator

First identified in *Xcc* (*Xanthomonas campestris* pv. *campestris*), HpaR1 was earlier known as a positive regulator of several genes responsible for pathogenicity, hypersensitive reaction, and auto-repress its expression. Su et al. [111] demonstrated that HpaR1 could also positively regulate xanthan production along with motility and stress tolerance. Through various mutational studies revealed that disruption of *hpar1* gene reduced the transcription of *gumB*. Many in vitro transcription assays confirmed that HpaR1 enables the binding of RNA polymerase with the *gumB* promoter, resulting in upregulation of *gum* cluster genes.

4.8 MetR—A LysR-Type Transcriptional Regulator

In a recent study on prodigiosin producing Gram-negative bacterium, *Serratia marcescens*, novel roles were established for highly conserved MetR protein. The results indicated a decreased expression of EPS biosynthesis pathway genes *wzx, wzy*, and *galU* and less biofilm production in *metR* mutant strain ZK66, thus proving that MetR positively regulates biofilm formation [76]. Genomic analysis of a *S. marcescens* strain that could sequester carbon dioxide and produce EPS was carried out by Kumar et al. [54]. The strain possessed in its genome eight putative EPS gene clusters accountable for the production of lipopolysaccharide (cluster 1), emulsan (cluster 4 and 6), stewartan (cluster 2), polysaccharide B (cluster 5), capsular polysaccharide (cluster 7) and fatty acid-saccharide (cluster 8). Among the wza, wzx, wzy, wzc, wzz, wca GTs and transporters were regulators. These included genes responsible for two-component system regulator, diguanylate cyclase, GalU, GalF, tyrosine kinase, sigma-54-dependent Fis family transcriptional regulator, and other transcriptional regulators. This indicates the presence of diverse regulatory mechanisms in the same species living in different environments.

4.9 RpfR—An Integral Biofilm Regulator Between Sessile and Motile Cycles of Bacteria

Many bacteria that survive in fluctuations of attachment and dispersal environments needing a well-regulated system of numerous external cues by several genetic networks. In a recent study, Cooper and colleagues performed four biofilm model evolution experiments on *Burkholderia cenocepacia*, an opportunist betaproteobacterium. The conserved gene *rpfR* product uniquely integrates quorum sensing and the motile–sessile switch. This integration is mediated by c-di-GMP through two domains that sense and respond to- and control the production of cis-2-dodecenoic acid (BDSF), an autoinducer. The resulting response of BDSF in turn controls the action of diguanylate cyclase and phosphodiesterase, thus regulating the levels of c-di-GMP, biofilm production, and EPS composition. This unique adaptation reveals the nexus of signaling networks and regulatory proteins [68].

4.10 EmmABC—The Three-Component Regulatory Circuit

Sinorhizobium meliloti, a soil-dwelling alphaproteobacterium forms nitrogen-fixing root nodules on certain plants. The bacterium's cell envelope comprises of five symbiotically vital surface polysaccharides including capsular polysaccharide, and cyclic-(1,2)-D-glucan, lipopolysaccharide, and acidic EPS-I and EPS-II. Among these, the EPS-I and EPS-II are synthesized the by-products of the *exp* genes- *exo* and *wgx*. ExoX interacts with ExoY and makes it unavailable for the EPS-I biosynthesis indicating post-translational regulation of EPS. The gene *ExpR* in *S. meliloti* positively regulates the expression of the biosynthetic genes for EPS-I and II and downregulates the motility genes. In another study by Barnett and Long [6] on EPS-I overproducing strains of *S. meliloti*, the authors discovered a novel gene *emmD*. This gene could be associated with the EmmABC three-component EPS-I regulation circuitry. EmmA is a small protein while EmmB and EmmC are a part of the putative two-component systems. A disruption in either of these genes results in an amplified yield of the symbiotically significant EPS-I succinoglycan. It was observed that a disruption in *emmB* or *emmC* leads to downregulated expression of *emmA*. The authors propose that the ExoR–ExoS-ChvI and EmmABC systems may function simultaneously but separately since they each control and regulate essential cellular processes, such as the biosynthesis of EPS-I (succinoglycan) and the microorganism's motility in response to external signals. (Morris and González [70].

4.11 The GacS/A System—A Regulator of Alginates Through SRNA

The GacS/A and Rsm posttranscriptional regulatory systems are a part of the two-component system, Gac-Rsm. In several enteric bacteria, the transcriptional regulation of the small regulatory RNAs (sRNAs) belonging to the Rsm regulatory system is controlled by the GacS/A. The Rsm system is composed of two or more sRNAs, and a repressor protein (RsmA/CsrA). Generally, these sRNAs possess GGA motif bearing stem-loop structures, that can counteract the activity of RsmA repressor protein. However, in the absence of Rsm-sRNAs, the repressor protein RsmA binds at or near the ribosomal binding site of its target mRNA. This action results in the blocking of translation, while endorsing its own degradation [55]. In *Azotobacter* vinelandii, a nitrogen-fixing soil bacterium, the alginate biosynthesis is controlled by the GacS/A-Rsm pathway. It regulates the expression of *algD* gene encoding the key enzyme GDP-mannose dehydrogenase of the biosynthesis pathway. RsmA repressor identifies the *algD* mRNA leader, and ensures blocking of translation. While GacA is observed to be necessary for the expression of Rsm-sRNAs, it counters the RsmA translational repression effect [61]. Another two-component system **CbrA/CbrBB**, studied in *A. vinelandii,* negatively regulates the production of alginate [84].

4.12 PcoR and RfiA-LuxR Regulators of Alginate Production

LuxR is a protein superfamily consisting of transcription regulatory proteins possessing a helix-turn-helix (HTH) motif in their C-terminal region. Two proteins PcoR and RfiA of this family were observed to coregulate various processes in *P. corrugate* such as quorum sensing, virulence, antimicrobial peptide production as well as positively regulate all the biosynthetic/structural alginate genes [58].

4.13 GlnA—Regulator of Curdlan Biosynthesis

The role of glutamine synthetase gene (*glnA*) was recently studied by Zhang et al. [135], in curdlan EPS-producing *Agrobacterium* spp was investigated. It was observed that after 96 h of fermentation, the Δ*glnA* strain exhibited 93% impaired curdlan biosynthesis in comparison with that of the wild-type strain. The mutated strain's fermentation profiles revealed an impairment in the uptake of carbon and nitrogen sources and cell growth. Among the several genes downregulated in Δ*glnA* were the genes for electron transport chain, and heme synthesis. An accumulation of adenosine diphosphate (ADP) and flavin adenine dinucleotide (FAD) was also observed. All these observations the critical role of *glnA* plays in curdlan biosynthesis in maintaining ATP levels via regulated expression of genes contributing in the electron transport chain and heme synthesis.

4.14 AmrZ and FleQ—Transcriptional Regulators of Cellulose Biosynthesis

The cellulose biosynthesis reported in *Pseudomonas syringae* pv. Tomato DC3000 (Pto DC3000) is transcriptionally regulated by AmrZ and FleQ, transcriptional regulators involved. The production of acetylated cellulose is triggered by elevated levels of c-di-GMP. Various studies are trying to decipher the regulatory circuits of cellulose production. Pérez-Mendoza et al. [80] demonstrated several interrelated roles of AmrZ and FleQ as well as the stimulus of c-di-GMP related to cellulose biosynthesis. When physiological levels of c-di-GMP are maintained by the cell, these regulatory proteins associate with the neighboring regions on the *wss* promoter of the *wssABCDEFGHI* operon of cellulose biosynthesis, thus preventing its transcription. In response to the elevated levels of c-di-GMP, the FleQ protein acts by releasing itself from the *wss* operator, thus changing into an activator from being a repressor of cellulose synthesis. It was observed that the two proteins behave as independent regulators, unlike those observed in other *Pseudomonas* sp.

4.15 Multicomponent Regulation of EPS1

Ralstonia (*Pseudomonas*) *solanacearum* is a phytopathogen that regulates the production of its virulent EPS via a complex interplay of about seven regulatory genes (*vsrA-vsrD, phcA, phcB, xpsR,* and *vsrB-vsrC*). These are arranged into three converging cascardes of signal transduction. Together they regulate the high-level expression of the *eps* operon. The operon is controlled by the members of the two-component system including xpsR and the vsrB-vsrC. It is observed that *eps* promoter is a key for the transcription activation by xpsR and vsrB-vsrC to stabilize the binding. This binding negatively regulates the transcription of the *eps* operon. Another putative negative regulator of the EPS-I synthesis is EpsR, which also binds to the *eps* promoter [34].

5 The Multi-faceted Role of c-di-GMP

As an allosteric regulator of cellulose EPS biosynthesis, bis(3′-5′)-cyclic dimeric GMP or c-di-GMP was first identified in *Gluconacetobacter xylinum* [92]. Ever since, its activity has been studied in *Streptomyces coelicolor, Listeria monocytogenes*, *Bacillus subtilis* and other Gram-positive bacteria, as well as firmicutes and actinobacteria. Its crucial role is to regulate the transition between distinct phenotypes, sessile and motile [57]. This feature is evident in integrating environmental inputs in biofilm formation and biofilm dispersal [109]. Regulation by c-di-GMP could occur at transcriptional and posttranscriptional levels. This also seems like a vastly conserved regulatory mechanism. In cellulose synthesis pathway, c-di-GMP brings about an allosteric change in the BcsA. This allows the active site accessible to UDP-Glc. The conformational change is brought about by a break in salt-bridges [69]. *Vibrio fischeri,* a marine bacterium produces a symbiosis polysaccharide (Syp) and cellulose. Bassis and Visick [7] reported the presence of a cyclic-di-GMP phosphodiesterase called BinA, which negatively regulates the cellulose biosynthesis genes. This activity could be a link between the biosynthesis regulation of the two polymers.

Poly-β-1,6-*N*-acetylglucosamine (PNAG), is an anionic liner homoglycan that is an essential component for attachment and biofilm formation for various bacteria [131]. In *E. coli* and *Acinetobacter baumannii,* PNAG is synthesized and exported by the *pgaABCD* operon. PgaD is the member protein os the biosynthesis pathway, which has transmembrane helices. Its glycosyltransferase activity is turned on with the binding of c-di-GMP at the interface of PgaD and PgaC. Thus c-di-GMP promotes or shuts the PNAG biosynthesis by acting like a switch [107].

In a human pathogen *Burkholderia cenocepacia*, Bcam1349, was reported to regulate biofilm formation [30]. It is a c-di-GMP responsive protein. Bcam1349 binds to the promoter region of the EPS gene cluster which is enhanced by c-di-GMP levels. It was also observed that high levels of Bcam1349 and c-di-GMP results in an amplified transcription of the EPS genes. Thus, it is evident that the two players Bcam1349 and c-di-GMP coregulate EPS production in *B. cenocepacia*.

There have been various studies to determine how the input signals are sensed by the c-di-GMP dependent pathways. Several studies have reported that the one-component or two-component systems are the usual sensors which carry sensory and c-di-GMP metabolizing domains [33, 114]. A combination of these systems and variation in c-di-GMP levels in multi-level signaling cascades enables bacteria to sense and respond to external inputs and thus regulate the making of EPS and secondary metabolites.

6 Genetic Engineering Strategies to Modify EPS

Enhancement of the physical properties of EPS using genetic engineering approaches has been widely researched upon. Due to the presence of common mechanisms of EPS biosynthesis, engineering of EPS is possible. The current evidence has demonstrated the necessary steps to carry out heterologous expression genes from different origins. The resulting system is functional that can produce modified polymers of desired properties.

Various studies have produced xanthan derivatives with modified acetylation, pyruvylation, and sidechain differences by targeting specific genes. The resulting polymers had either a lesser or higher viscosity as compared with the wild-type xanthan. Acetan, an EPS produced by *Acetobacter xylimum* was tailored using chemical mutagenesis. The resulting polymer composed of tetrasaccharide repeating units possessed higher viscosity.

6.1 Modifying Glycosyltransferase Genes

New polymers with different sugar compositions can be made by adding new or replacing the existing GT genes in an EPS-producing organism. Through total gene deletions, along with precise GT expression in *L. lactis*, Kleerebezem et al. [52], demonstrated the possibility of a system of GT expression that otherwise could be deleterious or lethal. Through random mutagenesis, an EPS-producing *L. sake* was obtained to produce a temperature-dependent production of EPS composed of different sugars. Overexpression of *epsD* by cloning it on a high copy number plasmid in *L. lactis* resulted in a higher yield of EPS. Rutering et al. [94] utilized CRISPR-Cas9 tool to disrupt the non-priming GT gene in a mesophilic strain *Paenibacillis polymyxa*. The EPS variant composed of altered monosaccharides possessed a different rheological behavior.

6.2 Modifying Central Metabolism Pathways

In an effort to maximize EPS production, researchers have taken up mutating certain genes so as to channel the resources into EPS biosynthesis. A high yielding colanic acid *E.coli* strain was generated by knocking out genes contributing in the biosynthesis and regulation of lipopolysaccharide. When strain (Δ*waaF*) was grown in optimized media composition, colonic acid overproduction was obtained [37]. Ren et al.'s [90] study reported that the cell envelop defect in ΔwaaF mutant could induce stress to the cells, impeding the normal cellular growth while triggering the CA biosynthesis.

6.3 Use of New Microbial Cell Factories

This strategy can change the production of current EPS, especially for food industry applications, thus reducing the cost and use of cheaper substrates. For example, EPS production by an obligate aerobic strain or to design a production of EPS by LAB, food-grade hosts. The same molecular weight and different composition in an EPS were accomplished by Stingele et al. [110]. The authors cloned the *eps* gene cluster from *S. thermophilus* SFi6 in *L. lactis* MG1363 on a plasmid. The new lactococcal EPS lacked sidechain sugar and contained galactose as a substitute for *N*-acetylgalactosamine, probably due to the inability of the new strain to produce UDP-*N*-acetylgalactosamine. This study proved that polymerization and secretion are fully functional for the new repeating residue with no exclusivity for the original repeat unit.

If the native EPS is a plasmid-encoded polysaccharide, the EPS plasmid can be added into new hosts through conjugal transfer, the EPS plasmid can be added into new host. This was done, for *L. lactis* associated EPS plasmid PnZ4000. To overcome the problem of limited-host range plasmid replicons, the *Streptococcal* Sfi6 associated *eps* gene cluster was cloned on a broad range plasmid. Heterologous synthesis of xanthan was accomplished by cloning the *gum* gene cluster into several strains of *Sphingomonas* [52, 100]. Even though the ease of production is significant, the low yield levels foil the commercialization of such novel strains.

Franken et al. [31] reported for the first time levan production in *S. cerevisiae* by co-expressing levansucrase M1FT and spinach sucrose transporter (SUT) genes. When grown in rich medium containing sucrose, this strain demonstrated a hyper synthesis and extracellular accumulation of levan EPS. Using an industrial microbe, *S. cerevisiae* opens potential avenues in EPS biotechnology.

A recent study, Williams et al. [122] demonstrated the heterologous cloning and expression of the heparosan synthase enzymes from *Pasteurella multocida* in *B. megaterium.* Using a fed-batch system, the new host could demonstrate heparosan production of 2.74 g/L. To further validate the photoautotrophic production of heparosan, the same group of researchers, introduced glucose phosphate uridyltransferase gene and heparosan synthase gene in the genome of *Synechococcus* Also, by overexpression of the transcriptional positive regulator, RfaH in the engineered strains, studies have demonstrated higher molecular weight fractions of heparosan EPS [99].

7 Conclusions

EPSs produced by microorganisms are a promising and sustainable platform for biotechnological and environmentally friendly biomaterials. Several studies have been carried out over the last decade to establish the biosynthetic pathways of EPS production. These pathways are conserved across species due to their ecological significance and survival strategy in adverse environments. Identifying gene clusters, proteins, and regulatory networks responsible for EPS synthesis is vital for EPS producers' bioprospecting. With the advantage of whole-genome sequencing and bioinformatics tools, the sequence of genes, and the metabolic pathways of the corresponding EPS, molecular structure of the EPS, have been decoded. However, information gaps exist in several areas including, secretory mechanisms, enzymes involved in synthesis and secretion, enzymatic structure, and regulation. There are many eps-producing cyanobacteria, fungi, extremophiles, LAB, whose genome needs to be sequenced to understand the genetic makeup of EPS production.

The challenges include:

(i) relationship between central metabolism (glycolysis, amino acid metabolism, etc.) and the synthesis of sugar/activated sugar repeats, (ii) the identification and characterization of critical members of the EPS production machinery, and (iii) elucidation of the intricate regulatory networks of the EPS production process.

Many optimization experiments have demonstrated the effects of growth conditions, including the type and concentration of sugars, pH, temperature, salinity, co-culture, etc. However, there is a need to decipher the mechanistic influence of these factors on EPS synthesis and excretion.

Various genetic engineering techniques have used cloning essential EPS genes into new hosts or overexpressing certain transferases. These approaches are being tailored further to increase EPS production, control of the properties and structure of EPSs, and have produced fermentation-friendly strains for future use.

By combining systems biology tools and structural studies on both EPS and associated proteins, researchers could address the crucial challenges. Moving beyond this, just biochemical functionalization of EPSs can significantly add to the reservoir of natural polymers. Nonetheless, the studies reported so far on several EPS and its biotechnological value presents a promising path forward.

Acknowledgements SK thanks UGC for providing her Senior Research Fellowship. SK is also thankful to Prof. C. Jayabaskaran, Department of Biochemistry, Indian Institute of Science, for his constant encouragement and support.

References

1. Alkhateeb RS, Vorhölter FJ, Rückert C et al (2016) Genome wide transcription start sites analysis of *Xanthomonas campestris* pv. *campestris* B100 with insights into the gum gene cluster directing the biosynthesis of the exopolysaccharide xanthan. J Biotechnol 225:18–28. https://doi.org/10.1016/j.jbiotec.2016.03.020
2. Almiron-Roig E, Mulholland F, Gasson MJ et al (2000) The complete cps gene cluster from *Streptococcus thermophilus* NCFB 2393 involved in the biosynthesis of a new exopolysaccharide. Microbiology 146:2793–2802. https://doi.org/10.1099/00221287-146-11-2793
3. Ayyash M, Abu-Jdayil B, Itsaranuwat P et al (2020) Characterization, bioactivities, and rheological properties of exopolysaccharide produced by novel probiotic *Lactobacillus plantarum* C70 isolated from camel milk. Int J Biol Macromol 144:938–946. https://doi.org/10.1016/j.ijbiomac.2019.09.171
4. Bamford NC, Le Mauff F, Van Loon JC et al (2020) Structural and biochemical characterization of the exopolysaccharide deacetylase Agd3 required for *Aspergillus fumigatus* biofilm formation. Nat commun 11(1):1–13. https://doi.org/10.1038/s41467-020-16144-5
5. Barnett MJ, Long SR (2018) Novel genes and regulators that influence production of cell surface exopolysaccharides in *Sinorhizobium meliloti*. J Bacteriol 200:e00501–e00517. https://doi.org/10.1128/JB.00501-17
6. Barnett MJ, Long SR (2015) The *Sinorhizobium meliloti* SyrM regulon: effects on global gene expression are mediated by *syrA* and *nodD3*. J Bacteriol 197:1792–1806. https://doi.org/10.1128/jb.02626-14
7. Bassis CM, Visick KL (2010) The cyclic-di-GMP phosphodiesterase BinA negatively regulates cellulose-containing biofilms in *Vibrio fischeri*. J Bacteriol 192(5):1269–1278. https://doi.org/10.1128/jb.01048-09
8. Becker A, Kuster H, Niehaus K et al (1995) Extension of the *Rhizobium meliloti* succinoglycan biosynthesis gene cluster: identification of the exsA gene encoding an ABC transporter protein, and the exsB gene which probably codes for a regulator of succinoglycan biosynthesis. Mol Gen Genet 249:487–497. https://doi.org/10.1007/bf00290574
9. Becker A (2015) Challenges and perspectives in combinatorial assembly of novel exopolysaccharide biosynthesis pathways. Front Microbiol 6:687. https://doi.org/10.3389/fmicb.2015.00687
10. Bhunia B, Prasad Uday US, Oinam G et al (2018) Characterization, genetic regulation and production of cyanobacterial exopolysaccharides and its applicability for heavy metal removal. Carbohydr Polym 179:228–243. https://doi.org/10.1016/j.carbpol.2017.09.091
11. Bianco MI, Jacobs M, Salinas SR et al (2014) Biophysical characterization of the outer membrane polysaccharide export protein and the polysaccharide co-polymerase protein from *Xanthomonas campestris*. Protein Expr Purif 101:42–53. https://doi.org/10.1016/j.pep.2014.06.002
12. Black WP, Wang L, Jing X et al (2017) The type IV pilus assembly ATPase PilB functions as a signalling protein to regulate exopolysaccharide production in *Myxococcus xanthus*. Sci Rep 7(1):1–13. https://doi.org/10.1038/s41598-017-07594-x
13. Brandt JU, Jakob F, Behr J et al (2016) Dissection of exopolysaccharide biosynthesis in *Kozakia baliensis*. Microb Cell Fact 15(1):170. https://doi.org/10.1186/s12934-016-0572-x
14. Breton C, Šnajdrová L, Jeanneau C et al (2006) Structures and mechanisms of glycosyltransferases. Glycobiology 16:29R–37R. https://doi.org/10.1093/glycob/cwj016
15. Bugert P, Geider K (1995) Molecular analysis of the ams operon required for exopolysaccharide synthesis of *Erwinia amylovora*. Mol Microbiol 15:917–933. https://doi.org/10.1111/j.1365-2958.1995.tb02361.x
16. Bundalovic-Torma C, Whitfield GB, Marmont LS et al (2020) A systematic pipeline for classifying bacterial operons reveals the evolutionary landscape of biofilm machineries. PLoS Comput Biol 16: https://doi.org/10.1371/journal.pcbi.1007721

17. Byrd MS, Sadovskaya I, Vinogradov E et al (2009) Genetic and biochemical analyses of the *Pseudomonas aeruginosa* Psl exopolysaccharide reveal overlapping roles for polysaccharide synthesis enzymes in Psl and LPS production. Mol Microbiol 73(4):622–638. https://doi.org/10.1111/j.1365-2958.2009.06795.x
18. Cantarel BL, Coutinho PM, Rancurel C et al (2009) The carbohydrate-active enzymes database (CAZy): an expert resource for glycogenomics. Nucleic Acids Res 37:233–238. https://doi.org/10.1093/nar/gkn663
19. Caro-Astorga J, Álvarez-Mena A, Hierrezuelo J et al (2020) Two genomic regions encoding exopolysaccharide production systems have complementary functions in *B. cereus* multicellularity and host interaction. Sci rep 10(1):1–15. https://doi.org/10.1038/s41598-020-57970-3
20. Chen X, Wang QQ, Liu NN et al (2017) A glycosyltransferase gene responsible for pullulan biosynthesis in Aureobasidium melanogenum P16. Int J Biol Macromol 95:539–549. https://doi.org/10.1016/j.ijbiomac.2016.11.081
21. Chiang P, Sampaleanu LM, Ayers M et al (2008) Functional role of conserved residues in the characteristic secretion NTPase motifs of the *Pseudomonas aeruginosa* type IV pilus motor proteins PilB, PilT and PilU. Microbiology 154(1):114–126. https://doi.org/10.1099/mic.0.2007/011320-0
22. Chong BF, Blank LM, Mclaughlin R et al (2005) Microbial hyaluronic acid production. Appl Microbiol Biotechnol 66:341–351. https://doi.org/10.1007/s00253-004-1774-4
23. Coleman RJ, Patel YN, Harding NE (2008) Identification and organization of genes for diutan polysaccharide synthesis from *Sphingomonas* sp. ATCC 53159. J Ind Microbiol Biotechnol 35:263–274. https://doi.org/10.1007/s10295-008-0303-3
24. Dabour N, LaPointe G (2005) Identification and molecular characterization of the chromosomal exopolysaccharide biosynthesis gene cluster from Lactococcus lactis subsp. cremoris SMQ-461. Appl Environ Microbiol 71:7414–7425. https://doi.org/10.1128/AEM.71.11.7414-7425.2005
25. Damron FH, Owings JP, Okkotsu Y et al (2012) Analysis of the *Pseudomonas aeruginosa* regulon controlled by the sensor kinase KinB and sigma factor RpoN. Journal of bacteriology 194(6):1317–1330. https://doi.org/10.1128/jb.06105-11
26. Delattre C, Pierre G, Laroche C et al (2016) Production, extraction and characterization of microalgal and cyanobacterial exopolysaccharides. Biotechnol Adv 34(7):1159–1179. https://doi.org/10.1016/j.biotechadv.2016.08.001
27. Deo D, Davray D, Kulkarni R (2019) A diverse repertoire of exopolysaccharide biosynthesis gene clusters in *Lactobacillus* revealed by comparative analysis in 106 sequenced genomes. Microorganisms 7(10):444. https://doi.org/10.3390/microorganisms7100444
28. Diaz M, Castro M, Copaja S et al (2018) Biofilm formation by the acidophile bacterium *Acidithiobacillus thiooxidans* involves c-di-GMP pathway and Pel exopolysaccharide. Genes 9(2):113. https://doi.org/10.3390/genes9020113
29. Donot F, Fontana A, Baccou JC et al (2012) Microbial exopolysaccharides: main examples of synthesis, excretion, genetics and extraction. Carbohydr Polym 87(2):951–962. https://doi.org/10.1016/j.carbpol.2011.08.083
30. Fazli M, O'Connell A, Nilsson M et al (2011) The CRP/FNR family protein Bcam1349 is ac-di-GMP effector that regulates biofilm formation in the respiratory pathogen *Burkholderia cenocepacia*. Mol Microbiol 82(2):327–341. https://doi.org/10.1111/j.1365-2958.2011.07814.x
31. Franken J, Brandt BA, Tai SL et al (2013) Biosynthesis of levan, a bacterial extracellular polysaccharide, in the yeast *Saccharomyces cerevisiae*. PLoS ONE 8(10): https://doi.org/10.1371/journal.pone.0077499
32. Fukao M, Zendo T, Inoue T et al (2019) Plasmid-encoded glycosyltransferase operon is responsible for exopolysaccharide production, cell aggregation, and bile resistance in a probiotic strain, *Lactobacillus brevis* KB290. J Biosci Bioeng 128(4):391–397. https://doi.org/10.1016/j.jbiosc.2019.04.008

33. Galperin MY, Nikolskaya AN, Koonin EV (2001) Novel domains of the prokaryotic two-component signal transduction systems. FEMS Microbiol Lett 203(1):11–21. https://doi.org/10.1111/j.1574-6968.2001.tb10814.x
34. Garg RP, Huang J, Yindeeyoungyeon W et al (2000) Multicomponent transcriptional regulation at the complex promoter of the exopolysaccharide I biosynthetic operon of *Ralstonia solanacearum* 182(23):6659–6666. https://doi.org/10.1128/jb.182.23.6659-6666.2000
35. Guo XP, Sun YC (2017) New insights into the non-orthodox two component Rcs phosphorelay system. Front Microbiol 8:2014. https://doi.org/10.3389/fmicb.2017.02014
36. Hamidi M, Kennedy JF, Khodaiyan F et al (2019) Production optimization, characterization and gene expression of pullulan from a new strain of *Aureobasidium pullulans*. Int J Biol Macromol 138:725–735. https://doi.org/10.1016/j.ijbiomac.2019.07.123
37. Han HM, Kim IJ, Yun EJ et al (2020) Overproduction of exopolysaccharide colanic acid by *Escherichia coli* by strain engineering and media optimization. Appl Biochem Biotechnol 1–17. https://doi.org/10.1007/s12010-020-03409-4
38. Hay ID, Ur Rehman Z, Moradali MF et al (2013) Microbial alginate production, modification and its applications. Microb Biotechnol 6:637–650. https://doi.org/10.1111/1751-7915.12076
39. Hay ID, Wang Y, Moradali MF et al (2014) Genetics and regulation of bacterial alginate production. Environ Microbiol 16:2997–3011. https://doi.org/10.1111/1462-2920.12389
40. Hay ID, Rehman ZU, Rehm BH (2010) Membrane topology of outer membrane protein AlgE, which is required for alginate productionin Pseudomonas aeruginosa. Appl Environ Microbiol 76(6):1806–12. https://doi.org/10.1128/AEM.02945-09
41. Hubbard C, McNamara JT, Azumaya C et al (2012) The hyaluronan synthase catalyzes the synthesis and membrane translocation of hyaluronan. J Mol Biol 418(1–2):21–31. https://doi.org/10.1016/j.jmb.2012.01.053
42. Islam ST, Lam JS (2014) Synthesis of bacterial polysaccharides via the Wzx/Wzy-dependent pathway. Can J Microbiol 60(11):697–716. https://doi.org/10.1139/cjm-2014-0595
43. Jacobs M, Salinas SR, Bianco MI et al (2012) Expression, purification and crystallization of the outer membrane lipoprotein GumB from *Xanthomonas campestris*. Acta Crystallogr Sect F Struct Biol Cryst Commun 68:1255–1258. https://doi.org/10.1107/s1744309112036597
44. Janecek S, Svensson B, Macgregor EA (2011) Structural and evolutionary aspects of two families of non-catalytic domains present in starch and glycogen binding proteins from microbes, plants and animals. Enzyme Microb Technol 49:429–440. https://doi.org/10.1016/j.enzmictec.2011.07.002
45. Jin LH, Lee JH (2014) Effect of uracil addition on proteomic profiles and 1,3-β-glucan production in *Agrobacterium* sp. Biotechnol Appl Biochem 61:280–288
46. Jones KM, Kobayashi H, Davies BW et al (2007) How rhizobial symbionts invade plants: the Sinorhizobium-Medicago model. Nat Rev Microbiol 5:619–633. https://doi.org/10.1016/j.enzmictec.2011.07.002
47. Kamat S, Kumari M, Sajna KV et al (2020) Endophytic fungus, *Chaetomium globosum,* associated with marine green alga, a new source of Chrysin. Sci Rep 10(1):1–17. https://doi.org/10.1038/s41598-020-72497-3
48. Kamat S, Kumari M, Taritla S et al (2020) Endophytic fungi of marine alga from Konkan coast, India—a rich source of bioactive material. Front Mar Sci 7:31. https://doi.org/10.3389/fmars.2020.00031
49. Kawano S, Tajima K, Uemori Y et al (2002) Cloning of cellulose synthesis related genes from *Acetobacter xylinum* ATCC23769 and ATCC53582: comparison of cellulose synthetic ability between strains. DNA Res 9(5):149–156. https://doi.org/10.1093/dnares/9.5.149
50. Keidan M, Broshy H, Van Moppes D et al (2006) Assimilation of sulphur into the cell-wall polysaccharide of the red microalga *Porphyridium* sp. (Rhodophyta). Phycologia 45(5):505–511. https://doi.org/10.2216/05-57.1

51. Kharadi RR, Castiblanco LF, Waters CM et al (2019) Phosphodiesterase genes regulate amylovoran production, biofilm formation, and virulence in *Erwinia amylovora*. Appl Environ Microbiol 1;85(1). https://doi.org/10.1128/aem.02233-18
52. Kim SY, Kim JG, Lee BM et al (2009) Mutational analysis of the gum gene cluster required for xanthan biosynthesis in *Xanthomonas oryzae* pv *oryzae*. Biotechnology letters 31(2):265. https://doi.org/10.1007/s10529-008-9858-3
53. Kleerebezem M, van Kranenburg R, Tuinier R et al (1999) Exopolysaccharides produced by *Lactococcus lactis*: from genetic engineering to improved rheological properties? In: Konings WN, Kuipers OP, In 't Veld JHJH (eds) Lactic acid bacteria: genetics, metabolism and applications. Springer, Dordrecht, pp 357–365. https://doi.org/10.1007/978-94-017-2027-4_21
54. Kubiak K, Kurzawa M, Jedrzejczak-Krzepkowska M et al (2014) Complete genome sequence of *Gluconacetobacter xylinus* E25 strain—valuable and effective producer of bacterial nanocellulose. J Biotechnol 176:18–19. https://doi.org/10.1016/j.jbiotec.2014.02.006
55. Kumar M, Kumar M, Pandey A et al (2019) Genomic analysis of carbon dioxide sequestering bacterium for exopolysaccharides production. Sci Rep 9(1):1–2. https://doi.org/10.1038/s41598-019-41052-0
56. Lapouge K, Schubert M, Allain FHT et al (2008) Gac/Rsm signal transduction pathway of γ-proteobacteria, from RNA recognition to regulation of social behaviour. Mol Microbiol 67 (2):241–253. https://doi.org/10.1111/j.1365-2958.2007.06042.x
57. Lee KY, Buldum G, Mantalaris A et al (2014) More than meets the eye in bacterial cellulose: biosynthesis, bioprocessing, and applications in advanced fiber composites. Macromol Biosci 14:10–32. https://doi.org/10.1002/mabi.201300298
58. Liang ZX (2015) The expanding roles of c-di-GMP in the biosynthesis of exopolysaccharides and secondary metabolites. Nat Product Rep 32(5):663–683. https://doi.org/10.1039/c4np00086b
59. Licciardello G, Caruso A, Bella P et al (2018) The LuxR regulators PcoR and RfiA co-regulate antimicrobial peptide and alginate production in *Pseudomonas corrugata*. Front Microbiol 9:521. https://doi.org/10.3389/fmicb.2018.00521
60. Liu W, Xie Y, Ma J et al (2015) IBS: an illustrator for the presentation and visualization of biological sequences. Bioinformatics 31(20):3359–3361. https://doi.org/10.1093/bioinformatics/btv362
61. Liu X, Zhao M, Xu Z et al (2020) Construction of a Robust *Sphingomonas* sp. Strain for Welan gum production via the expression of global transcriptional regulator IrrE. Front Bioeng Biotechnol 8:674. https://doi.org/10.3389/fbioe.2020.00674
62. Low KE, Howell PL (2018) Gram-negative synthase-dependent exopolysaccharide biosynthetic machines. Curr Opin Struct Biol 53:32–44. https://doi.org/10.1016/j.sbi.2018.05.001
63. López-Pliego L, García-Ramírez L, Cruz-Gómez EA et al (2018) Transcriptional study of the RsmZ-sRNAs and their relationship to the biosynthesis of alginate and alkylresorcinols in *Azotobacter vinelandii*. Mol Biotechnol 60:670–680. https://doi.org/10.1007/s12033-018-0102-7
64. Ma S, Wozniak DJ, Ohman DE (1997) Identification of the histidine protein kinase KinB in *Pseudomonas aeruginosa* and its phosphorylation of the alginate regulator AlgB. J Biol Chem 272(29):17952–17960. https://doi.org/10.1074/jbc.272.29.17952
65. Madhuri KV, Prabhakar KV (2014). Microbial exopolysaccharides: biosynthesis and potential applications. Orient J Chem 30(3):1401–1410. https://doi.org/10.13005/ojc/300362
66. Mayer MJ, D'Amato A, Colquhoun IJ et al (2020) Identification of genes required for glucan exopolysaccharide production in *Lactobacillus johnsonii* suggests a novel biosynthesis mechanism. Appl Environ Microbiol 86:e02808–e02819. https://doi.org/10.1128/AEM.02808-19

67. Meng J, Bai J, Chen J (2020) Transcriptomic analysis reveals the role of RcsB in suppressing bacterial chemotaxis, flagellar assembly and infection in *Yersinia enterocolitica.* Curr Genet 3:1–8. https://doi.org/10.1007/s00294-020-01083-x
68. Meredith TC, Mamat U, Kaczynski Z et al (2007) Modification of lipopolysaccharide with colanic acid (M-antigen) repeats in *Escherichia coli.* J Biol Chem 282:7790–7798. https://doi.org/10.1074/jbc.m611034200
69. Mhatre E, Snyder DJ, Sileo E et al (2008) One gene, multiple ecological strategies: a biofilm regulator is a capacitor for sustainable diversity. PNAS 17(35):21647–21657. https://doi.org/10.1073/pnas.2008540117
70. Morgan JL, McNamara JT, Zimmer J (2014) Mechanism of activation of bacterial cellulose synthase by cyclic di-GMP. Nature Struct Mol Biol 21(5):489–496. https://doi.org/10.1038/nsmb.2803
71. Morris J, González JE (2009) The novel genes emmABC are associated with exopolysaccharide production, motility, stress adaptation, and symbiosis in *Sinorhizobium meliloti.* J Bacteriol 191(19):5890–5900. https://doi.org/10.1128/jb.00760-09
72. Nachtigall C, Surber G, Herbi F et al (2020) Production and molecular structure of heteropolysaccharides from two lactic acid bacteria. Carbohyd Polym 236: https://doi.org/10.1016/j.carbpol.2020.116019
73. Niemeyer D, Becker A (2001) The molecular weight distribution of Succinoglycan produced by *Sinorhizobium meliloti* is influenced by specific tyrosine phosphorylation and ATPase activity of the cytoplasmic domain of the ExoP protein. J Bacteriol 183:5163–5170. https://doi.org/10.1128/jb.183.17.5163-5170.2001
74. Okkotsu Y, Pritchett CL, Schurr MJ (2012) Regulation of exopolysaccharide biosynthesis in *Pseudomonas aeruginosa.* Regulation of bacterial virulence ASM Press, pp 171–189. https://doi.org/10.1128/9781555818524.ch9
75. Okonkwo CC, Ujor V, Cornish K, Ezeji TC (2020) Inactivation of the levansucrase gene in *Paenibacillus polymyxa* DSM 365 diminishes exopolysaccharide biosynthesis during 2,3-butanediol fermentation. Appl Environ Microbiol 86:e00196-20. https://doi.org/10.1128/AEM.00196-20
76. Padmanabhan A, Tong Y, Wu Q et al (2020) Proteomic analysis reveals potential factors associated with enhanced EPS production in *Streptococcus thermophilus* ASCC 1275. Sci Rep 10:807. https://doi.org/10.1038/s41598-020-57665-9
77. Pan X, Sun C, Tang M et al (2020) LysR-type transcriptional regulator MetR controls prodigiosin production, methionine biosynthesis, cell motility, H2O2 tolerance, heat tolerance, and exopolysaccharide synthesis in *Serratia marcescens.* Appl Environ Microbiol 86:e02241-19. https://doi.org/10.1128/AEM.02241-19
78. Pando JM, Karlinsey JE, Lara JC et al (2017) The Rcs-regulated colanic acid capsule maintains membrane potential in *Salmonella enterica* serovar Typhimurium. MBio 8: e00808–e008017. https://doi.org/10.1128/mbio.00808-17
79. Patel KB, Toh E, Fernandez XB et al (2012) Functional characterization of UDP-glucose: undecaprenyl-phosphate glucose-1-phosphate transferases of *Escherichia coli* and caulobacter crescentus. J Bacteriol 194:2646–2657. https://doi.org/10.1128/jb.06052-11
80. Perez-Burgos M, Garcia-Romero I, Jung J et al (2020) Characterization of the exopolysaccharide biosynthesis pathway in *Myxococcus xanthus.* J Bacteriol 202:e00335-20. https://doi.org/10.1128/jb.00335-20
81. Periasamy A, Shadiac N, Amalraj A et al (2013) Cell-free protein synthesis of membrane (1,3)- β-d-glucan (curdlan) synthase: co-translational insertion in liposomes and reconstitution in nanodiscs. Biochim Biophys Acta 1828:743–757. https://doi.org/10.1016/j.bbamem.2012.10.003
82. Pollock TJ (2005). Sphingan group of exopolysaccharides (EPS). Biopolym Online 5. https://doi.org/10.1002/3527600035.bpol5010
83. Prajapati VD, Jani GK (2013) Khanda SM (2013) Pullulan: an exopolysaccharide and its various applications. Carbohyd Polym 95(1):540–549. https://doi.org/10.1016/j.carbpol.2013.02.082 (Jun 5)

84. Pérez-Mendoza D, Felipe A, Ferreiro MD et al (2019) AmrZ and FleQ Co-regulate cellulose production in *Pseudomonas syringae* pv. tomato DC3000. Front microbiol 10:746. https://doi.org/10.3389/fmicb.2019.00746
85. Quiroz-Rocha E, Bonilla-Badía F, García-Aguilar V et al (2017) Two-component system CbrA/CbrB controls alginate production in *Azotobacter vinelandii*. Microbiology 163 (7):1105–1115. https://doi.org/10.1099/mic.0.000457
86. Rana S, Upadhyay LS (2020). Microbial exopolysaccharides: synthesis pathways, types and their commercial applications. Int J Biol Macromol 157:577–583. https://doi.org/10.1016/j.ijbiomac.2020.04.084
87. Rasulov BA, Dai J, Pattaeva MA et al (2020) Gene expression abundance dictated exopolysaccharide modification in *Rhizobium radiobacter* SZ4S7S14 as the cell's response to salt stress. Int J Biol Macromol 164:4339–4347. https://doi.org/10.1016/j.ijbiomac.2020.09.038
88. Rehm BHA (ed) (2009) Alginate production: precursor biosynthesis, polymerization and secretion in Alginates: biology and applications. In: Microbiology monographs, vol 13. Springer, Berlin, pp 55–71
89. Rehm B (2010) Bacterial polymers: biosynthesis, modifications and applications. Nat Rev Microbiol 8(8):578–592. https://www.nature.com/articles/nrmicro2354.ris
90. Rehman ZU, Rehm BH (2013) Dual roles of *Pseudomonas aeruginosa* AlgE in secretion of the virulence factor alginate and formation of the secretion complex. Appl Environ Microbiol 79:2002–2011. https://doi.org/10.1128/aem.03960-12
91. Ren G, Wang Z, Li Y et al (2016) Effects of lipopolysaccharide core sugar deficiency on colanic acid biosynthesis in *Escherichia coli*. J Bacteriol 198(11):1576–1584. https://doi.org/10.1128/jb.00094-16
92. Robyt JF, Yoon SH, Mukerjea R (2008) Dextransucrase and the mechanism for dextran biosynthesis. Carbohydr Res 343:3039–3048. https://doi.org/10.1016/j.carres.2008.09.012
93. Ross P, Weinhouse H, Aloni Y et al (1987) Regulation of cellulose synthesis in *Acetobacter xylinum* by cyclic diguanylic acid. Nature 325(6101):279–281. https://doi.org/10.1038/325279a0
94. Ruffing AM, Chen RR (2012) Transcriptome profiling of a curdlan-producing *Agrobacterium* reveals conserved regulatory mechanisms of exopolysaccharide biosynthesis. Microb Cell Fact 11(1):17. https://doi.org/10.1186/1475-2859-11-17
95. Rutering M, Cress BF, Schilling M et al (2017) Tailor-made exopolysaccharides-CRISPR-Cas9 mediated genome editing in *Paenibacillus polymyxa*. Synth Biol 2:ysx007. https://doi.org/10.1093/synbio/ysx007
96. Saadat YR, Khosroushahi AY, Gargari BP (2019) A comprehensive review of anticancer, immunomodulatory and health beneficial effects of the lactic acid bacteria exopolysaccharides. Carbohydr Polym 217:79–89. https://doi.org/10.1016/j.carbpol.2019.04.025
97. Sajna KV, Kamat S, Jayabaskaran C (2020) Antiproliferative role of secondary metabolites from *Aspergillus unguis* AG 1.1 (G) isolated from marine macroalgae enteromorpha sp. by inducing intracellular ROS production and mitochondrial membrane potential loss leading to apoptosis. Front Mar Sci 7:543523:1–16. https://doi.org/10.3389/fmars.2020.543523
98. Schatschneider S, Persicke M, Watt SA et al (2013) Establishment, in silico analysis, and experimental verification of a large-scale metabolic network of the xanthan producing *Xanthomonas campestris* pv. *campestris* strain B100. J Biotechnol 167:123–134. https://doi.org/10.1016/j.jbiotec.2013.01.023
99. Schilling C, Badri A, Sieber V et al (2020) Metabolic engineering for production of functional polysaccharides. Curr Opin Biotechnol 66:44–51. https://doi.org/10.1016/j.copbio.2020.06.010
100. Schmid J, Fariña J, Rehm B et al (2016) Microbial exopolysaccharides: from genes to applications. Front Microbiol 7:308. https://doi.org/10.3389/fmicb.2016.00308
101. Schmid J, Sieber V, Rehm B (2015) Bacterial exopolysaccharides: biosynthesis pathways and engineering strategies. Front Microbiol 6:496. https://doi.org/10.3389/fmicb.2015.00496

102. Schmid J (2018) Recent insights in microbial exopolysaccharide biosynthesis and engineering strategies. Curr Opin Biotechnol 53:130–136. https://doi.org/10.1016/j.copbio.2018.01.005
103. Schulte F, Leβmeier L, Voss J et al (2019) Regulatory associations between the metabolism of sulfur-containing amino acids and xanthan biosynthesis in *Xanthomonas campestris* pv. *campestris* B100. FEMS Microbiol Lett 366(2):fnz005. https://doi.org/10.1093/femsle/fnz005
104. Song X, Xiong Z, Kong L, Wang G et al (2018) Relationship between putative eps genes and production of exopolysaccharide in *Lactobacillus casei* LC2W. Front Microbiol 9:1882. https://doi.org/10.3389/fmicb.2018.01882
105. Sreenivasan S, Kandasamy R (2017) Levan: a biocompatible homopolysaccharide excipient for stabilization of peptide drugs. Int J Pept Res Ther 23:305–311. https://doi.org/10.1007/s10989-016-9562-4
106. Srikanth R, Reddy CHSSS, Siddartha G et al (2015) Review on production, characterization and applications of microbial levan. Carbohydr Polym 120:102–114. https://doi.org/10.1016/j.carbpol.2014.12.00
107. Steiner S, Lori C, Boehm A et al (2013) Allosteric activation of exopolysaccharide synthesis through cyclic di-GMP-stimulated protein–protein interaction. EMBO J 32(3):354–368. https://doi.org/10.1038/emboj.2012.315
108. Stevenson G, Lan R, Reeves PR (2000) The colanic acid gene cluster of *Salmonella enterica* has a complex history. FEMS Microbiol Lett 191(1):11–16. https://doi.org/10.1111/j.1574-6968.2000.tb09312.x
109. Stewart PS, Costerton JW (2001) Antibiotic resistance of bacteria in biofilms. Lancet 358 (9276):135–138. https://doi.org/10.1016/S0140-6736(01)05321-1
110. Stingele F, Vincent SJ, Faber EJ et al (1999) Introduction of the exopolysaccharide gene cluster from *Streptococcus thermophilus* Sfi6 into *Lactococcus lactis* MG1363: production and characterization of an altered polysaccharide. Mol Microbiol 32(6):1287–1295. https://doi.org/10.1046/j.1365-2958.1999.01441.x
111. Su HZ, Wu L, Qi YH, Liu GF et al (2016) Characterization of the GntR family regulator HpaR1 of the crucifer black rot pathogen *Xanthomonas campestris* pathovar *campestris*. Sci Rep 6(1):1–13. https://doi.org/10.1038/srep19862
112. Sung GH, Poinar GO, Spatafora JW (2008) The oldest fossil evidence of animal parasitism by fungi supports a Cretaceous diversification of fungal-arthropod symbioses. Mol Phylogenet Evol 49(2):495–502. https://doi.org/10.1016/j.ympev.2008.08.028
113. Sá-Correia I, Fialho AM, Videira P et al (2002) Gellan gum biosynthesis in *Sphingomonas paucimobilis* ATCC 31461: genes, enzymes and exopolysaccharide production engineering. J Indust Microbiol Biotechnol 29(4):170–176. https://doi.org/10.1038/sj.jim.7000266
114. Temel DB, Dutta K, Alphonse S et al (2013) Regulatory interactions between a bacterial tyrosine kinase and its cognate phosphatase. J Biol Chem 288:15212–15228. https://doi.org/10.1074/jbc.m113.457804
115. Ulrich LE, Koonin EV, Zhulin IB (2005) One-component systems dominate signal transduction in prokaryotes. Trends Microbiol 13(2):52–56. https://doi.org/10.1016/j.tim.2004.12.006
116. Van Kranenburg R, Boels IC, Kleerebezem M et al (1999) Genetics and engineering of microbial exopolysaccharides for food: approaches for the production of existing and novel polysaccharides. Curr Opin Biotechnol 10(5):498–504. https://doi.org/10.1016/s0958-1669(99)00017-8
117. Vorholter FJ, Schneiker S, Goesmann A et al (2008) The genome of *Xanthomonas campestris* pv. *campestris* B100 and its use for the reconstruction of metabolic pathways involved in xanthan biosynthesis. J Biotechnol 134:33–45. https://doi.org/10.1016/j.jbiotec.2007.12.013

118. Wang J, Salem DR, Sani RK (2019) Extremophilic exopolysaccharides: a review and new perspectives on engineering strategies and applications. Carbohydr Polym 205:8–26. https://doi.org/10.1016/j.carbpol.2018.10.011
119. Wang X, Tao F, Gai Z et al (2012) Genome sequence of the welan gum-producing strain *Sphingomonas* sp. ATCC 31555. J Bacteriol 194:5989–5990. https://doi.org/10.1128/jb.01486-12
120. Wehland M, Bernhard F (2000) The RcsAB box characterization of a new operator essential for the regulation of exopolysaccharide biosynthesis in enteric bacteria. J Biol Chem 275 (10):7013–7020. https://doi.org/10.1074/jbc.275.10.7013
121. Whitfield GB, Marmont LS, Bundalovic-Torma C et al (2020) Discovery and characterization of a Gram-positive Pel polysaccharide biosynthetic gene cluster. PLoS Pathog 16(4): https://doi.org/10.1371/journal.ppat.1008281
122. Whitfield GB, Marmont LS, Ostaszewski A et al (2020a) Pel polysaccharide biosynthesis requires an inner membrane complex comprised of PelD, PelE, PelF, and PelG. J Bacteriol 202:e00684-19. https://doi.org/10.1128/JB.00684-19
123. Williams A, Gedeon KS, Vaidyanathan D et al (2020) Metabolic engineering of *Bacillus megaterium* for heparosan biosynthesis using *Pasteurella multocida* heparosan synthase, PmHS2. Microb Cell Fact 18(1):132. https://doi.org/10.1016/j.copbio.2020.06.010 (2019 Dec 1)
124. Willis LM, Whitfield C (2013) Structure, biosynthesis, and function of bacterial capsular polysaccharides synthesized by ABC transporter-dependent pathways. Carbohydr Res 378:35–44. https://doi.org/10.1016/j.carres.2013.05.007
125. Wong HC, Fear AL, Calhoon RD et al (1990) Genetic organization of the cellulose synthase operon in *Acetobacter xylinum*. Proc Natl Acad Sci USA 87:8130–8134. https://doi.org/10.1073/pnas.87.20.8130
126. Wu M, Huang H, Li G et al (2017) The evolutionary life cycle of the polysaccharide biosynthetic gene cluster based on the *Sphingomonadaceae*. Sci Rep 7:46484. https://doi.org/10.1038/srep46484
127. Wu D, Li A, Ma F et al (2016) Genetic control and regulatory mechanisms of succinoglycan and curdlan biosynthesis in genus *Agrobacterium*. Appl Microbiol Biotechnol 100 (14):6183–6192. https://doi.org/10.1007/s00253-016-7650-1
128. Wu X, Xu R, Ren Q et al (2012) Factors affecting extracellular and intracellular polysaccharide production in submerged cultivation of *Tricholoma mongolicum*. African J Microbiol Res 6:909–916. https://doi.org/10.5897/AJMR11.632
129. Xiong ZQ, Kong LH, Lai PF et al (2020) Genomic and phenotypic analyses of exopolysaccharide biosynthesis in *Streptococcus thermophilus* S-3. J Dairy Sci 102 (6):4925–4934. https://doi.org/10.3168/jds.2018-15572
130. Yang S, Yang X, Zhang H (2020) Extracellular polysaccharide biosynthesis in Cordyceps. Crit Rev Microbiol 25:1–22. https://doi.org/10.1080/1040841X.2020.1794788
131. Yang HD, Wu ZC, He DJ, et al (2017) Enzyme-assisted extraction and Pb2+ biosorption of polysaccharide from *Cordyceps militaris*. J Polym Environ 25(4):1033–1043. https://doi.org/10.1007/s10924-016-0882-4
132. Ye L, Zheng X, Zheng H (2014) Effect of sypQ gene on poly-N-acetylglucosamine biosynthesis in *Vibrio parahaemolyticus* and its role in infection process. Glycobiology 24 (4):351–358. https://doi.org/10.1093/glycob/cwu001
133. Yoshimura H, Kotake T, Aohara T et al (2012) The role of extracellular polysaccharides produced by the terrestrial cyanobacterium Nostoc sp. strain HK-01 in NaCl tolerance. J Appl Phycol 24(2):237–243. https://doi.org/10.1007/s10811-011-9672-5
134. Yoshimura H, Okamoto S, Tsumuraya Y et al (2007) Group 3 sigma factor gene, sigJ, a key regulator of desiccation tolerance, regulates the synthesis of extra- cellular polysaccharide in cyanobacterium Anabaena sp. strain PCC 7120. DNA Res 14(1):13–24. https://doi.org/10.1093/dnares/dsm003

135. Yuan L, Li X, Du L et al (2020) RcsAB and Fur coregulate the iron-acquisition system via entC in *Klebsiella pneumoniae* NTUH-K2044 in response to iron availability. Front Cell Infect Microbiol 10:282. https://doi.org/10.3389/fcimb.2020.00282
136. Zhang Z, Chen Y, Wang R et al (2015) The fate of marine bacterial exopolysaccharide in natural marine microbial communities. PLoS ONE 10(11): https://doi.org/10.1371/journal.pone.0142690
137. Zhang W, Gao H, Huang Y et al (2020) Glutamine synthetase gene glnA plays a vital role in curdlan biosynthesis of *Agrobacterium* sp. CGMCC 11546. Int J Biol Macromol 165:222–230. https://doi.org/10.1016/j.ijbiomac.2020.09.152

Molecular Engineering of Bacterial Exopolysaccharide for Improved Properties

Joyleen Fernandes, Dipti Deo, and Ram Kulkarni

Abstract Exopolysaccharides are highly diverse polymers secreted by microorganisms and have enormous potential in the food, medical and industrial sectors. Genetic engineering strategies have been applied to various bacterial strains in order to obtain exopolysaccharides with enhanced properties or the known exopolysaccharide at higher yields. Homopolysaccharides which are produced with the help of fewer genes are easier to produce at higher levels. On the other hand, heteropolysaccharides whose production is dependent on dedicated gene clusters as well as several housekeeping genes are relatively trickier to be modulated. Biochemical characterization of the least studied and highly diverse genes encoding the enzymes such as flippases and polymerases involved in the heteropolysaccharide can provide a toolbox for engineering the polysaccharide production. This chapter details various approaches of molecular engineering for improving the properties of the microbial polysaccharides.

Keywords EPS · Metabolic engineering · Structural alterations · Homopolysaccharides

1 Introduction

Exopolysaccharides (EPS) are the polysaccharides synthesized and then secreted by various bacteria into the surrounding environment. These complex polysaccharides can be categorized on the basis of their monomer composition as homopolysaccharides (e.g. cellulose, curdlan, dextran, levan, pullulan), which comprise of one

J. Fernandes · D. Deo · R. Kulkarni (✉)
Symbiosis School of Biological Sciences, Symbiosis International (Deemed University), Lavale, Pune 412115, India
e-mail: ram.kulkarni@ssbs.edu.in

A. K. Nadda et al. (eds.), *Microbial Exopolysaccharides as Novel and Significant Biomaterials*, Springer Series on Polymer and Composite Materials,
https://doi.org/10.1007/978-3-030-75289-7_4

type of monosaccharide and heteropolysaccharides (e.g. xanthan, gellan) which comprise of repeating units made up of two to seven different monosaccharides. Depending on the monomer composition and the glycosidic bond, homopolysaccharides are grouped into three major classes, viz. glucans, fructans and polygalactans. Heteropolysaccharides comprise of a complex combination of monosaccharides such as galactose, glucose, rhamnose, fucose, mannose and in certain cases their derivatives such as N-acetylglucosamine, N-acetylgalactosamine or glucuronic acid. Some of the EPS also have additional chemical moieties such as phosphate, pyruvate, acetate and succinate. Homopolysaccharides and heteropolysaccharides also differ in terms of the enzymes involved in their synthesis and the overall site of production [53]. Homopolysaccharides can be produced intracellularly or extracellularly depending on the mechanism involved, while heteropolysaccharides are restricted to intracellular biosynthesis pathways [74].

1.1 Biological Role of EPS

EPS are secreted into the environment during the microbial growth and mediate the microbial interactions with the surrounding environment [20]. EPS in general form a barrier to the translocation of solutes, and it is assigned a role in protection against attack by phages, phagocytosis, desiccation, toxic compounds, antibiotics, extreme temperatures, salinity and other stresses [1, 17, 46, 51]. EPS plays a role in cellular associations which include co-aggregation, adhesion to the eukaryotic cells or mucosa, cell recognition, biofilm production and might also aid in the uptake of the environmental DNA [66, 86]. Apart from the microenvironment, microbial EPS are known to have a great impact on the macroenvironment as well. EPS present in the soil enhances its water retention potential and soil aggregate stability [65]. Bacterial EPS has been studied to biologically detoxify areas contaminated with heavy metals [3, 29, 48]. Certain EPS have also shown to be involved in signalling pathways in plant-bacteria interaction and in virulence of pathogens of animal and plant.

1.2 Properties and Applications of EPS

Along with the biological role of EPS in the microbial ecology and physiology, many EPS also have unique physical properties, which can make them useful for certain technological applications. These include their chelation ability, flocculation and emulsification properties along with their nontoxic nature. The existing flocculating agents which include inorganic salts of aluminium have been shown to be harmful to human health [67]. Hence, the use of EPS can be promising in various processes of treatment of wastewater and contaminated groundwater [10, 23, 45]. Furthermore, EPS are being investigated for their ability to utilize petroleum hydrocarbons and facilitate in situ microbial enhanced oil recovery [25]. Biotechnological advancements have sparked the utilization of microbial biopolymers by humankind, opening the door to a myriad of potential medical and industrial applications. The synthesis of EPS in lactic acid bacteria (LAB) is now a topic of major focus and are studied to be associated with several health promoting benefits, especially with regard to its antioxidant, immunostimulating and immunomodulating activity [8, 37, 43, 44, 88, 91, 94]. EPS, especially, those from LAB, have been extensively shown to affect the rheological properties of the foods in which these bacteria are found. This, along with generally regarded as safe (GRAS) status of LAB make their EPS highly suitable for altering rheology and texture of various foods and beverages [9, 35, 39]. EPS has potential in its use as stabilizers and thickening agents in the cosmetic and food industry because of their ability to give rise to dense solutions in an aqueous environment [28]. Many LAB strains tend to produce EPS which help in improving the overall viscosity and texture of fermented foods. EPS can be effectively employed not only in the food industry but also in drug delivery and controlled release, as scaffolds in the process of tissue engineering, wound dressings, accelerated wound healing, pharmaceuticals and so on [4, 56, 62, 70, 71, 76].

1.3 Commercially Employed EPS

A large number of EPS have been reported from bacteria and several of them are commercially used. Bacterial exopolysaccharides date back to the nineteenth century with its discovery in wine, an EPS now infamously recognized as dextran produced by *Leuconostoc mesenteroides* [62]. Various applications of dextran were discovered which led to its commercialization. Dextran is widely employed in the food industry and is renowned for its inhibitory role in the development of ice crystals in ice creams and pudding preparations. It is also employed in confectionery items in order to enhance the viscosity and moisture retention properties [58]. Low molecular weight dextran has been suggested to act as a therapeutic agent in the restoration of blood volume and iron dextran given intravenously is employed to relieve anaemia [50]. Gellan is an EPS synthesized by *Sphingomonas elodea* during large-scale fermentation with a yield of approximately 40–50% from glucose [82]. It is a gelling agent which has been commercialized over several years and approved for its use also as a suspending and stabilizing agent in the food industry [69]. Gellan has been studied in different formulations for nasal and ophthalmic treatments. Few reports have also suggested the use of gellan-based materials in gene transfer and regenerative medicine [55]. Curdlan is an EPS produced by the *Agrobacterium* and *Alcaligenes* species. It is employed in the solidification of fine powders and liquids in food products. Porous particles formulated from a mixture of activated carbon and curdlan have been developed in order to selectively remove heavy metal containing compounds from aqueous solutions [49]. Both curdlan and dextran are FDA approved for their applications in the medical, cosmetic and food industry [38, 90]. Levan is a branched homofructan which is oil and water-soluble and possesses film forming, emulsifying and adhesive properties. In the biofilm of *Bacillus subtilis*, levan was observed to protect this bacteria from desiccation due to changes in the soil water levels [19]. Kefiran is an EPS synthesized by LAB such as *Lentilactobacillus kefiri*, *Lacticaseibacillus rhamnosus* and *Lactobacillus kefiranofaciens* obtained from kefir grains [47]. It is made up of glucose and galactose monomers and forms a yellow coloured water-soluble gel. Kefiran is employed in the production of traditional fermented milk and greatly increases the viscosity of milk gels [60]. Alginate is a polymer secreted by bacteria which belong to the genera *Azotobacter* and *Pseudomonas* comprising guluronic acid and mannuronic acid monomers. It is used in food products to enhance the texture resulting from its intrinsic gelling nature [34]. Since alginate possesses the ability to absorb large volumes of wound fluids and creates an exchange between the sodium and calcium ions resulting in the formation of a swelling, it has been widely employed in wound dressings to improve the quality of the healing process [79]. Hyaluronic acid is a linear polysaccharide produced by bacteria which include the genera *Streptococcus* and *Bacillus* and comprises glucuronic acid and *N*-acetylglucosamine residues. It also serves various medical applications in wound healing, treatment of osteoarthritis and as a vitreous substitute in eye surgery [15, 36]. *Acinetobacter calcoaceticus* is studied to produce emulsan which comprises fatty acids linked to the sugar backbone and has excellent emulsification property even at very low concentrations. Emulsan finds its application in industries such as soap, lotions, shampoo, creams, toothpaste and oil recovery [14]. Xanthan, a polysaccharide produced by *Xanthomonas campestris* is commercially employed in the food industry due to its beneficial attributes, viz. enhanced mouthfeel, improved texture and rapid release of flavour. Xanthan solutions possess pseudoplastic nature and are highly viscous even at lower concentrations. In addition to this, they do not show sensitivity towards varying pH, electrolyte concentrations and temperature. As a result, xanthan is widely employed in industries such as pharmaceutical, textile, paint, cosmetic, oil and paper where it finds its role as a suspending and flocculating agent [2, 77]. Although there is a great demand for microbial EPS in terms of their applications in industry and medicine, very few bacterial EPS such as gellan, dextran and xanthan are employed commercially due to extremely high cost of production [24, 42, 54]. Structures of few bacterial derived EPS have been illustrated in Fig. 1.

Cyanobacterial EPS possess water retention capacity and have the ability to form hydrated gels [59]. Exopolysaccharides from *Cyanobacteria* have been studied in degrading metal containing compounds such as chromium (VI) due to their chelating ability and strong anionic character [57]. Algal EPS in the marine environment tend to be in the form of ubiquitous molecules, wherein they participate in functions crucial in survival and adaptation. In terms of EPS produced by fungi, the major genera involved are *Candida*, *Cryptococcus*, *Aureobasidium* among many others. Pullulan is a well-studied EPS produced by *A. pullulans*; it is a linear polymer possessing adhesive properties [12]. The regular defined alternation between the ($\alpha 1 \rightarrow 4$) and ($\alpha 1 \rightarrow 6$) linkages present in this polymer results in its

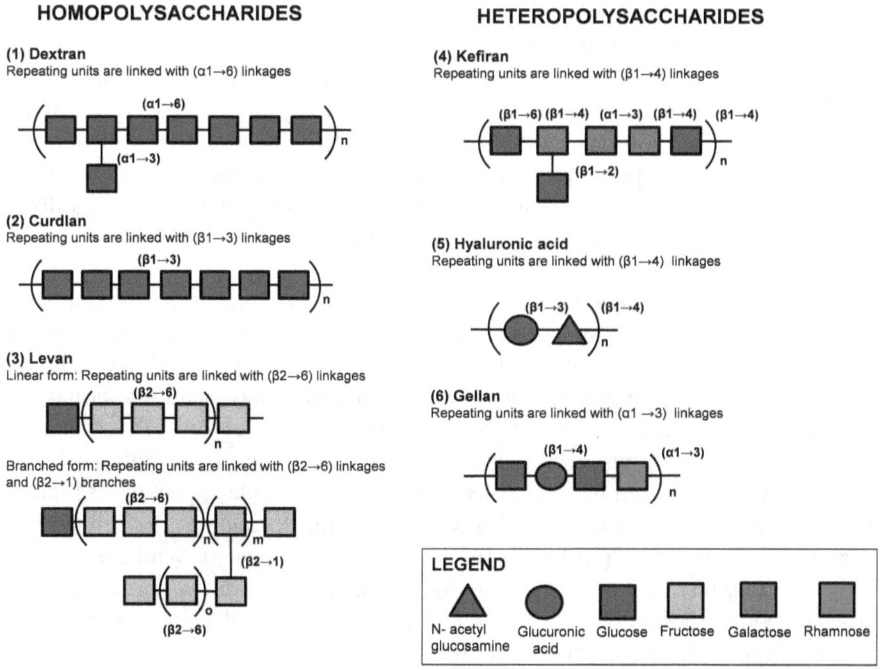

Fig. 1 Structures of few commercially employed bacterial EPS

enhanced properties in terms of flexibility and solubility. Pullulan holds economic value in its use in industries dealing with chemicals and pharmaceuticals [13].

2 Brief Pathways of EPS Production in Bacteria

Techniques such as enzymatic characterization, protein structure analysis, gene overexpression and knockouts, genome sequencing have been used to get molecular insights into the EPS biosynthesis pathways. Such studies carried out over the years on various bacteria have revealed the presence of four pathways, viz. Wzy-dependent pathway, the ATP-binding ABC transporter pathway, synthase-dependent pathway and extracellular synthesis by sucrase-dependent pathway [74, 92]. The requirement in terms of the production of activated sugar precursors and their assembly is common in the first three pathways. In extracellular synthesis, the EPS is synthesized by direct addition of the monomer formed by the cleavage of di- or trisaccharide. However, in a single bacterial species, there is a possible coexistence of two or more biosynthesis pathways to enable the production of different exopolysaccharides.

Wzy-dependent pathway: This pathway involves the assembly of individual repeating units via the action of several glycosyltransferases (GTs) which are then translocated by a flippase enzyme across the cytoplasmic membrane. In the periplasmic space, the polymerization of these repeating units is carried out by the enzyme known as polysaccharide polymerase. Additionally, the pathway also includes a few transcriptional regulators and modulatory proteins. Lactobacilli are well-known producers of EPS via Wzy-dependent pathway with enormous diversity in the Wzy gene clusters [18] and numerous genes encoded by the plasmids [16].

ATP-binding ABC transporter pathway: The polysaccharides produced by this pathway are generally attached to the cell surface and hence are called capsular polysaccharides (CPS). Similar to the Wzy-dependent pathway, GTs are involved in the assembly of repeating units. This pathway may lead to homo- or heteropolysaccharide production depending on the number of GTs associated. A complex of ABC transporters and periplasmic proteins of the polysaccharide co-polymerase (PCP) and the outer membrane polysaccharide export (OPX) family are involved in exporting the EPS followed by its translocation to the cell surface.

Synthase-dependent pathway: In this pathway, a single synthase enzyme carries out polymerization followed by translocation. This pathway generally produces homopolysaccharides (for example, curdlan, bacterial cellulose) in which only one type of sugar precursor is utilized.

Extracellular synthesis by sucrase-dependent pathway: This pathway also leads to homopolysaccharide production (e.g. dextran, levan) from sucrose by the activity of sucrase enzyme, which falls under the category of glycosylhydrolases.

3 Metabolic Engineering Strategies for EPS Production

There is a great demand for bacteria that can produce extracellular polysaccharides having the desired properties. This can be done by re-modelling the carbon metabolism and regulating the various pathways responsible for EPS biosynthesis [73]. Genetic engineering strategies have been devised, wherein the composition, molecular weight and the monomer sequence are the putative targets of manipulation (Fig. 2). There are various strategies, which have been explored by scientists based on metabolic engineering to increase the overall yield of the EPS and also to bring about changes in the structure and composition of the desired polysaccharide. A few of these strategies have been listed in Table 1.

3.1 Metabolic Engineering for Increase in Yield

Metabolic engineering of EPS poses a significant amount of challenges due to its naturally low production levels. Homologous and heterologous gene expression has

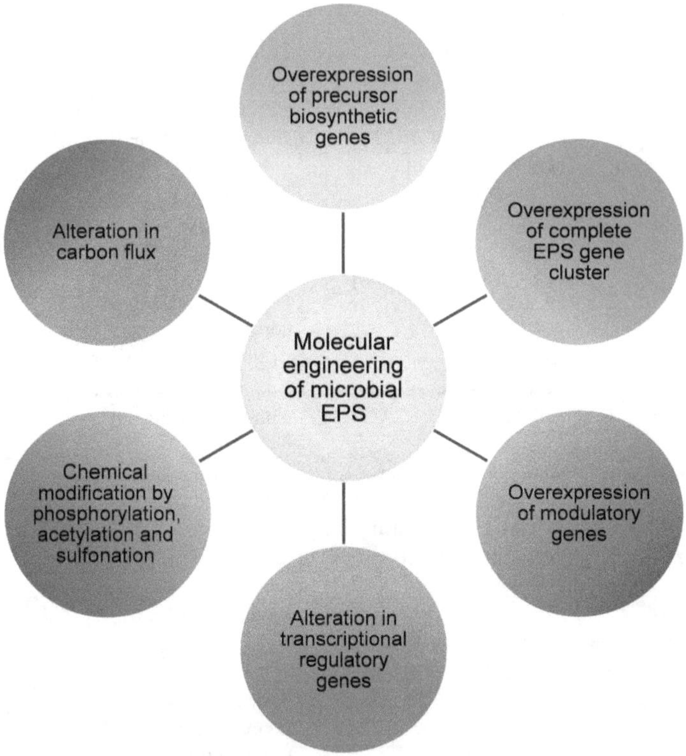

Fig. 2 Schematic representation of major strategies employed for improved EPS production

been most commonly used for enhancing the EPS yield. As sugar precursors for the EPS biosynthesis are provided by the central metabolic pathway, one strategy is overexpressing genes from such a pathway which would result in an increase in the availability of sugar precursors and in turn increase in the yield of EPS. *Streptococcus thermophilus* utilizes the Leloir pathway to form UDP-galactose and UDP-glucose when lactose is a carbon source. A study conducted on *S. thermophilus* showed that overexpression of the genes encoding the Leloir enzymes, viz. UDP-glucose-4-epimerase, galactose-phosphate uridyl transferase and galactokinase (*galE*, *galT* and *galK*) by introducing mutations in its promoter resulted in about 3.3 fold increase in EPS production [78].

Phosphoglucomutase (PGM) is an enzyme which interconnects glycolysis and the Leloir pathway by catalysing the conversion of the substrate glucose-6-phosphate into glucose-1-phosphate. Overexpression of both the genes *galU* and *pgmA* encoding UDP-glucose pyrophosphorylase and phosphoglucomutase, respectively, in *S. thermophilus* resulted in an overall increase in the EPS yield from 0.17 to 0.31 g/mol of carbon from lactose. However, the overexpression of *pgmA* gene alone had no effect on the EPS production levels [40]. Overexpression

Table 1 Various metabolic engineering experiments for EPS production

Experimental bacteria	Strategy employed	Property changed	Reference
Lactococcus lactis	Heterologous overexpression of *EPS* gene cluster from *S. thermophilus*	EPS deficient in galactose side chain, unlike that obtained in the native producer	[75]
Sphingomonas sp. S7	Homologous overexpression of normal *pgm* gene and *sps* gene cluster	Increase by 20% in EPS yield along with increased culture viscosity and change in the sugar composition	[80]
Streptococcus thermophilus	Homologous overexpression of *galU* and *pgmA*	EPS yield increased from 0.17 to 0.31 g/mol of carbon from the sugar lactose	[40]
Lactococcus lactis	Homologous overexpression of the EPS gene cluster using a high-copy number plasmid	Fourfold increase in the EPS yield	[5]
Streptococcus thermophilus	Homologous overexpression of genes encoding Leloir enzymes	Increase in EPS by 3.3 fold	[78]
Lacticaseibacillus rhamnosus	Deletion of *welE* gene by homologous recombination and transposon mutagenesis	Shorter EPS due to premature chain termination	[6]
Sphingomonas strain PPD3 and PPD6	Homologous overexpression of genes encoding enzymes involved in diutan biosynthesis	Enhanced intrinsic viscosity (> 150 dL/g) of the EPS and deficient in polyhydroxybutyrate (PHB) production	[31]
Azotobacter vinelandii	Homologous overexpression of *alg8* gene	Increase in molecular weight of alginate	[21]
Sinorhizobium meliloti	Homologous overexpression of *exoY* gene	Increase in the yield of succinoglycan by 2.5 fold	[33]
Sphingomonas sanxanigenens	Homologous overexpression of *pgmG* gene	Increase in total production of sphingan up to 12.5 g/L	[32]
Xanthomonas campestris	Homologous overexpression of genes *gumB* and *gumC*	Increase in the length of the xanthan chain and viscosity	[26]
Lacticaseibacillus casei LC2W	Homologous overexpression of *rfbB*, *galT*, *pfk* and *nox* genes	Increase in EPS yield by 17.4, 19.6, 20 and 46%, respectively	[41]
Ganoderma lucidum	Homologous overexpression of *pgm* gene	Increase in EPS yield by 44.3%	[89]

(continued)

Table 1 (continued)

Experimental bacteria	Strategy employed	Property changed	Reference
Limosilactobacillus reuteri	Heterologous expression in *E. coli* and site directed mutagenesis of *gtfA* gene	Reuteran with varying glycosidic bonds and molecular weight	[11]
Paenibacillus polymyxa	Homologous directed deletions of genes *pepF, pepJ, pepC, ugdH1* and *manC* using CRISPR-Cas9 followed by integration	EPS deficient mutant with yields ranging from 16 to 50% and an altered monomer composition compared to that of the parent strain	[68]
Xanthomonas campestris CGMCC 15155	Generation of knockout strains lacking genes *gumF, gumG, gumL* and *gumI* involved in xanthan gum synthesis; same genes employed for homologous overexpression	Xanthan lacking the terminal mannose residue had low zero-shear viscosity, whereas that having acetyl groups resulted in stabilization of the xanthan helix structure	[87]
Escherichia coli	Homologous expression and deletion of *waaF* gene	12-fold increase in colanic acid from 177 to 2052.8 mg/L	[30]

of a *pgmG* gene which encodes a bifunctional protein having both phosphomannomutase and phosphoglucomutase activities in *Sphingomonas sanxanigenens* resulted in 17% increase in the total sphingan production resulting in a yield of 12.5 g/L [32]. A study conducted to improve EPS production via the overexpression of *pgm* in *Ganoderma lucidum* reported maximum yield of EPS (1.76 g/L), which was 44.3% greater than that produced by the parent strain. In addition to the EPS yield, the effects of overexpression of the gene *pgm* were analysed on the transcription levels of the genes encoding the enzymes UGP (UDP-glucose pyrophosphorylase) and GLS (β-1,3-glucan synthase). An upregulation of these genes was observed by 1.71 and 1.53 fold, respectively, which suggest that the overexpression of *pgm* could contribute to increased production of polysaccharide via the upregulation of other genes involved in the EPS biosynthesis [89].

Lacticaseibacillus casei LC2W which produced an EPS with lower efficiency was genetically manipulated by overexpressing eight genes which were related to its central metabolism, cofactor engineering, supply of sugar nucleotides and glycosyltransferase. The results showed that the *rfbB* gene encoding dTDP glucose-4,6 dehydratase which participates in dTDP-rhamnose synthesis, *galT* encoding a uridyl transferase responsible for the synthesis of UDP-galactose, *pfk* encoding phosphofructokinase which catalyses a rate limiting step of glycolysis and *nox* encoding NADH oxidase which determines the intracellular redox level greatly contributed to the biosynthesis of EPS and their overexpression increased the yields by 17.4, 19.6, 20 and 46%, respectively. Furthermore, the overexpression of NADH oxidase showed reduced growth rate and lactate production in *L. casei* which resulted in the availability of carbon source for EPS production. This work provided a newer approach for manipulating the metabolic flux which could be exploited for increased EPS yield [41].

In *Lactococcus lactis*, a higher EPS yield was obtained after the whole *EPS* gene cluster was homologously overexpressed using a high-copy plasmid [5]. Single gene knockouts have also been studied to enhance the titre of the EPS. *Azotobacter vinelandii* is a well-studied soil bacterium renowned for its ability to fix atmospheric nitrogen and for the production of alginate. An experiment involving *A. vinelandii* lacking the gene *nqrE* encoding the sodium translocating NADH quinone oxidoreductase (Na^+ NQR) showed overproduction of alginate. Na^+ NQR was suggested to be the major sodium pump for this bacterium as no Na^{+}/H^+ antiport function was observed upon *nqrE* gene deletion, and it further led to the absence of a Na^+ transmembrane gradient which might be a specific signal that results in high production of alginate [27, 52]. Colanic acid is a negatively charged EPS produced during the formation of a biofilm by members of the *Enterobacteriaceae* family and has great potential in the industry owing to its significantly high fucose content. *Escherichia coli* strains possessing defects in their cell membrane can be made use of to increase the production of colanic acid. Such knockout strains lacking the genes *waaF, waaP, waaB* and *waaG* were found to produce mucoid colonies indicating enhanced polysaccharide production [63]. A study focussed on the overproduction of the EPS colanic acid constructed an *E. coli* mutant which lacked the *waaF* gene, and hence resulted in a 12 fold increase in the EPS content from 177 to 2052.8 mg/L [30]. A mutation in *frp* (pleiotropic regulator of multiple cellular functions) in *Streptococcus mutans* resulted in decreased transcriptional efficiency of the genes encoding glucosyltransferases (*gtfB, gtfC*) and frucotosyltransferase (*ftfF*) via a cascade of events which affected EPS production, development of competence and formation of a sucrose-dependent biofilm [84].

In certain cases, the manipulation of a particular gene can have pleiotropic effects. The overexpression of *exoY* encoding a phosphotransferase in the *Sinorhizobium meliloti* resulted in increased production of succinoglycan and the symbiotic productivity with *Medicago truncatula*, its host plant, was enhanced [33]. Biological soil crusts have been studied to harbour various microbial communities producing exopolysaccharides [7]. Cyanobacteria produce EPS which are crucial components for the formation of biofilm in the biological soil crust via the association of various microbes. Metabolic engineering has been employed to overproduce the cyanobacterial EPS which are further used in applications like enabling the generation of biological soil crusts that are artificially induced [81].

3.2 Metabolic Engineering for the Structural Changes

The industrial applications of EPS are largely dependent on their ability to impart certain texture to the medium which is in turn because of their specific chemical structures. Thus, it is possible to enhance the utility of EPS or produce EPS with novel applications by altering its structure. Most often, techniques involving single gene knockouts are employed so as to modify the chemical composition of the EPS. LAB produce a wide variety of EPS differing in their structural and chemical composition, which can further be engineered depending on the requirement [64]. The glycosyltransferase genes which play an important role in the assembly of EPS can be a target for engineering EPS. In *Lacticaseibacillus rhamnosus*, the *welE* gene encoding a priming glycosyltransferase responsible for the attachment of the first monosaccharide unit to the lipid carrier was manipulated by the use of antisense RNA which altered the molecular mass of the EPS produced with no considerable effect on its yield. The suppressed glycosyltransferase activity resulted in shorter EPS due to premature chain termination [6]. *Acetobacter xylinum* comprises a pathway for acetan biosynthesis. A study focused on genetically manipulating this pathway developed an *Acetobacter* strain CKE5 which was better suited for gene transfer. An *aceP* gene homologous to the β-D-glucosyltransferases was identified in the acetan biosynthetic pathway and its disruption in *A. xylinum* CKE5 strain provided confirmation regarding its role and resulted in the production of a truncated acetan with a pentasaccharide repeating unit instead of heptameric unit as seen in the wild type. This experiment also reported the effect of *aceP* disruption only on the assembly of acetan and not on the yield [22].

Xanthan gum is a natural polysaccharide which is employed as an effective stabilizing agent and has diverse functions ranging from oil drilling to that of the food industry. A metabolic network was formulated and curated for *Xanthomonas campestris* pv. campestris (Xcc) B100 in order to analyse and predict the total biomass generated and the impact of production of xanthan [72]. Xanthan biosynthesis comprises of an operon containing a set of twelve genes, viz. *gumB* to *gumM.* It was found that the strains of *X. campestris* lacking the UDP-glucose phosphorylase activity had changes in the xanthan chain length. The researchers analysed the proteins which were encoded by the genes *gumB* and *gumC.* GumB was an outer membrane protein while GumC was linked to the inner membrane. The GumB-GumC protein levels were also observed to modulate the chain length, viscosity and composition of the EPS xanthan [26]. In a latest study, eight strains of *X. campestris* CGMCC 15155 producing xanthan having the same repeating unit but differing with respect to the presence or absence of the acetyl or pyruvyl groups or the terminal mannose residue were engineered. This was achieved by knocking-out or overexpressing the genes *gumF*, *gumG*, *gumL* and *gumI* in various combinations. It was further found that the xanthan lacking the terminal mannose residue had low zero-shear viscosity, whereas that having acetyl groups resulted in stabilization of the xanthan helix structure [87].

A relatively modern approach for obtaining tailor-made EPS includes genome editing tools such as CRISPR-Cas9 which not only enable the engineering of strains, but also plays a major role in improving its economic production process. Paenibacilli are a group of gram-positive bacteria, many of which have been described as beneficial for their role in solubilization of phosphates and nitrogen fixation. They are known to produce EPS with enhanced rheological and antioxidant characteristics [61]. A study focussed on *Paenibacillus polymyxa* made use of a single plasmid system to bring about homologous deletions of three genes *pepF, pepJ, pepC* encoding putative GTs and two genes *ugdH1* and *manC* which might be involved in precursor biosynthesis [68]. This resulted in an EPS deficient mutant with yields ranging from 16 to 50% of that obtained from the parent strain. This formed the first evidence of the function of EPS biosynthetic gene cluster in *P. polymyxa* DSM 365. Furthermore, it was observed that the deletion of the genes *pepF* and *pepJ* resulted in the EPS having an altered monomer composition. In the case of the *pepJ* mutant, the total mannose and glucose concentrations were reduced by 15 and 7.5%, respectively; whereas, in the *pepF* mutant, the concentration of galactose was decreased by 50% [68].

Heterologous expression of *EPS* gene clusters can result in a polymer with varied composition from that of the native species. In a study conducted on a non-EPS producing *Lactococcus lactis*, an *EPS* gene cluster obtained from *S. thermophilus* was expressed heterologously. This resulted in the production of an EPS which was deficient in a galactose side chain, contrary to that obtained from the native EPS producer [75]. Mutagenesis is yet another technique by which the EPS produced can be manipulated. Site directed mutagenesis of *gtfA* gene encoding reuteransucrase enzyme in *Limosilactobacillus reuteri* resulted in an EPS reuteran with varying glycosidic bonds and molecular weights which was superior to the unmodified reuteran, thereby finding its potential in the bread-baking industry [11]. A study focused on increasing the productivity and in turn change in the viscosity and monomer composition of the EPS produced by *Sphingomonas* sp S7 involved the augmentation of the normal copy of the *pgm* gene (phosphoglucomutase) and also the gene cluster *sps* which is crucial for the assembly of the repeating units. A sixfold increase in the activity of the *pgm* genes resulted in a small percentage increase in the overall production of polysaccharide from glucose. However, the overexpression of the *sps* genes led to a 20% increase in the total yield of the EPS obtained from glucose and directly resulted in increased culture viscosity and a change in the sugar composition present in the polymer [80].

Alginates are EPS which are employed as stabilizers and thickeners in various industries. Such a biopolymer is produced in large-scale fermenters using *Azotobacter* as the host organism with minimum understanding of its biosynthetic processes. Overexpression of the gene *alg8* encoding the catalytic subunit of the enzyme alginate polymerase in *A. vinelandii* resulted in increase in the mean molecular weight of alginate [21]. In general, gums are most often employed in pharmacy, oil recovery and the food industry. Since these gums are usually in the form of a gel or a solution, their properties such as viscosity and fluidity are highly studied. A diutan polysaccharide with enhanced properties of intrinsic viscosity was obtained when genes encoding the biosynthesis of diutan which are polysaccharide secretory proteins, proteins involved in rhamnose sugar precursors and glycosyl transferase were overexpressed in a PHB deficient strain of *Sphingomonas* resulting in more viscous diutan gum which was further patented [31].

EPS obtained from LAB have been chemically modified by phosphorylation, acetylation and sulphonation in order to obtain a tailor-made polymer. Phosphorylation enhances the affinity of EPS to immune cells and is also studied to hinder the growth of cancerous cells [83, 85]. Sulphonation has been observed to enhance the bioactivity of EPS due to the existence of the functional groups. A study conducted on the EPS produced by *Lactiplantibacillus plantarum* showed that sulphonation enhanced its antibacterial effect and antagonistic potential against the enterotoxins produced by *Bacillus cereus* [93].

4 Conclusion

An ever-increasing interest has been observed in isolating and identifying newer microbial-derived polysaccharides which possess novel properties, thereby rendering them useful for a wide variety of applications. For a few strains, the genes encoding EPS production and their biosynthesis pathway have been identified and elucidated. It is quite fascinating to note that when the whole EPS gene cluster encoding wzy-dependent pathway is overexpressed, the bacterial growth is drastically reduced. This is not unexpected as the production of EPS is an energy-intensive process. Since the biosynthesis of heteropolysaccharides is dependent on the concerted action of many genes in addition to those partaking in the synthesis of sugar nucleotide precursors, engineering their production is challenging. This situation is exacerbated by the fact that very few flippases and polymerases, which play a crucial role in EPS biosynthesis, have been characterized in bacteria. Thus, effective engineering for the production of the desired EPS at high levels requires holistic understanding of the system biology of EPS biosynthesis.

Acknowledgements We are grateful to the Symbiosis Centre for Research and Innovation (SCRI), Symbiosis International (Deemed University), Science and Engineering Research Board (Government of India), Department of Biotechnology (Government of India), Board of Research for Nuclear Sciences (Government of India), and DAAD (Germany) financial support.

References

1. Angelin J, Kavitha M (2020) Exopolysaccharides from probiotic bacteria and their health potential. Int J Biol Macromol 162:853–865
2. Becker A, Katzen F, Pühler A, Ielpi L (1998) Xanthan gum biosynthesis and application: a biochemical/genetic perspective. Appl Microbiol Biotechnol 50:145–152

3. Bhunia B, Prasad Uday US, Oinam G et al (2018) Characterization, genetic regulation and production of cyanobacterial exopolysaccharides and its applicability for heavy metal removal. Carbohydr Polym 179:228–243
4. Biswas Majee S, Avlani D, Roy Biswas G (2017) Rheological behavior and pharmaceutical applications of bacterial exopolysaccharides. J Appl Pharm Sci 7:224–232. https://doi.org/10.7324/JAPS.2017.70931
5. Boels IC, Van Kranenburg Richard, Kanning MW et al (2003) Increased exopolysaccharide production in *Lactococcus lactis* due to increased levels of expression of the NIZO B40 eps gene cluster. Appl Environ Microbiol 69:5029–5031. https://doi.org/10.1128/AEM.69.8.5029-5031.2003
6. Bouazzaoui K, LaPointe G (2006) Use of antisense RNA to modulate glycosyltransferase gene expression and exopolysaccharide molecular mass in *Lactobacillus rhamnosus*. J Microbiol Methods 65:216–225. https://doi.org/10.1016/j.mimet.2005.07.011
7. Cania B, Vestergaard G, Kublik S et al (2020) Biological soil crusts from different soil substrates harbor distinct bacterial groups with the potential to produce exopolysaccharides and lipopolysaccharides. Microb Ecol 79:326–341. https://doi.org/10.1007/s00248-019-01415-6
8. Cao C, Liu Y, Li Y et al (2020) Structural characterization and antioxidant potential of a novel exopolysaccharide produced by *Bacillus velezensis* SN-1 from spontaneously fermented Da-Jiang. Glycoconj J 37:307–317. https://doi.org/10.1007/s10719-020-09923-1
9. Casas JA, Santos VE, García-Ochoa F (2000) Xanthan gum production under several operational conditions: molecular structure and rheological properties. Enzyme Microb Technol 26:282–291. https://doi.org/10.1016/S0141-0229(99)00160-X
10. Casentini B, Gallo M, Baldi F (2019) Arsenate and arsenite removal from contaminated water by iron oxides nanoparticles formed inside a bacterial exopolysaccharide. J Environ Chem Eng 7: https://doi.org/10.1016/j.jece.2019.102908
11. Chen XY, Levy C, Gänzle MG (2016) Structure-function relationships of bacterial and enzymatically produced reuterans and dextran in sourdough bread baking application. Int J Food Microbiol 239:95–102. https://doi.org/10.1016/j.ijfoodmicro.2016.06.010
12. Cheng KC, Demirci A, Catchmark JM (2011) Pullulan: biosynthesis, production, and applications. Appl Microbiol Biotechnol 92:29–44
13. Chi Z, Wang F, Chi Z et al (2009) Bioproducts from *Aureobasidium pullulans*, a biotechnologically important yeast. Appl Microbiol Biotechnol 82:793–804
14. Choi JW, Choi HG, Lee WH (1996) Effects of ethanol and phosphate on emulsan production by *Acinetobacter calcoaceticus* RAG-1. J Biotechnol 45:217–225. https://doi.org/10.1016/0168-1656(95)00175-1
15. Chong BF, Blank LM, Mclaughlin R, Nielsen LK (2005) Microbial hyaluronic acid production. Appl Microbiol Biotechnol 66:341–351
16. Davray D, Deo D, Kulkarni R (2020) Plasmids encode niche-specific traits in Lactobacillaceae. bioRxiv 2020.08.20.258673. https://doi.org/10.1101/2020.08.20.258673
17. Deming JW, Young JN (2017) The role of exopolysaccharides in microbial adaptation to cold habitats. In: Psychrophiles: from biodiversity to biotechnology, 2nd Edn. Springer, Berlin, pp 259–284
18. Deo D, Davray D, Kulkarni R (2019) A diverse repertoire of exopolysaccharide biosynthesis gene clusters in *Lactobacillus* revealed by comparative analysis in 106 sequenced genomes. Microorganisms 7:444. https://doi.org/10.3390/microorganisms7100444
19. Dogsa I, Brloznik M, Stopar D, Mandic-Mulec I (2013) Exopolymer diversity and the role of levan in *Bacillus subtilis* biofilms. PLoS ONE 8:e62044. https://doi.org/10.1371/journal.pone.0062044
20. Donot F, Fontana A, Baccou JC, Schorr-Galindo S (2012) Microbial exopolysaccharides: main examples of synthesis, excretion, genetics and extraction. Carbohydr Polym 87:951–962. https://doi.org/10.1016/j.carbpol.2011.08.083

21. Díaz-Barrera A, Soto E, Altamirano C (2012) Alginate production and alg8 gene expression by *Azotobacter vinelandii* in continuous cultures. J Ind Microbiol Biotechnol 39:613–621. https://doi.org/10.1007/s10295-011-1055-z
22. Edwards KJ, Jay AJ, Colquhoun IJ et al (1999) Generation of a novel polysaccharide by inactivation of the aceP gene from the acetan biosynthetic pathway in *Acetobacter xylinum*. Microbiology 145:1499–1506. https://doi.org/10.1099/13500872-145-6-1499
23. Elkady MF, Farag S, Zaki S et al (2011) *Bacillus mojavensis* strain 32A, a bioflocculant-producing bacterium isolated from an Egyptian salt production pond. Bioresour Technol 102:8143–8151. https://doi.org/10.1016/j.biortech.2011.05.090
24. Escárcega-González CE, Garza-Cervantes JA, Vázquez-Rodríguez A, Morones-Ramírez JR (2018) Bacterial exopolysaccharides as reducing and/or stabilizing agents during synthesis of metal nanoparticles with biomedical applications. Int J Polym Sci 2018. https://doi.org/10.1155/2018/7045852
25. Fan Y, Wang J, Gao C et al (2020) A novel exopolysaccharide-producing and long-chain n-alkane degrading bacterium *Bacillus licheniformis* strain DM-1 with potential application for in-situ enhanced oil recovery. Sci Rep 10:1–10. https://doi.org/10.1038/s41598-020-65432-z
26. Galván EM, Ielmini MV, Patel YN et al (2013) Xanthan chain length is modulated by increasing the availability of the polysaccharide copolymerase protein GumC and the outer membrane polysaccharide export protein GumB. Glycobiology 23:259–272. https://doi.org/10.1093/glycob/cws146
27. Gaytán I, Peña C, Núñez C et al (2012) *Azotobacter vinelandii* lacking the Na + -NQR activity: a potential source for producing alginates with improved properties and at high yield. World J Microbiol Biotechnol 28:2731–2740. https://doi.org/10.1007/s11274-012-1084-4
28. Gientka I, Błażejak S, Stasiak-Różańska L et al (2015) Undefined exopolysaccharides from yeast: insight into optimal conditions for biosynthesis, chemical composition and functional properties-review. Acta Sci, food.actapol.net. https://doi.org/10.17306/J.AFS.2015.4.29
29. Gupta P, Diwan B (2017) Bacterial exopolysaccharide mediated heavy metal removal: a review on biosynthesis, mechanism and remediation strategies. Biotechnol Rep 13:58–71
30. Han HM, Kim IJ, Yun EJ et al (2020) Overproduction of exopolysaccharide colanic acid by *Escherichia coli* by strain engineering and media optimization. Appl Biochem Biotechnol 193:1–17. https://doi.org/10.1007/s12010-020-03409-4
31. Harding NE, Talashek TA, Patel YN, inventors; CP Kelco US Inc, assignee (2011) *Sphingomonas* Strains Producing Greatly Increased Yield Of PHB-Deficient Sphingan (Diutan). United States patent application US 12/533,649, 2011 Feb 3
32. Huang H, Li X, Wu M et al (2013) Cloning, expression and characterization of a phosphoglucomutase/phosphomannomutase from sphingan-producing *Sphingomonas sanxanigenens*. Biotechnol Lett 35:1265–1270. https://doi.org/10.1007/s10529-013-1193-7
33. Jones KM (2012) Increased production of the exopolysaccharide succinoglycan enhances *Sinorhizobium meliloti* 1021 symbiosis with the host plant *Medicago truncatula*. J Bacteriol 194:4322–4331. https://doi.org/10.1128/JB.00751-12
34. Kim HS, Lee CG, Lee EY (2011) Alginate lyase: structure, property, and application. Biotechnol Bioprocess Eng 16:843–851
35. Kleerebezem M, van Kranenburg R, Tuinier R et al (1999) Exopolysaccharides produced by *Lactococcus lactis*: from genetic engineering to improved rheological properties? Lactic acid bacteria: genetics metabolism and applications. Springer, Netherlands, pp 357–365
36. Kogan G, Šoltés L, Stern R, Gemeiner P (2007) Hyaluronic acid: a natural biopolymer with a broad range of biomedical and industrial applications. Biotechnol Lett 29:17–25
37. Korcz E, Kerényi Z, Varga L (2018) Dietary fibers, prebiotics, and exopolysaccharides produced by lactic acid bacteria: potential health benefits with special regard to cholesterol-lowering effects. Food Funct 9:3057–3068

38. Lakshmi Bhavani A, Nisha J (2016) Int J Pharma Bio Sci Dextran Polysaccharide Versatile Uses. Citeseer
39. Laws AP, Marshall VM (2001) The relevance of exopolysaccharides to the rheological properties in milk fermented with ropy strains of lactic acid bacteria. Int Dairy J 11:709–721, Elsevier
40. Levander F, Svensson M, Rådström P (2002) Enhanced exopolysaccharide production by metabolic engineering of *Streptococcus thermophilus*. Appl Environ Microbiol 68:784–790. https://doi.org/10.1128/AEM.68.2.784-790.2002
41. Li N, Huang Y, Liu Z et al (2015) Regulation of EPS production in *Lactobacillus casei* LC2W through metabolic engineering. Lett Appl Microbiol 61:555–561. https://doi.org/10.1111/lam.12492
42. Llamas I, Amjres H, Mata JA et al (2012) The potential biotechnological applications of the exopolysaccharide produced by the halophilic bacterium *Halomonas almeriensis*. Molecules 17:7103–7120. https://doi.org/10.3390/molecules17067103
43. Lobo RE, Gómez MI, Font de Valdez G, Torino MI (2019) Physicochemical and antioxidant properties of a gastroprotective exopolysaccharide produced by *Streptococcus thermophilus* CRL1190. Food Hydrocoll 96:625–633. https://doi.org/10.1016/j.foodhyd.2019.05.036
44. Lynch KM, Coffey A, Arendt EK (2018) Exopolysaccharide producing lactic acid bacteria: their techno-functional role and potential application in gluten-free bread products. Food Res Int 110:52–61. https://doi.org/10.1016/j.foodres.2017.03.012
45. Mabinya LV, Cosa S, Nwodo U, Okoh AI (2012) Studies on bioflocculant production by *Arthrobacter* sp. Raats, a freshwater bacteria isolated from Tyume River, South Africa. Int J Mol Sci 13:1054–1065. https://doi.org/10.3390/ijms13011054
46. Mancuso Nichols CA, Guezennec J, Bowman JP (2005) Bacterial exopolysaccharides from extreme marine environments with special consideration of the Southern Ocean, sea ice, and deep-sea hydrothermal vents: a review. Mar Biotechnol 7:253–271
47. Micheli L, Uccelletti D, Palleschi C, Crescenzi V (1999) Isolation and characterisation of a ropy *Lactobacillus* strain producing the exopolysaccharide kefiran. Appl Microbiol Biotechnol 53:69–74. https://doi.org/10.1007/s002530051616
48. Mohite BV, Koli SH, Narkhede CP et al (2017) Prospective of microbial exopolysaccharide for heavy metal exclusion. Appl Biochem Biotechnol 183:582–600. https://doi.org/10.1007/s12010-017-2591-4
49. Moon CJ, Lee JH (2005) Use of curdlan and activated carbon composed adsorbents for heavy metal removal. Process Biochem 40:1279–1283. https://doi.org/10.1016/j.procbio.2004.05.009
50. Naessens M, Cerdobbel A, Soetaert W, Vandamme EJ (2005) *Leuconostoc* dextransucrase and dextran: Production, properties and applications. J Chem Technol Biotechnol 80:845–860
51. Nichols CM, Lardière SG, Bowman JP et al (2005) Chemical characterization of exopolysaccharides from antarctic marine bacteria. Microb Ecol 49:578–589. https://doi.org/10.1007/s00248-004-0093-8
52. Núñez C, Bogachev AV, Guzmán G et al (2009) The Na + -translocating NADH: ubiquinone oxidoreductase of *Azotobacter vinelandii* negatively regulates alginate synthesis. Microbiology 155:249–256. https://doi.org/10.1099/mic.0.022533-0
53. Nwodo U, Green E, Okoh A (2012) Bacterial exopolysaccharides: functionality and prospects. Int J Mol Sci 13:14002–14015. https://doi.org/10.3390/ijms131114002
54. Oleksy M, Klewicka E (2018) Exopolysaccharides produced by *Lactobacillus* sp.: biosynthesis and applications. Crit Rev Food Sci Nutr 58:450–462. https://doi.org/10.1080/10408398.2016.1187112
55. Osmałek T, Froelich A, Tasarek S (2014) Application of gellan gum in pharmacy and medicine. Int J Pharm 466:328–340

56. Otero A, Vincenzini M (2003) Extracellular polysaccharide synthesis by *Nostoc* strains as affected by N source and light intensity. J Biotechnol 102:143–152. https://doi.org/10.1016/S0168-1656(03)00022-1
57. Ozturk S, Aslim B, Suludere Z (2009) Evaluation of chromium(VI) removal behaviour by two isolates of *Synechocystis* sp. in terms of exopolysaccharide (EPS) production and monomer composition. Bioresour Technol 100:5588–5593. https://doi.org/10.1016/j.biortech.2009.06.001
58. Paniagua-Michel J de J, Olmos-Soto J, Morales-Guerrero ER (2014) Algal and microbial exopolysaccharides: new insights as biosurfactants and bioemulsifiers. In: Advances in food and nutrition research. Academic Press, pp 221–257
59. Pereira SB, Sousa A, Santos M et al (2019) Strategies to obtain designer polymers based on cyanobacterial extracellular polymeric substances (EPS). Int J Mol Sci 20:5693. https://doi.org/10.3390/ijms20225693
60. Piermaria JA, Pinotti A, Garcia MA, Abraham AG (2009) Films based on kefiran, an exopolysaccharide obtained from kefir grain: development and characterization. Food Hydrocoll 23:684–690. https://doi.org/10.1016/j.foodhyd.2008.05.003
61. Raza W, Makeen K, Wang Y et al (2011) Optimization, purification, characterization and antioxidant activity of an extracellular polysaccharide produced by *Paenibacillus polymyxa* SQR-21. Bioresour Technol 102:6095–6103. https://doi.org/10.1016/j.biortech.2011.02.033
62. Rehm BHA (2010) Bacterial polymers: biosynthesis, modifications and applications. Nat Rev Microbiol 8:578–592
63. Ren G, Wang Z, Li Y et al (2016) Effects of lipopolysaccharide core sugar deficiency on colanic acid biosynthesis in *Escherichia coli*. J Bacteriol 198:1576–1584. https://doi.org/10.1128/JB.00094-16
64. Riaz Rajoka MS, Wu Y, Mehwish HM et al (2020) *Lactobacillus* exopolysaccharides: new perspectives on engineering strategies, physiochemical functions, and immunomodulatory effects on host health. Trends Food Sci Technol 103:36–48
65. Roberson EB, Shennan C, Firestone MK, Sarig S (1995) Nutritional management of microbial polysaccharide production and aggregation in an agricultural soil. Soil Sci Soc Am J 59:1587–1594. https://doi.org/10.2136/sssaj1995.03615995005900060012x
66. Ruas-Madiedo P, Hugenholtz J, Zoon P (2002) An overview of the functionality of exopolysaccharides produced by lactic acid bacteria. Int Dairy J 12: 163–171 (Elsevier)
67. Rudén C (2004) Acrylamide and cancer risk—expert risk assessments and the public debate. Food Chem Toxicol 42:335–349
68. Rütering M, Cress BF, Schilling M et al (2017) Tailor-made exopolysaccharides—CRISPR-Cas9 mediated genome editing in *Paenibacillus polymyxa*. Synth Biol 2:ysx007. https://doi.org/10.1093/synbio/ysx007
69. Sá-Correia I, Fialho AM, Videira P et al (2002) Gellan gum biosynthesis in *Sphingomonas paucimobilis* ATCC 31461: genes, enzymes and exopolysaccharide production engineering. J Ind Microbiol Biotechnol 29:170–176
70. Sahana TG, Rekha RD (2020) A novel exopolysaccharide from marine bacterium *Pantoea* sp. YU16-S3 accelerates cutaneous wound healing through Wnt/β-catenin pathway. Carbohydr Polym 238:116191. https://doi.org/10.1016/j.carbpol.2020.116191
71. Sahana TG, Rekha PD (2019) A bioactive exopolysaccharide from marine bacteria *Alteromonas* sp. PRIM-28 and its role in cell proliferation and wound healing in vitro. Int J Biol Macromol 131:10–18. https://doi.org/10.1016/j.ijbiomac.2019.03.048
72. Schatschneider S, Persicke M, Watt SA et al (2013) Establishment, in silico analysis, and experimental verification of a large-scale metabolic network of the xanthan producing *Xanthomonas campestris* pv. campestris strain B100. J Biotechnol 167:123–134. https://doi.org/10.1016/j.jbiotec.2013.01.023

73. Schilling C, Badri A, Sieber V et al (2020) Metabolic engineering for production of functional polysaccharides. Curr Opin Biotechnol 66:44–51
74. Schmid J, Sieber V, Rehm B (2015) Bacterial exopolysaccharides: biosynthesis pathways and engineering strategies. Front Microbiol 6:496. https://doi.org/10.3389/fmicb.2015.00496
75. Stingele F, Vincent SJF, Faber EJ et al (1999) Introduction of the exopolysaccharide gene cluster from *Streptococcus thermophilus* Sfi6 into *Lactococcus lactis* MG1363: production and characterization of an altered polysaccharide. Mol Microbiol 32:1287–1295. https://doi.org/10.1046/j.1365-2958.1999.01441.x
76. Sun ML, Zhao F, Chen XL, et al (2020) Promotion of wound healing and prevention of frostbite injury in rat skin by exopolysaccharide from the Arctic marine bacterium *Polaribacter* sp. SM1127. Mar Drugs 18. https://doi.org/10.3390/md18010048
77. Sutherland IW (1998) Novel and established applications of microbial polysaccharides. Trends Biotechnol 16:41–46
78. Svensson M, Waak E, Svensson U, Rådström P (2005) Metabolically improved exopolysaccharide production by *Streptococcus thermophilus* and its influence on the rheological properties of fermented milk. Appl Environ Microbiol 71:6398–6400. https://doi.org/10.1128/AEM.71.10.6398-6400.2005
79. Tabernero A, Cardea S (2020) Microbial exopolysaccharides as drug carriers. Polymers (Basel) 12:2142. https://doi.org/10.3390/polym12092142
80. Thorne L, Mikolajczak MJ, Armentrout RW, Pollock TJ (2000) Increasing the yield and viscosity of exopolysaccharides secreted by *Sphingomonas* by augmentation of chromosomal genes with multiple copies of cloned biosynthetic genes. J Ind Microbiol Biotechnol 25:49–57. https://doi.org/10.1038/sj.jim.7000019
81. Tiwari ON, Bhunia B, Mondal A et al (2019) System metabolic engineering of exopolysaccharide-producing cyanobacteria in soil rehabilitation by inducing the formation of biological soil crusts: a review. J Clean Prod 211:70–82
82. Vartak NB, Lin CC, Cleary JM et al (1995) Glucose metabolism in "*Sphingomonas elodea*": pathway engineering via construction of a glucose-6-phosphate dehydrogenase insertion mutant. Microbiology 141:2339–2350. https://doi.org/10.1099/13500872-141-9-2339
83. Verkhnyatskaya S, Ferrari M, De Vos P, Walvoort MTC (2019) Shaping the infant microbiome with non-digestible carbohydrates. Front Microbiol 10:343
84. Wang B, Kuramitsu HK (2006) A pleiotropic regulator, Frp, affects exopolysaccharide synthesis, biofilm formation, and competence development in *Streptococcus mutans*. Infect Immun 74:4581–4589. https://doi.org/10.1128/IAI.00001-06
85. Wang J, Zhao X, Yang Y et al (2015) Characterization and bioactivities of an exopolysaccharide produced by *Lactobacillus plantarum* YW32. Int J Biol Macromol 74:119–126. https://doi.org/10.1016/j.ijbiomac.2014.12.006
86. Wolfaardt GM, Lawrence JR, Korber DR (1999) Function of EPS. Microbial extracellular polymeric substances. Springer, Berlin, Heidelberg, pp 171–200
87. Wu M, Qu J, Tian X et al (2019) Tailor-made polysaccharides containing uniformly distributed repeating units based on the xanthan gum skeleton. Int J Biol Macromol 131:646–653. https://doi.org/10.1016/j.ijbiomac.2019.03.130
88. Xu Y, Cui Y, Yue F et al (2019) Exopolysaccharides produced by lactic acid bacteria and Bifidobacteria: structures, physiochemical functions and applications in the food industry. Food Hydrocoll 94:475–499
89. Xu JW, Ji SL, Li HJ et al (2015) Increased polysaccharide production and biosynthetic gene expressions in a submerged culture of *Ganoderma lucidum* by the overexpression of the homologous α-phosphoglucomutase gene. Bioprocess Biosyst Eng 38:399–405. https://doi.org/10.1007/s00449-014-1279-1

90. Yang M, Zhu Y, Li Y et al (2016) Production and optimization of curdlan produced by *Pseudomonas* sp. QL212. Int J Biol Macromol 89:25–34. https://doi.org/10.1016/j.ijbiomac.2016.04.027
91. Zannini E, Waters DM, Coffey A, Arendt EK (2016) Production, properties, and industrial food application of lactic acid bacteria-derived exopolysaccharides. Appl Microbiol Biotechnol 100:1121–1135
92. Zeidan AA, Poulsen VK, Janzen T et al (2017) Polysaccharide production by lactic acid bacteria: from genes to industrial applications. FEMS Microbiol Rev 41:S168–S200. https://doi.org/10.1093/femsre/fux017
93. Zhang Z, Liu Z, Tao X, Wei H (2016) Characterization and sulfated modification of an exopolysaccharide from *Lactobacillus plantarum* ZDY2013 and its biological activities. Carbohydr Polym 153:25–33. https://doi.org/10.1016/j.carbpol.2016.07.084
94. Zhu Y, Wang X, Pan W et al (2019) Exopolysaccharides produced by yogurt-texture improving *Lactobacillus plantarum* RS20D and the immunoregulatory activity. Int J Biol Macromol 121:342–349. https://doi.org/10.1016/j.ijbiomac.2018.09.201

Extremophiles: A Versatile Source of Exopolysaccharide

Monalisa Padhan

Abstract Prokaryotic organisms are omnipresent, inhabiting even extremely harsh ecological niches. Many microorganisms living in extreme environments produce exopolysaccharides (EPSs) like a major approach to live in adverse situations. The EPSs hold unique features because of the diverse unfavorable condition that stimulate the microbes to secret these biopolymers. Extremophilic microbes are also involved in the formation of biofilms. The biofilm-derived EPS has appliances in the pharmaceutical, cosmetics along food industries as stabilizers, thickeners, coagulants, gelling agents, and emulsifiers. This chapter reports the treasured insights related to the features of different sorts of EPSs secreted by diverse groups of extremophilic microbes and additionally discusses their relevance in the field of biotechnology in the modern era.

Keywords Extremophiles · Exopolysaccharide · Biofilm · Biotechnological relevance

1 Introduction

Extremophiles are living creatures that thrive in extreme environments and are capable of surviving in unfavorable climates that were earlier considered to be adverse or deadly for life [66]. Extremophiles are divided into two broad categories: extremophiles, which need one or more extreme conditions to survive, while extremotolerant organisms, grow optimally at more 'normal' conditions but are still able to survive one or more extreme physiochemical values [12, 66].

A substance produced by the extremophiles called EPS enables them to function in such forbidding environments. A large variety of microbes secret exopolysaccharides which are principally made up of polysaccharides, DNA, and proteins

M. Padhan (✉)
Microbiology, School of Life Sciences, Sambalpur University, Jyoti Vihar, Burla, Sambalpur Odisha-768019, India

A. K. Nadda et al. (eds.), *Microbial Exopolysaccharides as Novel and Significant Biomaterials*, Springer Series on Polymer and Composite Materials,
https://doi.org/10.1007/978-3-030-75289-7_5

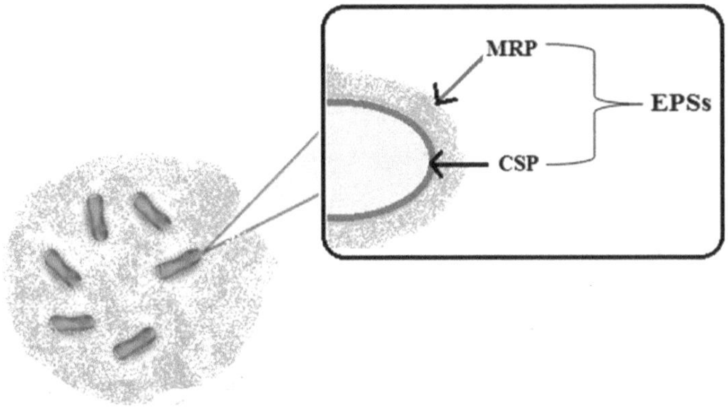

Fig. 1 Topology of exopolysaccharides (EPSs) (N.B. MRPs: Medium released exopolysaccharides; CPSs: Capsular polysaccharides)

[22]. Their structures, function, physical and chemical characteristics that build up their primary conformation vary from one bacterial species to another [5].

These are high molecular weight extracellular polymers surrounding most microbial cells in the environment [70, 76, 86]. Generally, the EPS produced by bacteria occur in two forms, namely medium released exopolysaccharides (MRPs) and capsular polysaccharides (CPSs). The polymers are covalently bound to the cell surface in CPSs, whereas in case of MRPs they are loosely bound to the cell surface or are found in the extracellular medium as the amorphous matrix or embedded into biofilms [17, 70] (Fig. 1).

EPSs are principally made up of carbohydrates (various sugar residues) and noncarbohydrate constituents (like pyruvate, acetate). EPSs have led to significant interest among all the products from extremophiles, because of the rising demand for natural polymers in various sectors like food, pharmaceutical, and other industries [85]. Different EPS produced by diverse groups of extremophiles and their applications in the field of biotechnology are summarized in this chapter. Exopolysaccharides constitute a considerable element of extracellular polymers encompassing mainly the microbial cells in the extreme environment. The extremophiles have build up several adaptations, allowing them to compensate for the detrimental effects of extreme conditions, like high temperatures, low pH or temperature, salt, high radiation [63].

Extreme habitats harbor a host of extremophilic microbes (extremophiles), such as thermophiles, halophiles, acidophiles, alkaliphiles, piezophiles, and psychrophiles [23, 57, 66]. The general classification of extremophiles is given in Table 1.

Table 1 General classification of extremophiles based on the environmental parameters [57, 66]

Environmental parameter	Group of extremophile	Growth range
Temperature	Psychrophile	<10 °C
	Thermophile	45–80 °C
	Hyperthermophile	>80 °C
pH	Hyperacidophile	<pH3
	Acidophile	<pH5
	Alkaliphile	>pH9
	Hyperalkaliphile	>pH11
Pressure	Peizotolerant/Barotolerant	0.1–10 MPa
	Peizophile/Barophile	10–50 MPa
	Hyperpeizophile/ Hyperbarophile	>50 MPa
Salinity[a]	Halotolerant	(1.2–2.9%; tolerate 14.6%)
	Halophile	(>8.8%)
	Extreme halophile	(>14.6%, cannot grow < 8.8%)
Water activity	Xerophile	aw < 0.7
UV radiation	Radiotolerant	40–400 nm
Multiple parameters	Polyextremophile	Tolerance for combined multiple parameters

[a]Salinity expressed as a percent of NaCl (W/V)

2 Significance of EPS Production by Extremophiles

Extremophiles should adjust to adverse conditions through distinct mechanisms, and the production of EPS is one of the important mechanisms for survival. EPS provides a truly ideal environment for the nutrient supplement, chemical reactions, and protect against environmental stresses like drought and salinity [22]. Biofilm producing microbes are about one thousand times more resistant to antimicrobial chemicals than the planktonic cells [25]. The physiological function of EPS relies upon the indigenous habitat from which microorganisms have been isolated. Additionally, EPS can take away nutrients from the nearby environment [26, 28], encourages biofilm development, and prevents the entry of antimicrobial agents into the biofilms [34]. Some of the functions are demonstrated as follows (Fig. 2).

3 Biofilm Production and EPS

EPS is considered the most significant factor affecting the process of microbial adhesion. EPS molecules fortify the connections linking the microbes, and accordingly, they decide the process of formation of cell aggregates on the solid

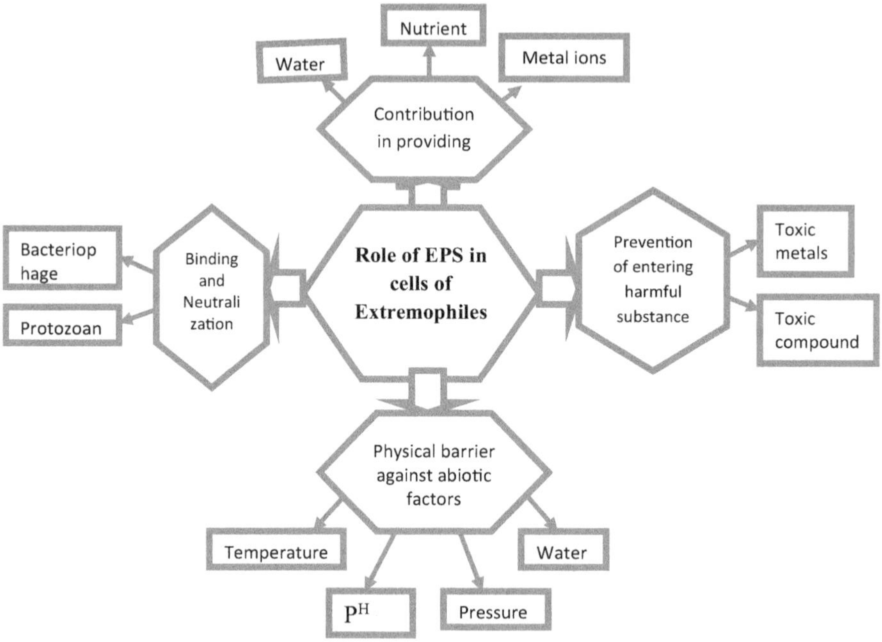

Fig. 2 Role of EPS in microbial cells of Extremophiles inhabiting the extreme environment

surface [10]. EPSs play a significant job in the biofilm matrix concerning the biochemical interactions among microbes and encompassing cells [27]. The process of formation of biofilm and EPS matrix is described in Fig. 3.

4 EPS Produced by a Diverse Group of Extremophiles

Extremophiles (thermophiles, psychrophiles, acidophiles, alkaliphiles, and halophiles) are well known for the productions of EPS [46]. These EPSs have many novel characteristics that may be useful in the field of industrial biotechnology. A huge number of EPSs from extremophiles were described over the last decades, and various researchers have extensively studied their biosynthesis and functional properties. EPS produced by microbes which are generally has a broad range of industrial applications in the recovery of oil, textile, cosmetics, pharmaceutical, and food industries.

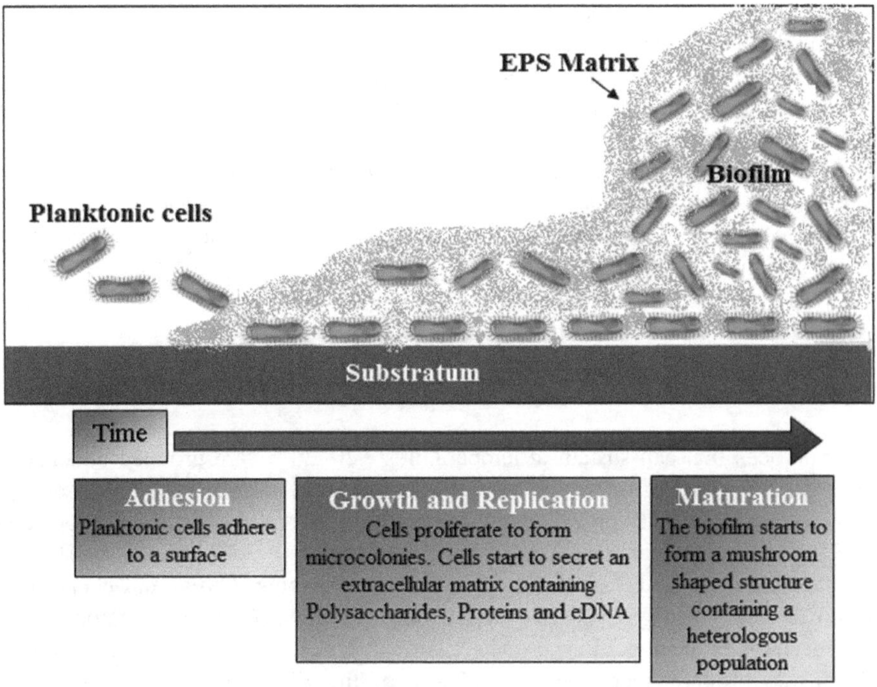

Fig. 3 Process of formation of biofilm and EPS matrix

4.1 Production of EPS by Thermophilic Microorganisms

Thermophiles are mostly prokaryotic organisms with optimum growth temperatures of >44 °C, and hyperthermophiles are thermophiles that grow at temperatures >80 °C [66]. Both the thermophiles and hyperthermophiles [75] are found in various natural ecosystems, for example, marine and terrestrial hot springs, geothermal waters, deep-sea hydrothermal vents, fumaroles, and volcanoes [42, 66]. Generally, in the case of thermophilic microorganisms, the rate of chemical reactions is increased at high temperature that finally increases the production in a stipulated period, and for this reason, thermophiles are industrially important in the production of various compounds [44, 45, 64, 85]. Thermophiles utilize branched-chain fatty acids and polyamines for the stabilization of the membranes. Though the production of EPS is lower as compared to the mesophiles, due to their short fermentation period, thermophiles are commercially significant [71].

Many thermophilic microbes inhabiting hot sea vents and deep-sea hot springs are observed to produce EPSs [59, 87, 88]. They can survive in these types of environments commonly described by their elevated temperature and pressure with high concentrations of toxic metals also. The production of EPS by thermophiles protects themselves from the unfavorable condition. The important factor to improve the maximum production of EPSs is the sugar, which acts as a source of carbon and energy. Thermostability is a significant criterion of the EPSs produced by thermophiles besides their molecular weight and chemical composition. The EPSs produced by thermophilic microbes comprised of a wide range of intriguing features for modern applications in industries [35, 43, 63, 65]. Some thermophilic bacterial strains, such as *Bacillus thermantarcticus*, *Bacillus licheniformis*, *Geobacillus thermodenitrificans*, *Geobacillus tepidamans*, *Thermotoga maritima*, *Methanococcus jannaschii*, can produce an unusually thermostable EPS that starts to decompose at about 280 °C [68, 70].

4.2 Production of EPS by Psychrophilic Microorganisms

The microbes that can sustain at temperatures −20–20 °C are grouped under psychrophiles, but their growth is optimum at temperature <15°C [20]. This group of microorganisms usually prevail in marine environments and psychrophilic bacteria have varied molecular mechanisms to sustain at low temperatures [14]. The mechanisms include the synthesis of cold-shock proteins (CSPs) and chaperones which protect the synthesis of RNA and proteins. Generally, carboxylated polysaccharides are the EPSs produced from the psychrophilic marine microbes. The acidic properties and negative charge of the EPSs are mainly due to the presence of carboxyl groups in them [15, 16].

In the cold oceanic environment, the EPS produced by bacteria is important for supplying nutrients, the formation of biofilm, and protecting from the low temperature as well as from the high salinity. The exopolysaccharides secreted by psychrophiles can accumulate cations due to their polyanionic nature. In this way, the EPSs produced by psychrophiles act as a cryoprotectant as well as a buffering agent.

For the inhibition of recrystallization of ice, the pseudohelicoidal structure might be beneficial as evident from the molecular structural analysis of EPS produced from psychrophiles. The length of the polymer chain can also affect the physical and chemical properties of EPSs. EPS plays the role of a cryoprotectant [55] in the high salinity, the low-temperature habitat of sea ice brine channels [47, 61, 62]. The biopolymers produced by psychrophilic microbes can be utilized for the stabilization of many important enzymes used in industries during unfavorable situations [37, 40].

4.3 Production of EPS by Halophilic Microorganisms

The salt-loving microorganisms are known as halophiles. For their growth, halophiles require high salt concentration, and also they possess advanced physiological characters to live in high salt concentration [10]. Halophilic microbes require more than 0.5 M NaCl for their optimal growth. Generally, halophiles are observed to be found in salt lakes and lagoons [49].

Generally, the EPSs produced by halophilic microorganisms are heteropolysaccharides and they vary in their physical and chemical properties. The most widely recognized monosaccharide moieties are mannose and glucose. The EPSs secreted by halophiles have greater significance because of their properties like gelling properties, metal binding [14, 48], and emulsifying capacity. It has a broad range of appliances in various sectors of industries, for example, in the food industry, it is utilized as an alternative for xanthan gum [72, 85]. The members of the genus *Haloferax* and *Halomonas* have been recognized as the most potent producers of EPSs. The biosynthesis of EPS in halophilic microbes is affected by the alteration in salinity, mainly the ratio for each type of monosaccharide composition. To protect the microbes from rising salinity, some monosaccharide components are present in the EPS may require to be changed to retain their functions. The halophilic EPSs generally contain large amounts of uronic acids. The high viscosity of the EPS solution at acidic pH and the gelification capacity may be due to the high uronic acid content [9].

The halophilic strains *Halomonas maura*, *Halomonas ventosae*, *Halomonas alkaliantarctica*, *Hahellache juensis* and the archaeal halophilic *Haloferax mediterranei* isolated from hypersaline environments were also shown to produce EPS [29]. Levan, an EPS produced by *Halomonas smyrnensis,* has many applications in foods, cosmetics, chemical, and pharmaceutical industries [31]. Further, *Halomonas almeriensis*, a halophilic bacterium synthesized significant quantities of EPS which is capable of emulsifying several hydrophobic substrates [52].

The most interesting halophilic EPS is mauran, produced by *Halomonas maura*, which has a viscosifying activity and polymers from *Halomonas eurihalina* is well known for its emulsifying activity and jellifying properties at acidic pH [53]. Aqueous solutions of mauran are highly viscous and display pseudoplastic, viscoelastic, and thixotropic behavior. Its viscosity is stable over a broad pH range of 3–11, after freezing–thawing processes, and in the presence of sucrose, salts, and surfactants and alpha-hydroxyl acids. It has a high capacity for binding lead and other cations [4]. Further, the polymers secreted from *Salipiger mucescens* is rich in fucose and used in the fields of cosmetics, pharmaceuticals, food, and cosmetics industries. Fucose is used in cosmetics as an anti-aging agent supporting skin elasticity. Besides this, other halophilic EPS producers include *Idiomarina, Alteromonas*, and *Palleronia* [32, 54].

4.4 Production of EPS by Acidophilic Microorganisms

Acidophiles are defined as the group of microorganisms that grow at pH less than 3 [7]. Acidophilic microorganisms are observed to be found in natural environments like sulphuric pools, sulfidic mines, and man-made environments like acid mine drainage [80]. Generally, sulfur, sulfide, and sulfur oxide are present in the acidic environment. The acidophilic exopolysaccharides protect the cells against the stress conditions of low pH and the presence of metals.

During the process of bioleaching, acidophilic EPSs are produced with the extracellular polymeric substances by mixed cultures of microbes. The exopolymeric substance is composed mainly of polysaccharides (polysaccharide is made up of glucose, mannose, xylose, rhamnose, and uronic acids), fatty acids, protein, and ferric ion.

In the process of bioleaching, the oxidation capability of bacteria is utilized to break down metal sulfides that may assist the extraction of valuable metals. Mostly the acidophilic, sulfur-oxidizing, and autotrophic iron-oxidizing bacteria are involved in this process [24, 58]. *Acidithiobacillus caldus* and *Leptospirillum ferriphilum* are the dominant group of bacteria in the mixed culture that can produce EPSs during the process of bioleaching [89].

4.5 Production of EPS by Alkaliphilic Microorganisms

Alkaliphilic microorganisms generally consist of two main physiological groups of microorganisms namely alkaliphiles and haloalkaliphiles. Alkaliphiles are the microbes that grow well at pH values exceeding pH 9, often in the 10–13 range of pH (optimal growth pH of around 10), but unable to grow or grow gradually at near-neutral pH values, while haloalkaliphiles require both an alkaline pH (>pH 9) and high salinity (up to 33% [wt/vol] NaCl) [39]. The most common stable alkaline environments are soda lakes and deserts.

The alkaliphilic microorganisms produce EPS as their metabolic products like the other group of extremophiles. The EPS produced by alkaliphiles is useful for the adhesion of microbial cells to matrix or substratum [67]. *Halomonas alkaliantartica* a haloalkaliphilic microbe was found to produce EPS that is proposed to have a role as an emulsifying agent in oil recovery [30]. Further, *Halomonas alkaliantartica* isolated from a saline lake in Antarctica produced maximum EPS when acetate was used as a carbon source [69]. *Vagococcus carniphilus*, a gram positive alkaliphilic microbe isolated from Lonar Lake, was also observed to produce EPS. The flocculating activity of the polysaccharide of *V. carniphilus* was found to be comparable to that of xanthan gum, suggesting that the polysaccharide has potential as a flocculating agent [33, 41].

5 Biotechnological Application of Extremophilic EPS

Biotechnological appliances of EPS commence from conventional areas as food, drug, and cosmetic industries to novel biomedicine zones [21, 81]. Extremophilic microorganisms give non-pathogenic EPS which are appropriate for applications in the drug, food, and cosmetics industries as emulsifiers, stabilizers, gel agents,

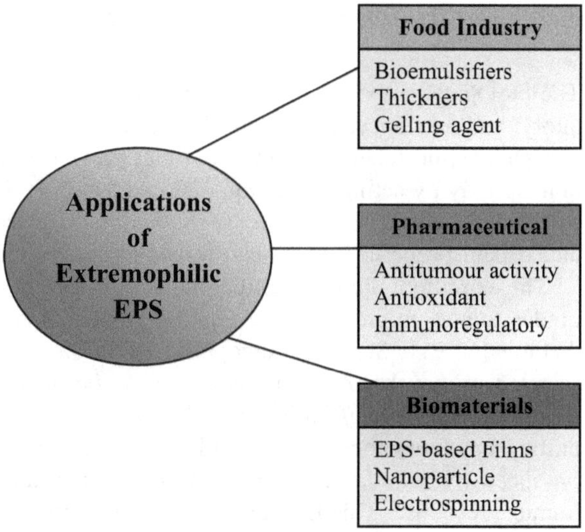

Fig. 4 Schematic representation of applications of extremophilic EPS in different fields

coagulants, thickeners, and suspending agents [63, 83]. The schematic representation of the application of EPS produced from extremophilic microorganisms is given below (Fig. 4).

5.1 Application in Food Industry

The EPSs produced by extremophiles are used as an emulsifier in the food industries. They are heteropolysaccharides in nature and can stabilize emulsions over a broad range of pH, temperature, and salinity [90].

As per the report of [4], the factors like changes in pH, the presence of salts or sugars, lactic acid, or freezing and thawing will not affect the viscosity and the pseudo-plasticity of halophilic EPS mauran. But, the protein content of the extremophilic exopolysaccharide may be necessary for emulsifying activity. Additionally, the acetyl group and uronic acid present in EPS also play a key role in its emulsifying activity [56].

One attractive characteristic of the EPSs is high pseudo-plasticity in various types of food products like pudding, syrup, cake, sauce, and salad dressing [38]. This pseudoplastic nature of EPS is beneficial to create comfortable sensory properties for instance flavor release. Cheap and renewable substrates like molasses should be utilized as a substitute for sucrose for the production of EPS [77].

5.2 Application in Pharmaceutical Industry

The biological activities of extremophilic EPSs include antioxidant, immunoregulatory, and anticancer properties [3, 60]. The EPSs produced by extremophiles are non-pathogenic to human beings. It can prevent oxidative injuries caused in the human body by acting as a natural antioxidant [84]. Furthermore, the EPS secreted by *Halomonas nitroreducens* (a halophilic bacterium) is well known for its antioxidant properties which can forage hydroxyl and DPPH radicals [19].

The extremophilic EPSs could play a vital role in the treatment of cancer. It can hinder tumor progression and can be applied in macrophage-mediated immune therapy [6, 91]. Some extremophilic EPSs are non-cytotoxic such as the EPS secreted by psychrophilic bacteria *Pseudoalteromonas* sp., halothermophilic bacteria *Bacillus licheniformis,* and *Geobacillus thermodenitrificans.* The extremophilic EPSs have potential to treat herpes virus infection [36, 82]. The exopolymer produced by *Bacillus licheniformis* acts as immunomodulator and stimulate the immune response. This EPS finally contributes to the antiviral immune defense and inhibit herpes simplex virus type 2 (HSV-2) replication, improving the immune surveillance performed by human peripheral blood mononuclear cells (PBMC) toward HSV-2 infection by triggering the production of Th1-type cytokines (IFN-γ, IFN-α, TFN-α, IL-12, and IL-18) [36].

5.3 Application in Productions of Biomaterials

The extremophilic EPSs have several implications in the production of biomaterials [1] which includes EPS-based films, nanoparticles, and manufacturing of synthetic and natural polymers through electrospinning.

The inborn purpose of EPSs is to give adhesion and provide security to the microbes inhabiting unfavorable conditions. The surface of EPS-based films have better biocompatibility, and the films give cohesive strength. For example, in the formation of the hybrid film, levan (a halophilic EPS) was combined with polyethylene oxide (PEO) and chitosan, then the mixed film indicated improved biocompatibility in comparison with chitosan-PEO film [11].

In the field of nanotechnology, the EPSs secreted by different extremophiles can be applied in two different manners. In one way, the EPS is directly used to form nanoparticles, and in the other way, EPS is used for the encapsulation of nanoparticles produced from other material. Levan, one type of halophilic EPS secreted by *Halomonas smyrnensis*, is used as a carrier of nanoparticles. EPS-coated nanoparticles have reduced cellular toxicity in comparison with the uncoated ones and therefore can be used in the drug industries [8].

The materials electrospun from extremophilic EPSs is biocompatible, biodegradable, and non-hazardous. EPS-based electrospun biomaterials are used in the biomedical sector as EPSs are highly water-soluble and it avoids toxic solvents

Table 2 Some applications of EPS in biotechnology in the modern era reported by many researchers

Name and group of EPS	Produced by bacteria	Application details	References
Thermophilic EPS	*Geobacillus* sp.	Inhibits the proliferation of hepatoma carcinoma cell	[85]
Thermophilic EPS	*Thermus aquaticus*	Acts as an immunomodulator that stimulates macrophage cells to secret TNFα and IL6	[51]
EPSs from halothermophilic strain and thermophile	*Bacillus licheniformis* & *Geobacillus thermodenitrificans*	Used in the treatment of viral infection (herpes)	[2]
Halophilic EPS levan	*Halomonas smyrnensis*	Used as nanocarrier	[79]
Halophilic EPS	*Halomonas smyrnensis*	Enhances the antitumor efficacy	[78]
Halophilic EPS mauran	*Halomonas maura*	Electrospinning	[74]
Psychrophilic EPS	*Pseudoalteromonas* sp	applied in cancer treatment	[18]
Psychrophilic EPS	*Pseudomonas* sp.	Used as emulsifier	[13]

and chemicals during the process of electrospinning. Mauran, a halophilic EPS mixed with PVA (poly vinyl alcohol) and subjected to electrospinning to produce nanofibers which will enhance cell adhesion and proliferation. A tremendous property of mauran is the maintenance of stable viscosity under extreme pH, and this property of mauran is highly favorable in getting steady electrospinning conditions [73].

Many EPSs secreted by extremophiles function as bioflocculant [50]. Chitosan and algin are examples of naturally occurring biopolymer flocculants (bioflocculants). Some of the applications of EPSs, especially in the field of biotechnology in this modern era are given in Table 2.

6 Conclusion

Extremophiles have developed numerous policies to sustain the harsh environmental conditions. Production of EPS is an important strategy for the supply of nutrients, promoting chemical reactions, and safeguard the cells against unfavorable abiotic factors as well as from toxic materials. The utilization of extremophilic EPS in food, biomedical, and biomaterial products is determined by biocompatibility, biodegradability, and non-toxicity. EPSs are industrially significant due to their thickening, gelling, and emulsion capacities. The combination of classic microbiological techniques with modern high-throughput methods will improve efforts to explore novel EPS from extreme environments produced by extremophiles and their functions in the field of biotechnology.

References

1. Aravamudhan A, Ramos DM, Nada AA, Kumbar SG (2014) Natural polymers: polysaccharides and their derivatives for biomedical applications. In: Kumbar S, Laurencin C, Deng M (eds) Natural and synthetic biomedical polymers. Elsevier, Brazil, pp 67–89
2. Arena A, Gugliandolo C, Stassi G et al (2009) An exopolysaccharide produced by *Geobacillus thermodenitrificans* strain B3–72: antiviral activity on immunocompetent cells. Immunol Lett 123(2):132–137
3. Arena A, Maugeri TL, Pavone B et al (2006) Antiviral and immunoregulatory effect of a novel exopolysaccharide from a marine thermotolerant *Bacillus licheniformis*. Int Immunopharmacol 6(1):8–13
4. Arias S, del Moral A, Ferrer MR et al (2003) Mauran, an exopolysaccharide produced by the halophilic bacterium *Halomonas maura*, with a novel composition and interesting properties for biotechnology. Extremophiles 7(4):319–326
5. Ates O (2015) Systems biology of microbial exopolysaccharides production. Front Bioeng Biotechnol 3:200
6. Bai Y, Zhang P, Chen G et al (2012) Macrophage immunomodulatory activity of extracellular polysaccharide (PEP) of Antarctic bacterium *Pseudoaltermonas* sp.S-5. Int Immunopharmacol 12(4):611–617
7. Baker-Austin C, Dopson M (2007) Life in acid: pH homeostasis in acidophiles. Trends Microbiol 15(4):165–171
8. Banerjee A, Bandopadhyay R (2016) Use of dextran nanoparticle: a paradigm shift in bacterial exopolysaccharide based biomedical applications. Int J Biol Macromol 87:295–301
9. Bejar V, Llamas I, Calvo C et al (1998) Characterization of exopolysaccharides produced by 19 halophilic strains of the species *Halomonas eurihalina*. J Biotechnol 61(2):135–141
10. Biswas J, Paul AK (2017) Diversity and production of extracellular polysaccharide by halophilic microorganisms. Biodivers Int J 1(2):00006
11. Bostan MS, Mutlu EC, Kazak H et al (2014) Comprehensive characterization of chitosan/PEO/levan ternary blend films. Carbohyd Polym 102:993–1000
12. Canganella F, Wiegel J (2011) Extremophiles: from abyssal to terrestrial ecosystems and possibly beyond. Naturwissenschaften 98:253–279. https://doi.org/10.1007/s00114-011-0775-2
13. Carrion O, Delgado L, Mercade E (2015) New emulsifying and cryoprotective exopolysaccharide from Antarctic *Pseudomonas* sp. ID1. Carbohyd Polym 117:1028–1034
14. Caruso C, Rizzo C, Mangano S et al (2018) Extracellular polymeric substances with metal adsorption capacity produced by *Pseudoalteromonas* sp. MER144 from Antarctic seawater. Environ Sci Pollut Res 25(5):4667–4677
15. Casillo A, Parrilli E, Sannino F et al (2017) Structure-activity relationship of the exopolysaccharide from a psychrophilic bacterium: a strategy for cryoprotection. Carbohyd Polym 156:364–371
16. Casillo A, Stahle J, Parrilli E et al (2017) Structural characterization of an all-aminosugar-containing capsular polysaccharide from *Colwellia psychrerythraea* 34H. Antonie Van Leeuwenhoek 110(11):1377–1387
17. Casillo A, Lanzetta R, Parrilli M et al (2018) Exopolysaccharides from marine and marine extremophilic bacteria: structures, properties, ecological roles and applications. Mar Drugs 16

18. Chen M, Li Y, Liu Z et al (2018) Exopolysaccharides from a Codonopsispilosula endophyte activate macrophages and inhibit cancer cell proliferation and migration. Thoracic Cancer 9 (5):630–639
19. Chikkanna A, Ghosh D, Kishore A (2018) Expression and characterization of a potential exopolysaccharide from a newly isolated halophilic thermotolerant bacteria *Halomonas nitroreducens* strain WB1. PeerJ 6:e4684
20. Clarke A, Morris GJ, Fonseca F et al (2013) A low temperature limit for life on Earth. PLoS ONE 8:e66207. https://doi.org/10.1371/journal.pone.0066207
21. Costa RR, Neto AI, Calgeris I et al (2013) Adhesive nanostructured multilayer films using a bacterial exopolysaccharide for biomedical applications. J Mater Chem B 1(18):2367–2374
22. Costa OY, Raaijmakers J, Kuramae E (2018) Microbial extracellular polymeric substances: ecological function and impact on soil aggregation. Front Microbiol 9:1636
23. Cragg G, Newman D (2007) 1.08—natural product sources of drugs: plants, microbes, marine organisms, and animals. Compr Med Chem 1:355–403
24. d'Hugues P, Joulian C, Spolaore P et al (2008) Continuous bioleaching of a pyrite concentrate in stirred reactors: population dynamics and exopolysaccharide production vs. bioleaching performance. Hydrometallurgy 94(1–4):34–41
25. Davey ME, O'toole GA (2000) Microbial biofilms: from ecology to molecular genetics. Microbiol Mol Biol Rev 64(4):847–867. https://doi.org/10.1128/mmbr.64.4.847-867.2000
26. Decho AW (2000) Microbial biofilms in intertidal systems: an overview. Cont Shelf Res 20 (11):1257–1273
27. Decho AW (1990) Microbial exopolymer secretions in ocean environments: their role(s) in food webs and marine processes. In: Barnes M (ed) Oceanography and marine biology: an annual review. Aberdeen University Press, Aberdeen, UK, pp 73–153
28. Decho AW, Hernd GJ (1995) Microbial activities and the transformation of organic matter within mucilaginous material. Sci Total Environ 65:33–42
29. Delbarre-Ladrat C, Sinquin C, Lebellenger L et al (2014) Exopolysaccharides produced by marine bacteria and their applications as glycosaminoglycan-like molecules. Front Chem 2:85. https://doi.org/10.3389/fchem.2014.00085
30. Deshmukh S, Kanekar P, Bhadekar R (2017) Production and characterization of an exopolysaccharide from marine moderately halophilic bacterium *Halomonas smyrnensis* SVD III. Int J Pharm Pharm Sci 9(10):146–151
31. Diken E, Ozer T, Arıkan M, Emrence Z et al (2015) Genomic analysis reveals the biotechnological and industrial potential of levan producing halophilic extremophile, *Halomonas smyrnensis* AAD6T. Springerplus 4:393
32. Fernando MC, Emilia QM, Jose M, Inmaculada L et al (2005) Palleroniamarisminoris gen. a moderately halophilic, exopolysaccharide producing bacterium belonging to the Alpha proteo bacteria, isolated from a saline soil. Int J Sys EvolMicr 55:2525–2530
33. Gao J, Bao H, Xin M, Liu Y et al (2006) Characterization of a bioflocculant from a newly isolated *Vagococcus* sp. W31*. J Zhejiang Univ Sci B 7(3):186–192
34. Gilbert P, Das J, Foley I (1997) Biofilm susceptibility to antimicrobials. Adv Dent Res 11 (1):160–167
35. Gomes E, Souza AR, Orjuela G, Silva R et al (2016) Applications and benefits of thermophilic microorganisms and their enzymes for industrial biotechnology in book gene expression systems in fungi: advancements and applications, pp 459–492. https://doi.org/10.1007/978-3-319-27951-0_21
36. Gugliandolo C, Spano A, Maugeri T, Poli A et al (2015) Role of bacterial exopolysaccharides as agents in counteracting immune disorders induced by Herpes Virus. Microorganisms 3 (3):464–483
37. Hamdan A (2018) Psychrophiles: ecological significance and potential industrial application. S Afr J Sci 114

38. Han P, Sun Y, Wu X et al (2014) Emulsifying, flocculating, and physicochemical properties of exopolysaccharide produced by cyanobacterium *Nostocflagelliforme*. Appl Biochem Biotechnol 172(1):36–49
39. Horikoshi K (1999) Alkaliphiles: some applications of their products for biotechnology. Microbiol Mol Biol Rev 63(4):735–750
40. Huston AL, Methe B, Deming JW (2004) Purification, characterization, and sequencing of an extracellular cold-active aminopeptidase produced by marine psychrophile *Colwellia psychrerythraea* strain 34H. Appl Environ Microbiol 70(6):3321–3328
41. Joshi AA, Kanekar PP (2011) Production of exopolysaccharide by *Vagococcus carniphilus* MCM B-1018 isolated from alkaline Lonar lake, India. Ann Microbiol 61:733–740
42. Kambourova M (2018) Thermostable enzymes and polysaccharides produced by thermophilic bacteria isolated from Bulgarian hot springs. Eng Life Sci 18(11):758–767
43. Kambourova M, Derekova A (2013) Developments in thermostable gellan lyase. In: Satyanarayana T, Littlechild J, Kawarabayasi Y (eds) Thermophilic microbes in environmental and industrial biotechnology: biotechnology of thermophiles. Springer, Berlin, pp 711–730
44. Kambourova M, Mandeva R, Dimova D, Poli A et al (2009) Production and characterization of a microbial glucan, synthesized by *Geobacillus tepidamans* V264 isolated from Bulgarian hot spring. Carbohyd Polym 77:338–343
45. Kambourova M, Radchenkova N, Tomova I et al (2016) Thermophiles as a promising source of exopolysaccharides with interesting properties. In: Rampelotto PH (ed), Biotechnology of extremophiles, grand challenges in biology and biotechnology 1, Bulgarian Academy of Sciences. Springer International Publishing, Switzerland. https://doi.org/10.1007/978-3-319-13521-2_4
46. Kazak H, Oner E, Dekker RH (2010) Extremophiles as sources of exopolysaccharides (Chapter 19)
47. Kristiansen E, Zachariassen KE (2005) The mechanism by which fish antifreeze proteins cause thermal hysteresis. Cryobiology 5:262–280. https://doi.org/10.1016/j.cryobiol.2005.07.007
48. Kulichevskaya IS, Milekhina EI, Borzenkov IA, Zvyagintseva IS et al (1992) Oxidation of petroleum hydrocarbons by extremely halophilic archaebacteria. Microbiology 60:596–601
49. Kunte HJ, Truper HG, Stan- H (2002) Halophilic microorganisms. In: Horneck G, Baumstark C (eds) Astrobiology: the quest for the conditions of life. Springer, Berlin, pp 185–200
50. Li WW, Zhou WZ, Zhang YZ, Wang J et al (2008) Flocculation behavior and mechanism of an exopolysaccharide from the deep-sea psychrophilic bacterium *Pseudoalteromonas* sp. SM9913. Bioresour Technol 99:6893–6899
51. Lin M, Yang Y, Chen Y et al (2011) A novel exopolysaccharide from the biofilm of *Thermus aquaticus* YT-1 induces the immune response through Toll-like receptor 2. J Biol Chem 286 (20):17736–17745
52. Llamas I, Amjres H, Mata JA, Quesada E et al (2012) The potential biotechnological applications of the exopolysaccharide produced by the Halophilic Bacterium *Halomonas almeriensis*. Molecules 17(6):7103–7120
53. Llamas I, Mata JA, Tallon R, Bressollier P et al (2010) Characterization of the exopolysaccharide produced by *Salipiger mucosus* A3T, a halophilic species belonging to the Alphaproteobacteria, isolated on the Spanish Mediterranean Seaboard. Mar Drugs 8:2240–2251
54. Lulu W, Hong Z, Long Y, Xinle L et al (2017) Structural characterization and bioactivity of exopolysaccharide synthesized by *Geobacillus* sp. TS3–9 isolated from radioactive radon hot spring. Adv Biotech Micro 4(2):555634. https://doi.org/10.19080/AIBM.2017.04.555635
55. Marx JG, Carpenter SD, Deming JW (2009) Production of cryoprotectant extracellular polysaccharide substances (EPS) by the marine psychrophilic bacterium *Colwellia psychrerythraea* strain 34H under extreme conditions. Can J Microbiol 55(1):63–72

56. Mata JA, Bejar V, Llamas I et al (2006) Exopolysaccharides produced by the recently described halophilic bacteria *Halomonas ventosae* and *Halomonas anticariensis*. Res Microbiol 157(9):827–835
57. Merino N, Aronson HS, Bojanova DP et al (2019) Living at the extremes: extremophiles and the limits of life in a planetary context. Front Microbiol 10:780. https://doi.org/10.3389/fmicb.2019.00780
58. Michel C, Beny C, Delorme F et al (2009) New protocol for the rapid quantification of exopolysaccharides in continuous culture systems of acidophilic bioleaching bacteria. Appl Microbiol Biotechnol 82(2):371–378
59. Moriello SM, Lama L, Poli A, Gugliandolo C et al (2003) Production of exopolysaccharides from a thermophilic microorganism isolated from a marine hot spring in flegrean areas. J Microbiol Biotech 30(2):95–101
60. Moscovici M (2015) Present and future medical applications of microbial exopolysaccharides. Front Microbiol 6:1012. https://doi.org/10.3389/fmicb.2015.01012
61. Nichols CM, Bowman JP, Guezennec J (2005) Effects of incubation temperature on growth and production of exopolysaccharides by an Antarctic sea ice bacterium grown in batch culture. Appl Environ Microbiol 71(7):3519–3523
62. Nichols CM, Guezennec J, Bowman JP (2005) Bacterial exopolysaccharides from extreme marine environments with special consideration of the southern ocean, sea ice, and deep-sea hydrothermal vents: a review. Mar Biotechnol 7(4):253–271. https://doi.org/10.1007/s10126-004-5118-2
63. Nicolaus B, Kambourova M, Oner ET (2010) Exopolysaccharides from extremophiles: from fundamentals to biotechnology. Environ Technol 31(10):1145–1158
64. Nicolaus B, Manca MC, Romano I, Lama L (1993) Production of an exopolysaccharide from two thermophilic archaea belonging to the genus Sulfolobus. FEMS Microbiol Lett 109(2–3):203–206
65. Nicolaus B, Moriello VS, Lama L et al (2004) Polysaccharides from extremophilic microorganisms. Orig Life Evol Biosph 34(1):159–169
66. Orellana R, Macaya C, Bravo G et al (2018) Living at the frontiers of life: extremophiles in chile and their potential for bioremediation. Front Microbiol 9:2309. https://doi.org/10.3389/fmicb.2018.02309
67. Perry TD, Klepac-Ceraj V, Zhang XV et al (2005) Binding of harvested bacterial exopolymers to the surface of calcite. Environ Sci Technol 39(22):8770–8775
68. Poli A, Anzelmo G, Nicolaus B (2010) Bacterial exopolysaccharides from extreme marine habitats: production, characterization and biological activities. Mar Drugs 8(6):1779–1802
69. Poli A, Esposito E, Oriando P, Lama L et al (2007) *Halomonas alkaliantarctica* sp. nov. isolated from saline lake cape russell in antarctica, an alkalophilic moderately halophilic, exopolysaccharide-producing bacterium. Syst Appl Microbiol 30:31–38
70. Poli A, Donato PD, Abbamondi GR, Nicolaus B (2011) Synthesis, production, and biotechnological applications of exopolysaccharides and polyhydroxyalkanoates by Archaea. Archaea
71. Radchenkova N, Vassilev S, Panchev I, Anzelmo G et al (2013) Production and properties of two novel exopolysaccharides synthesized by a thermophilic bacterium *Aeribacillus pallidus* 418. Appl Biochem Biotechnol 171(1):31–43
72. Rau DC, Parsegian VA (1990) Direct measurement of forces between linear polysaccharides xanthan and schizophyllan. Science 249:1278–1281
73. Raveendran S, Dhandayuthapani B, Nagaoka Y et al (2013) Biocompatible nanofibers based on extremophilic bacterial polysaccharide Mauran from *Halomonas maura*. Carbohyd Polym 92(2):1225–1233
74. Raveendran S, Poulose AC, Yoshida Y et al (2013) Bacterial exopolysaccharide based nanoparticles for sustained drug delivery, cancer chemotherapy and bioimaging. Carbohyd Polym 91(1):22–32

75. Rinker KD, Kelly RM (1996) Growth physiology of the hyperthermophilic archaeon *Thermococcus litoralis*: development of a sulfur-free defined medium, characterization of an exopolysaccharide, and evidence of biofilm formation. Appl Environ Microbiol 62(12):4478–4485
76. Roberts IS (1996) The biochemistry and genetics of capsular polysaccharide production in bacteria. Ann Rev Microbiol 50:285–315
77. Sam S, Kucukasik F, Yenigun O et al (2011) Flocculating performances of exopolysaccharides produced by a halophilic bacterial strain cultivated on agro-industrial waste. Bioresour Technol 102(2):1788–1794
78. Sarilmiser HK, Oner ET (2014) Investigation of anti-cancer activity of linear and aldehyde-activated levan from *Halomonas smyrnensis* AAD6T. Biochem Eng J 92:28–34
79. Sezer AD, Kazak H, Oner ET et al (2011) Levan-based nanocarrier system for peptide and protein drug delivery: Optimization and influence of experimental parameters on the nanoparticle characteristics. Carbohyd Polym 84(1):358–363
80. Sharma A, Parashar D, Satyanarayana T (2016) Acidophilic microbes: biology and applications in biotechnology of extremophiles, pp 215–241
81. Singha T (2012) Microbial extracellular polymeric substances: production, isolation and applications. IOSR J Pharm 2(2):276–281
82. Spano A, Lagana P, Visalli G et al (2016) In vitro antibiofilm activity of an exopolysaccharide from the marine thermophilic *Bacillus licheniformis* T14. Curr Microbiol 72(5):518–528
83. Tango MSA, Islam MR (2002) Potential of extremophiles for biotechnological and petroleum applications. Energy Sources 24:543–559. https://doi.org/10.1080/00908310290086555
84. Wang H, Liu YM, Qi ZM et al (2013) An overview on natural polysaccharides with antioxidant properties. Curr Med Chem 20(23):2899–2913
85. Wang J, Salem D, Sani R (2019) Extremophilic exopolysaccharides: a review and new perspectives on engineering strategies and applications. Carbohyd Polym 205:8–26
86. Wolfaardt GM, Lawrence JR, Korbe DR (1999) Function of EPS. In: Wingender J, Neu TR, Flemming H-C (eds) Microbial extracellular polymeric substances: characterization, structure and function. Springer-Verlag, New York, NY, USA, pp 171–200
87. Xu L, Wu YH, Zhou P et al (2018) Investigation of the thermophilic mechanism in the genus Porphyrobacter by comparative genomic analysis. BMC Genom 19:385. https://doi.org/10.1186/s12864-018-4789-4
88. Yildiz SY (2019) Exopolysaccharide production by thermophilic microorganisms. In: ÖzlemAteşDuru (ed) Microbial exopolysaccharides: current research and developments. Caister Academic Press, U.K, pp 57–82. https://doi.org/10.21775/9781912530267.03
89. Zeng W, Qiu G, Zhou H et al (2010) Characterization of extracellular polymeric substances extracted during the bioleaching of chalcopyrite concentrate. Hydrometallurgy 100(3–4):177–180
90. Zheng C, Li Z, Su J et al (2012) Characterization and emulsifying property of a novel bioemulsifier by *Aeribacillus pallidus* YM-1. J Appl Microbiol 113(1):44–51
91. Zong A, Cao H, Wang F (2012) Anticancer polysaccharides from natural resources: a review of recent research. Carbohyd Polym 90(4):1395–1410

Pullulan: Biosynthesis, Production and Applications

Supriya Pandey, Ishita Shreshtha, and Shashwati Ghosh Sachan

Abstract Pullulan, a exopolysaccharide polymer derived mainly from *Aureobasidium pullulan*, consists of alpha-1,4 linked maltotriose units which are connected by alpha-1,6 linkages between the terminal glycosidic residues of the trisaccharide and provides resistance against cell desiccation and predation. This chapter discusses pullulan chemistry and organisms responsible for its production, with an emphasis on fungus *Aureobasidium pullulan*. Further, the chapter focuses on pullulan biosynthesis, factors affecting the pathway, summarizing the state-of-the-art of production, upstream and downstream process, and its applications. Mechanism of biosynthesis intends to illustrate the key features involved in the intracellular pullulan synthesis using various sources such as glucose and sucrose. Pullulan production via fermentation process is widely accepted and this biopolymer shows large range of applications in various fields such as drug delivery, environment, medical science, food technology and nanotechnology.

Keywords *Aureobasidium pullulan* · Biosynthesis · Drug delivery · Fermentation · Food technology · Medical science · Nanotechnology · Production · Pullulan

1 Introduction

Polysaccharides are present in all life forms and exhibit multifarious biochemical structure. The microbial polysaccharides are found in different forms such as capsular polysaccharides (CPS), structural polysaccharides (SPS) and exopolysaccharide (EPS). Microbial EPS are omnipresent, consist of repeated sugar moiety units and show important industrial and biotechnological applications. Accumulation of EPS on the microbial cell surface stabilizes the membrane

S. Pandey · I. Shreshtha · S. G. Sachan (✉)
Department of Bio-Engineering, Birla Institute of Technology, Mesra 835215, Jharkhand, India
e-mail: ssachan@bitmesra.ac.in

A. K. Nadda et al. (eds.), *Microbial Exopolysaccharides as Novel and Significant Biomaterials*, Springer Series on Polymer and Composite Materials,
https://doi.org/10.1007/978-3-030-75289-7_6

structure and thus, provides protection to the cells against the unfavourable external environment. Number of microorganisms such as bacteria and cyanobacteria [21, 84, 83], marine microalga such as *Chroomonas* sp. [6] and *Dunaliella salina* [74], yeast [25], basidiomycetes [22, 69] such as the medicinal mushroom *Phellinus linteus* [124], ascomycetes such as *Aspergillus japonicus* [73], have been reported that produce various kinds of EPS. EPS such as xanthan, gellan, pullulan, dextran, alginate, acetan, welan and cellulose are being commercially used from food processing to pharmaceuticals further extended to detoxification, bioremediation, petrochemicals, and other biotechnological applications.

2 About Pullulan

Among the various exopolysaccharides, pullulan is an important EPS firstly reported by Bauer [5]. Pullulan is mainly produced by *Aureobasidium pullulan,* a ubiquitous fungus mainly present in humid climates such as land-soil, water bodies, air and plant materials such as crop and wood [36, 56]. Pullulan is colourless, odourless and a slow digesting macromolecule used as low calorific additives. Since it is not obtained from any animal source, it can be used by all consumer groups [75]. According to the FDA, pullulan consumption should be 10 g per day [105]. Overall production cost of pullulan is the major obstacle with its utilization as it is 3 times more expensive than any other polysaccharide [75]. The structure and properties of pullulan, mark their role for a range of industrial applications such as pharmaceutical, food processing and cosmetics. Their role is further extended in areas such as bioremediation, health and medical sciences.

2.1 Structure and Properties

Pullulan is a linear [18], non-ionic, hydrophilic, exopolysaccharide obtained as amorphous slime matter from a polymorphic fungus, *A. pulluan* [20, 105]. Aerobic fermentation broth of *A. pullulan* yields pullulan as a highly water-soluble homo polysaccharide with linear α-D-glucan monomer units that constitute of regularly occurring α-(1→4) maltotriose units connected by glycosidic linkages. Moving further the respective maltotriose units are linked by α-(1→6) linkages [35]. This unique linkage pattern of pullulan structure helps in its characteristic physical features such as hydrophilicity and elasticity. Production parameter of pullulan is affected by its molecular weight which range in between $4.5 \times 10^{4-6} \times 10^{5}$ Da [55]. This polymer exhibits, distinct chemical reactions such as:

1. Produces Cu^{2+} complexes
2. Shows no colour change on treatment with I_2 [113]

It exhibits unique linkage pattern due to which it shows peculiar physical properties such as adhesiveness, capacity to form fibres and thin bio-degradable films. These bio-degradable films formed are impermeable to oxygen, oil resistant and have transparency. Because of this oxygen barrier property pullulan is widely used in food packaging industries and also for food preservation [122]. When dissolved in H_2O pullulan produces viscous solution and remains stable in presence of metal ions. The viscosity of pullulan solution formed is independent of factors such as pH, temperature and salt concentration [41]. It is highly hygroscopic, thermally stable and decomposes, at higher temperatures (250–280 °C). Pullulan are crystalline, dry white powders, amorphous, absorb moisture and dissolve easily in hot or cold H_2O [75]. It has excellent mechanical strength, foam retention properties, capacity to form films, adhesive ability and can be broken down by enzymes during the production process. The potential characteristics of pullulan has made it a good biopolymer which is edible, non-mutagenic, odourless, non-carcinogenic, tasteless and non-toxic in nature. It dissolves in water but cannot dissolve in organic solvents except for dimethyl formamide and dimethyl sulfoxide [52, 112]. During the process of drying and dissolution in water, two characteristics properties of pullulan, the adhesive and foam preservation property, can be observed respectively [47]. The adhesive property of pullulan makes it suitable for cosmetic production and ease the use of active or viable particles on the skin [52, 92, 106]. The degradation index (temperature at which polymer properties such as molecular weight, colour, shape, tensile strength changes and it starts to degrade) of pullulan is 0.7 with intervals of 48 h of incubation whereas for dextran, it is 0.05 given the same condition, therefore, the degradation rate of pullulan in blood serum is much faster as compared to dextran [7]. Pullulan resists mammalian amylases therefore cannot be broken down into simple units to obtain energy. Thus, its partial breakdown releases small amount of energy and provides less calories so it act as a dietary fibre. It can undergo susceptible chemical modification and can synthesize a range of derivatives such as pullulan having cholesterol, acetate, carboxymethyl, succinylate and amine groups. This process is facilitated by two pullulan properties, chain flexibility and hydroxyl functional groups present in pullulan [103]. Further, pullulan is being used as texturizing and glazing component in both, chewing and bubble gum. It is also used to create foam in milk-desserts. It is promoted as a part of the drug delivery system because it is mouldable (powdered form of pullulan can be compressed and moulded into different shape and size), spinnable (fibres and filaments like structures can be obtained by spinning pullulan), and at the same time, it is non-toxic, bio-degradable, bio-compatible [35], non-immunogenic and inert in nature [75].

2.1.1 Pullulan Degradation and Derivatization

Pullulan degradation into monosaccharide units takes place by the action of series of enzymes. The classification of enzymes degrading pullulan [90] are as follows:

(1) Pullulanases: This group of enzymes (α-dextrin, 6-glucannohydrolase) leads to hydrolysation of α-(1–6)-glucosidic linkage of pullulan to produce maltotriose unit.
(2) Glucoamylases: This enzymatic group produces glucose by hydrolysis of pullulan from the non-reducing end.
(3) Neopullulanases: They produce panose by hydrolysis of α-(1–4)-glucosidic linkage.
(4) Isopullulanases: They are responsible for the production of isopanose by hydrolysis of α-(1–4)-glucosidic linkage.

To increase the activity and its application range we can change pullulan into its derivate (derivatization). The process of derivatization is carried out via several chemical reactions such as esterification, etherification, sulfation, oxidation and copolymerization. The hydroxyl groups (9-OH groups) found on the repeating units of pullulan can be used to carry out the substitution reaction using different chemical groups. Pullulan attains its distinctive properties on account of its distinctive α-1-4 and α-1-6 glycosidic linkages [90].

3 Biosynthesis of Pullulan

The biosynthesis of pullulan synthesizes two different polysaccharides including pullulan and a water-insoluble material which is jelly-like in nature. A study of cell walls of *A. pullulans* using an electron microscope was performed by Simon et al. [101] which revealed that both these polysaccharides are localized on an outer surface of the *chlamydospores*, which are the cells considered as the main producers of polysaccharide [100]. Most microorganisms prefer to cover themselves with a highly hydrated EPS layer, which may protect them against predation by protozoans or from desiccation [49].

Pullulan synthesis occurs intracellularly at the cell wall membrane. It is secreted out to the cell surface forming a loose, slimy layer [101]. Though the synthesis of pullulan is not very specifically defined, we will discuss in brief the biosynthesis of pullulan from the following two microorganisms: *Aureobasidium pullulans* (Fig. 1) and *Aureobasidium melanogenum.*

Mechanism of biosynthesis from *A. pullulans*

The cells of *A. pullulans* synthesize ATP via the pentose monophosphate route [100]. Polysaccharides break down into simpler monosaccharides which in turn change into fructose 6-phosphate following the glycolysis pathway upon the action of various enzymes like isomerase and hexokinase. Fructose 6-phosphate changes into Glucose 6-Phosphate via a reversible reaction. It breaks down into Glucose 3-Phosphate subsequently transforming into Uridine Diphosphate Glucose (UDPG), also known as the pullulan precursor. UDGP is an important medium and the first step for synthesis of pullulan, [11] as it triggers the attachment of D-glucose

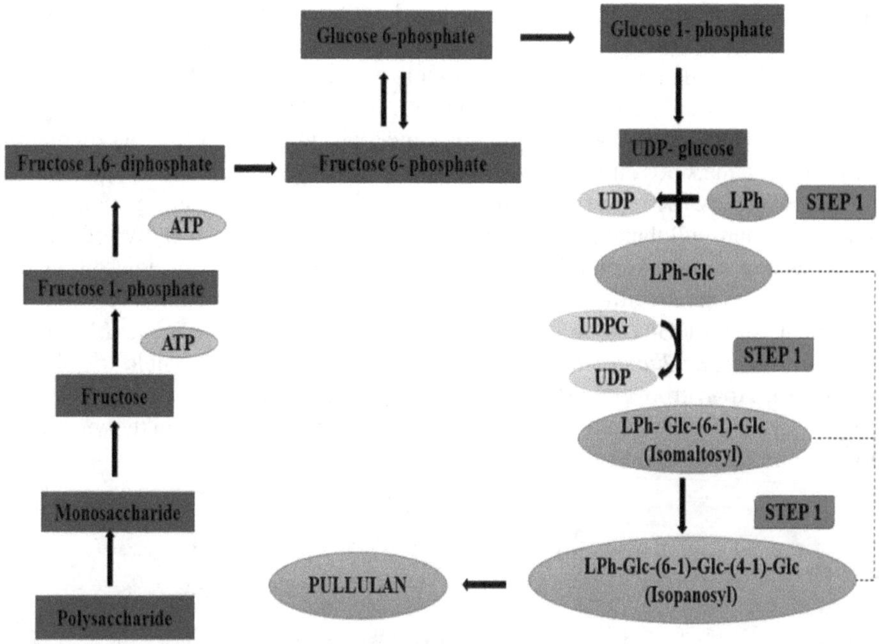

Fig. 1 Mechanism of pullulan production from *A. pullulan*

residues to the lipid molecule (lipid-hydroperoxides) with the help of phosphoester linkage. In the next step, lipid-linked isomaltose is synthesized from the further transfer of D-glucose residue from UDGP. Subsequently, participation of isomaltosyl with lipid-linked glucose yields an isopanosyl residue that act as precursor. Finally, the isopanosyl residue undergo polymerization to form the pullulan chain. Three key enzymes, α-phosphoglucose mutase, UDPG-pyrophosphorylase, and glucosyltransferase, should be present to convert the obtained glucose units into pullulan [19]. Apart from glucose, *A. pullulans* also utilizes sucrose, mannose, galactose, maltose and fructose. It also utilizes agricultural and agro-industrial wastes such as corn residue, olive oil waste, sugarcane bagasses to obtain carbon source from them [10, 52, 68].

A new yeast-like fungal strain related to the *Aureobasidium pullulans* family was obtained from leaf surfaces collected from the Kheyroodkenar forest, Iran [37]. It showed the same morphology, DNA molecular and other properties as the *A. pullulans* family and produces 37.55 ± 0.45 g/l of pullulan.

Mechanism of biosynthesis from *A. melanogenum*

Biosynthesis of pullulan from *A. melanogenum* is not well defined. However, in a study it was found that alternative primers for pullulan biosynthesis are also required in *Aureobasidium melanogenum* P16 [13]. Complementation of genes encoding glycogenins Glg1, Glg2, Gcs1(β-glucosyltransferase) and Sgt1(sterol glucosyltransferase) in the mutants restored pullulan biosynthesis in the strain P16 [13].

Although the detailed mechanisms of biosynthesis of pullulan from *A. melanogenum* P16 is unknown, it was found in a study that completely removing the CreA gene encoding a glucose repressor played an important role in its regulation [13]. In the same study, it was concluded that pullulan production from glucose can be improved by de-repression on pullulan biosynthesis. Also, glucoamylase, α-amylase, and pullulanase affect the biosynthesis of pullulan, thereby influencing the molecular weight of pullulan. So, to produce high molecular weight pullulan it is important to remove them (α-amylase, pullulanase and glucoamylase) from the disruptant strain-DG41.A multidomain α-glucan synthetase 2 gene (AmAGS2 gene) in *Aureobasidium melanogeum* P16 is the key gene responsible for biosynthesis [14]. Apart from *A. pullulan* and *A. melanogenum* few other fungi have also been reported to show pullulan production but their mechanism is unknown and are not commercially used (Table 1).

4 Production

Pullulan production process is regulated by series of factors (Table 2) including reactor condition and design and availability of raw materials.

Pullulan production can be achieved by using agro-waste or other sources such as starch, as the raw material. Sucrose act as the main substrate for industrial production of pullulan [45, 108] but production is limited at industrial scale due to comparative shortage of sucrose sources and has high cost. Further, accumulation of fructo-oligosaccharides which is a by-product of pullulan production from *A. melanogenum* P16 using sucrose as substrate is a matter of concern [59]. In order to make the production process cost- efficient, cost of carbon source has been reduced by using substrates such as glucose, hydrolysed potato starch, molasses and inulin instead of sucrose [45, 33, 66, 107]. An important carbon source is lignocellulosic biomass which is the most abundant sustainable carbon source and various methods have been employed for its bioconversion (Table 3). It has complex and recalcitrant structure with the potential to provide carbon source for industrial fermentation

Table 1 Other microbial sources reported to show pullulan biosynthesis

Microbial sources	References
Dematium pullulans	[8]
Aureobasidium mousonni	Chi et al. 2020
Cytaria harioti	[114]
Cytaria darwinii	
Termella mesenterica	[29]
Cryphonectria parasitica	[24, 28]
Teloschistes flavicans	[91]

Table 2 Factors affecting pullulan production

Factors	Description	References
Carbon	Glucose, sucrose, mannose, galactose, fructose and agricultural waste such as corn residue, sugarcane bagasses, olive oil waste, provides carbon source	[25, 104]
	Presence of xylose and lactose may result in low pullulan production activity and low cell growth	[20, 25]
	Excess carbon source inhibits pullulan production, which is because of suppression effect of sugars on enzymes that produce pullulan	[25, 48, 98]
Nitrogen	Nitrogen depletion act as signal for EPS formation of *A. pullulan*	[8, 31]
	NH4+ might control the carbon flow within the cell by acting as an effector for pullulan degradation activity	[9]
	Excess of nitrogen reduces polysaccharide production but contributes to increased biomass production	[82]
pH	For pullulan production-optimum pH range has to be 5.5–7.5	[16, 55, 58, 99]
	Under reduced pH conditions insoluble glucan production is stimulated and pullulan synthesis is suppressed	[67]
Temperature	Vary according to the strain, but is mainly between 25 and 30 °C	[26]
	Two-stage temperature have been suggested for optimum pullulan production that is- 25–26 °C—1st stage and 30–32°C—2nd stage (Preferring 26 C and 32 C)	[118]
Light intensity	Low density blue light (470 nm) ranging between 100- to 200-lux intensity increases pullulan production	[12]
	Fermentation carried out in dark conditions leads to the formation of melanin	
Oxygen profile	In absence of oxygen neither biomass increases nor pullulan is produced	[54]
	Aeration helps to promote the microbial growth. So, extreme aeration increases the number of pullulan synthesizing cells, thus increasing the production	[95]
	Under well-aerated conditions, the α-amylase activity of the fermentation broth changes leading to reduced molecular weight of pullulan produced (α-amylase is the key enzyme that regulates molecular weight of pullulan produced)	[3, 53]
Incubation time and Agitation speed	Incubation time depends on the type of the strain that is used. Production is influenced by constant optimization of both the parameters, incubation time and agitation speed	[125]
	Agitation at 250 rpm and fermentation time of 7.9 days was reported to give optimum pullulan production.	[102]

processes. One of the major advantages of using lignocellulosic biomass as substrate is that it is cost- effective and facilitates the production process.

In order to design a bioprocess method for pullulan production from lignocellulose, establishing compatibility between pullulan production and CBS process is important. Under various sugar ratios the *Aureobasidium* strains should together ferment C6/C5 sugars, at the same time have tolerance for high salt conditions. In other words, to obtain high yield pullulan producing strains must be able to together ferment, both glucose and the xylose. CBS process generates high osmotic pressure because of increased salinity in the system, therefore the pullulan producers should have tolerance to this system. *A. melanogenum* TN2-1-2, a new fungus had been reported to assimilate glucose and xylose simultaneously. It produces pigment free pullulan as it does not undergo melanin synthesis. The production was found to be 55.1 g/l with 0.5 g pullulan being produced per gram of lignocellulosic biomass (CBS hydrolysate) [63, 61]. Agriculture-based sources can be utilized to lower the cost and shortage of source material used for pullulan production (Table 4).

Further, liquefied starch along with non-toxic and non-pathogenic strains under particular conditions are used to produce pullulan at industrial scale. Both the cultivation parameters (such as raw materials, pH, agitation speed, incubation time and temperature) and bioreactor design (type of impeller/ mixing plate) contribute in deciding the concentration of the pullulan produced by *A. pullulan* [16–20].

Batch, fed-batch, continuous, immobilized and biofilm systems have been reportedly used to increase pullulan production by either optimizing the parameters (culture conditions) or the bioreactor. Response surface methodology had been used to study and evaluate cultivation parameters such as sucrose, ammonium sulphate, yeast extract, dipotassium hydrogen phosphate and sodium chloride, and it was found that at optimal conditions the pullulan produced by *A. pullulan* ATCC201253 was 44.2 g/l [19]. Fed-batch fermentation was performed to overcome the

Table 3 Methods for bioconversion of lignocellulose for production of pullulan

Process	Features	References
SHF and SSF$_a$	Depends on fungal cellulases and are off-site saccharification strategies	[64, 110]
	Process is limited by enzyme cost	
	Reduces enzyme cost	
	One step for lignocellulosic bioconversion	
CBP$_a$	Combination of fermentation, production of enzyme and hydrolysis of cellulose	[65]
	Helps in bioconversion of lignocellulose	
CBS$_a$	Separates fermentation process from CBP	[63, 61]
	Produces hydrolysates which can be used as carbon source to ease the downstream fermentation of pullulan	[60–61]

$_a$Separate enzymatic hydrolysis and fermentation (SHF), Simultaneous saccharification and fermentation (SSF)
Consolidated bioprocessing (CBP), Consolidated bio-saccharification (CBS)

Table 4 Agriculture-based substrates used for pullulan production

Substrates	Microbial sources	References
Deproteinized whey	*A. pullulan P56*	[94]
Beet molasses	*A. Pullulan* P56	[51]
Cassava starch residue	*A. pullulan* MTCC 1991	[89]
Sweet potato	*A. pullulan* AP329	[119]
Potatoes starch	*A pullulan P56*	[33]
De-oiled jatropha seed cake	*A. pullulan RBF 4A3*	[23]
Rice hull hydrolysate	*A pullulan CCTCC M 2012259*	[115]
Asian palm kernel	*A. pullulan* MTCC 2670	[109]
Potato starch hydrolysate	*A. pullulan* 201253	[1]
Sesame seed oil	*A. pullulan* KY767024	[72]
De-oiled rice bran	*A. pullulan* MTCC 6994	[102]

suppression effect of sucrose for pullulan production. However, the process could not determine the exact content of pullulan produced, the amount of exopolysaccharide obtained from *A. pullulan* was 58 g/l [98]. Using fed-batch process pullulan production increases with the time till the tenth day with no remarkable increase in the rate of production on incorporation of sucrose [19].

Further, trials have been made to increase the pullulan production via continuous fermentation methods and Schuster et al. [97], first synthesized pullulan in continuous system, with rate of synthesis varying between 0.16–0.35 g/l/h. Later, it was found that on optimizing cultivation parameters the production could be achieved to 78.4 g/l in continuous fermentation process. Production of pullulan mainly deals with parameters such as strain and inoculum preparation, media type and its development, fermentation process and its parameter optimization. Submerged fermentation method has been successfully executed for commercial production of pullulan [122]. Submerged fermentation has inherent advantages such as low operation cost, easily available, simplicity of use, low water output and higher productivity rate. These properties make submerged fermentation better than the solid-state fermentation [32, 72, 38]. In submerged fermentation during the production process the characteristics of the liquid medium changes drastically. Initially, the liquid medium shows Newtonian behaviour but as the polymer starts getting synthesized from the producer microorganism (*A. pullulan* P56) the liquid medium property changes to non- Newtonian behaviour. This shift is mainly because of the viscosity of the medium [51]. However, it has been reported that with *A. pullulan* KY767024 (melanin-deficient strain), on optimizing the cultivation conditions pullulan solution predominantly shows Newtonian behaviour which later changes to pseudoplastic behaviour [72].

To enhance the pullulan production cell immobilization techniques had been carried out using agar, calcium alginate and carrageenan in batch system and it was found that it helped in increasing the total biomass in the bioreactor. The drawback of using immobilized cells is that the production yield is very low with 0.43 mg of polysaccharide per gram of cells per hour with a purity of 36% using *A. pullulan* ATCC 42023 as the producer strain [116, 117]. On optimizing fermentation parameters (such as pH, agitation speed and incubation time) it was found that the pullulan produced from *A. pullulan* P56 increased to 19.5 g/l [125]. It was found that instead of artificial immobilization if biofilms are grown on a solid support it acts as natural cell immobilization. Adoption of biofilm reactors and use of plastic composite support (PCS) as a solid support, further increased the pullulan production to 60 g/l. Plastic composite support provides a perfect surface for biofilm formation as they are mixture of polypropylene (functions as matrix) and nutritious compounds (microbial nutrients such as peptone, mineral salts and bovine serum albumin) [17, 86].

Bioreactor design also determines pullulan productivity. Implementation of helical ribbon impeller- based reactors show highest viscosity but extremely low productivity. Reciprocating plate bioreactors are used for fermentation processes that give high viscous broth. The bioreactor design consists of a mixing plate that is made up of six perforated stainless-steel plates. The working volume of the bioreactor is 13 litres with cylinder of height 420 nm and 206 mm diameter. It was found that reciprocating plate bioreactors give the highest pullulan yield [2]. Mutagenic studies were carried out with respect to *Aureobasidium* strain using ethidium bromide as the chemical mutagen. It was found that high molecular weight pullulan was produced when compared to parent strain [85]. New strains of *A. pullulan* have been isolated from tree leaves that show 51.4 ± 0.50 g/L productivity with an average molecular weight of 2.07×10^5 g mol^{-1}. Further, it was found that the resulting pullulan had Newtonian behaviour and the decomposition temperature was approximately 300 C [36].

4.1 Downstream Processing

Downstream processing determines the product purity and quality obtained from the fermentation system. Biomass is separated from the fermentation broth, melanin pigments and proteins are removed. The pullulan is finally precipitated with appropriate organic solvent. The biomass is removed from the fermentation broth using cross-flow membrane filtration. Melanin production along with pullulan is one of the major problems with pullulan purification. The broth appears dark green to black in colour [46].

Further, to overcome the problem pigment free or pigment reduced pullulan producers are being investigated. Once the pigment is removed, product is precipitated with suitable organic solvent and two phases are formed. Molecular weight of the polymer (pullulan) determines the amount of the precipitated pullulan in solvent phase. Precipitation is usually carried out using combination of 95% ethanol and fermentation supernatant in the ratio 2:1 [121]. Ultrafiltration and ion exchange resins helps in pullulan purification [46] and the harvest is dried and mechanically grounded into powder. Therefore, after pullulan production downstream processing is crucial for obtaining the target pullulan produced by the producer microorganisms from the fermentation broth (Fig. 2).

4.2 Quality

Quality check helps in determining the exact amount of pullulan produced in the process. Several techniques are used to check for pullulan properties and to confirm the content of pullulan in the produced product (Table 5).

5 Applications of Pullulan

Pullulan is a neutral linear exopolysaccharide. Its properties make it suitable for various industrial applications. These properties include water solubility, its bio-degradable nature, impermeability to oxygen, non-hygroscopic and non-reducing nature, and its thermal stability [35].

5.1 Food

Pullulan has been used for various applications in food industry. Its properties including reduced calorific value, resistance to oil, very low oxygen permeability, tasteless, colourless, heat sealable properties, give it an advantage over other

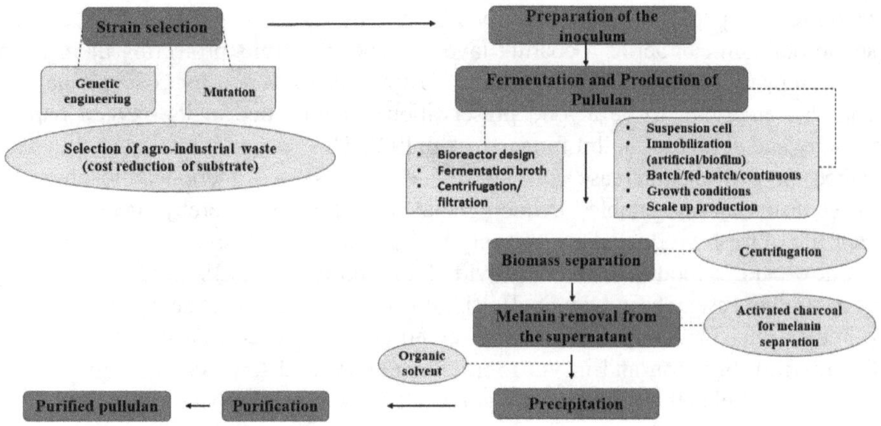

Fig. 2 Pullulan production and downstream processing

Table 5 Methods used for determining the quality of pullulan produced

Property	Analysis	References
Content	Assay carried out to check product sensitivity for pullulanase (digests α-1,6 bond and releases maltotriose units)	[54]
	The maltotriose units are further hydrolysed and content is determined using dinitrosalicyclic acid method and HPLC (High Performance Liquid Chromatography)	[71]
Molecular weight	Size exclusion chromatography	[72]
	The weight-average molecular weight of pullulan is 362–480 kDa	[80]
Surface	Scanning electron microscope (SEM) helps in determining the smooth surface of pullulan	[102]
Viscosity	Flow behaviour analysis carried out using rotational programmable viscometer shows Newtonian to pseudoplastic behaviour	[72]
Structural determination	Linkages are further confirmed using Fourier transform infrared spectroscopy (FT-IR)	[72, 102]
Thermal analysis	Thermo gravimetric analysis helps in determining the thermal degradation (approx. 300 °C but varies depending upon cultivation parameters)	[72]
Nature	X-ray diffraction techniques determines the amorphous characteristic of pullulan	

polysaccharides for the formation of edible films [122] or pullulan ester films. Both these films can be used for extending the shelf life of fruits and delaying their ripening. Pullulan ester films showed lower tensile strength for pullulan ester films and higher water vapour permeability ($3.718 \times 10{-}11$ g/m s Pa) for pure pullulan film. Hence, pullulan ester films are more preferred for packaging of fruits [79].

Another work shows the use of pullulan and chitosan-based multilayer coatings for protecting papayas post-harvest using layer-by-layer technique [123]. A whey isolate pullulan can act as a coating layer for chestnut fruits protecting them from loss of moisture [34]. Pullulan is used as coating due to its non-hygroscopic nature. It is also emerging to be a food preservation essential due to its oxygen barrier property and ability to inhibit fungal growth [92]. This water-soluble saccharide can reduce oxygen and increase carbon-dioxide levels in the internal atmosphere of coated fruits and vegetables, reducing rate of respiration, thereby increasing the shelf life of fresh produce. Due to its slow digestion and tasteless features, it is used as a low-calorie food-additive to provide bulk and texture [92].

In another research by Oku et al. [81], it was found that because of its resistance to human intestinal enzymes, pullulan served as dietary food as a starch substitute. The in vitro digestion and in vitro fermentation showed low degree of hydrolysis (DOH) of pullulan than xanthan and gellan along with its improved triggering effect on the growth of probiotics. The Prebiotic Activity Score (PAS) for pullulan was also assessed to be higher than other polysaccharides. Probiotics and prebiotics are main functional food ingredients that help in improving and restoring health [78]. Other applications that have also been reported include lower viscosity filler for beverage and sauce, stabilizing agent for mayonnaise, binder and stabilizing agent for food pastes, anti-sticking substance, cookies, and tobaccos [39, 41, 70, 76].

5.2 Biomedical

Biopolymers have become of great value in biomedical sectors considering their easy availability, non-toxicity, biocompatibility. Pullulan is one such biocompatible polymer having diverse biomedical applications (Table 6).

5.3 Cosmetics

Pullulan has a refined role in cosmetic applications. Photoallergic reactions in skin occur due to cosmetic components of molecular weight <500 Da that can diffuse through the skin and cause systemic effects. These reactions can be minimized with the use of components with high molecular weight. An organic sunscreen compound, dioxybenzone, upon conjugation with pullulan, showed reduced hypochromic effect and photoabsorption intensity as well as good epidermal retention ability. The synthesized pullulan was also able to absorb both UVA and UVB [40]. A cosmetic hydrogel composed of modified starch and pullulan smoothened wrinkles, firmed the skin and restored elasticity to the facial skin. Another cosmetic film invented by Halloran and Zolotarsky, contained 3–15% pullulan by weight. It was in semi-liquid gel form and non-sticky and easy to apply on skin. Pullulan was not just restricted in the domain of sunscreen and cosmetic films. Fogg and group developed a natural formulation for a mascara containing pullulan containing eye fixative system which did not smudge or form flakes [27].

5.4 Environmental

Pullulan has been explored for its applications in environmental remediation in the past decade. Pullulan producing strains on account of their adhesive nature possess heavy metal removal ability from aqueous solutions. It was also reported that as outcome of pullulan production, the content of metals like Cu, Mn, Cd, Pb, Ni, Fe, Zn and Cr decreased [88]. Pullulan exhibits promising outcomes for conduction of microbial-mediated oil recovery from the polluted field due to its high viscosity [44].

Table 6 Application of pullulan in biomedical sector

Biomedical sector	Applications	References
Cancer and tumour targeting	pH sensitive pullulan-doxorubicin conjugates can target the hepatic tissue	[56]
	Pullulan-based copolymer micelle added with folate was developed for tumour targeting. This helped in co-delivery of doxorubicin and shRNA of Beclin1 which proved effective for anti-tumour activity	[15]
	Pullulan nanoparticles formulated by reversible cross-linking are used for delivery of anticancer drug—paclitaxel	[42]
	Pullulan shell-based delivery system is used for chemotherapy and photodynamic therapy	Wang et al. 2014
	Core-shell nanoparticle synthesized by pullulan conjugate is used for simultaneous delivery of gene and chemotherapeutic agent	[62]
Stem cell therapy	Pullulan-based formulations are being used in Mesenchymal Stem Cells (MSC) therapy for rapid wound closure	[30]
	Pullulan-based hydrogels can be potentially used as carrier for stem cell delivery to develop musculoskeletal tissue engineering	[57]
	Spermine-pullulan complex can result in significant reduction in liver fibrosis	[43]
	Porous frames of pullulan and dextran are potential lines for vascular disease therapy using stem cell delivery	[50]
Photodynamic therapy	Pullulan acetate conjugated with pheophorbide A is used in photodynamic therapy for regression in tumour that is to reduce the tumour size and number, apoptosis of cancer cells and in treatment of cholangiocarcinoma in pigs	[4]
	Development of protoporphyrin-conjugated pullulan as nanocarrier system serve as reduction-triggered photodynamic therapy less toxic to normal cells but lead to apoptosis of tumour cells	[120]
Vaccine delivery	It uses Cholesteryl group bearing pullulan (CHP) This vaccine transport system (CHP used to produce non-toxic recombinant portion of Botulinum Type A neurotoxin bearing nanogel) evoke active immune responses and high titers of serum and secretory antibodies were obtained. CHP had been used to synthesize a nanogel bearing the non-toxic recombinant portion of Botulinum—Type A neurotoxin	[90]
Gene Delivery	Nanoplexes made from sulphide, linked cationically improved pullulan used for gene delivery	[90]

(continued)

Table 6 (continued)

Biomedical sector	Applications	References
Tissue Engineering	Pullulan-based scaffold is used in healing bone defects	[87]
	Pullulan and dextran-based biomaterial show bone regeneration. New blood vessels can be synthesized in rats with a femoral bone-defect using this biomaterial	[93]
	Pullulan used with dextran to prepare a frame for delivery of astaxanthin to treat ischemic-reperfusion injury	[111]
Imaging	Pullulan has an important application as a contrast agent for magnetic resonance imaging (MRI), in which parts of the body are imaged through the application of strong magnetic fields	[90]
Molecular Chaperones	Molecular chaperones are proteins that help other proteins in folding or unfolding and assembly. The nanogels systems based on cholesteryl group bearing pullulan (CHP) have shown artificial chaperone activity in cell-free protein formation systems	[96]

6 Conclusion

Pullulan is a linear exopolysaccharide produced mainly from *A. pullulans*, a polymorphic fungus. It is a biopolymer that shows wide range of characteristic features such as adhesive property, non-ionic, neutral and hydrophilicity. The biosynthesis of pullulan can occur by some fungal species including *A. pullulans* and *A. melanogenum*. Its production process involves batch, fed-batch, continuous and immobilized bioreactor systems which is influenced by cultivation parameters such as temperature, pH, agitation, and aeration that can be optimized to increase the production rate. New strains and production processes are being investigated to increase the rate of pullulan production. Further new agro-based sources or raw materials are also being explored to reduce its production cost. Pullulan is an amorphous slime matter possessing unique traits and properties which make it a promising biopolymer suitable for use in various applications ranging from pharmaceuticals to environmental remediation further extended to food technology, cosmetics, and drug delivery system.

References

1. An C, Ma SJ, Chang F, Xue WJ (2017) Efficient production of pullulan by *Aureobasidium pullulans* grown on mixtures of potato starch hydrolysate and sucrose. Braz J Microbiol 43:180–185
2. Audet J, Gagnon H, Lounes M, Thibault J (1998) Polysaccharide production: experimental comparison of the performance of four mixing devices. Bioprocess Eng 19:45–52

3. Audet J, Lounes M, Thibault J (1996) Pullulan fermentation in a reciprocating plate bioreactor. Bioprocess Eng 15:209–214
4. Bae B, Yang S-G, Jeong S, Lee D-H, Na K, Kim JM, Costamagna G, Kozarek RA, Isayama H, Deviere H (2014) Polymeric photosensitizer-embedded self-expanding metal stent for repeatable endoscopic photodynamic therapy of cholangiocarcinoma. Biomaterials 35:8487–8495
5. Bauer R (1938) Physiology of dematium pullulans de Bary. Zentralbl Bacteriol Parasitenkd Infektionskr Hyg Abt2 98:133–167
6. Bermu´dez J, Rosales N, Loreto C, Bricen˜o B, Morales E (2004) Exopolysaccharide, pigment and protein production by the marine microalga *Chroomonas* sp. in semicontinuous cultures. World J Microbiol Biotechnol 20:179–183
7. Bruneel D, Schacht E (1993) Chemical modification of pullulan: 1. Periodate oxidation. Polymer 34:2628–2632
8. Bulmer MA, Catley BJ, Kelly PJ (1987) The effect of ammonium ions and pH on the elaboration of the fungal extracellular polysaccharide, pullulan, by *Aureobasidium pullulans*. Appl Microbiol Biotechnol 25:362–365
9. Campbell SB, McDougall MB, Seviour JR (2003) Why do exopolysaccharide yields from the fungus *Aureobasidium pullulans* fall during batch culture fermentation? Enzym Microb Technol 33:104–112
10. Catley BJ (1971) Utilization of carbon sources by *Pullularia pullulans* for the elaboration of extracellular polysaccharides. Appl Microbiol 22:641–649
11. Catley BJ, McDowell W (1982) Lipid-linked saccharides formed during pullulan biosynthesis in *Aureobasidium*. Carbohydr Res 103:65–75
12. Chang YH (2009) The effect of light on the production of the fungal extracellular polysaccharide by *Aureobasidium pullulans*. Masteral thesis, Taoyen, Taiwan
13. Chen TJ, Liu GL, Chen L, Yang G, Hu Z, Chi Z-M, Chi Z (2020) Alternative primers are required for pullulan biosynthesis in *Aureobasidium melanogenum* P16. Int J Biol Macromol 147:10–17
14. Chen TJ, Liu GL, Wei X, Wang K, Hu Z, Chi Z, Chi ZM (2019) A multidomain α- glucan synthase 2 (AmAgs2) is the key enzyme for pullulan biosynthesis in *Aureobasidium melanogenum* P16. Int J Biol Macromol 150:1037–1045
15. Chen L, Qian M, Zhang L, Xia J, Bao Y, Wang J, Guo L, Li Y (2018) Co-delivery of doxorubicin and shRNA of Beclin1 by folate receptor targeted pullulan-based multifunctional nanomicelles for combinational cancer therapy. RSC Adv 8:17710–17722
16. Cheng KC, Demirci A, Catchmark MJ (2009) Effects of plastic composite support and pH profiles on pullulan production in a biofilm reactor. Appl Microbiol Biotechnol 86:853–861
17. Cheng KC, Demirci A, Catchmark MJ (2010) Advances in biofilm reactors for production of value-added products. Appl Microbiol Biotechnol 87:445–456
18. Cheng KC, Demirci A, Catchmark JM (2011) Continuous pullulan fermentation in a biofilm reactor. Appl Microbiol Biotechnol 90:921–927
19. Cheng KC, Demirci A, Catchmark JM (2011) Evaluation of medium composition and cultivation parameters on pullulan production by *Aureobasidium pullulans*. Food Sci Technol Int 17:99–109
20. Cheng KC, Demirci A, Catchmark JM (2011) Pullulan: biosynthesis, production, and applications. Appl Microbiol Biotechnol 92:29–44
21. Chi Z, Su CD, Lu WD (2007) A new exopolysaccharide produced by marine *Cyanothece* sp. 113. Bioresour Technol 98:1329–1332
22. Chi Z, Zhao S (2003) Optimization of medium and cultivation conditions for pullulan production by a new pullulan-producing yeast. Enzyme Microb Technol 33:206–211
23. Choudhury AR, Sharma N, Prasad G (2012) De-oiled jatropha seed cake is a useful nutrient for pullulan production. Microb Cell Fact 11:39
24. Delben F, Forabosco A, Guerrini M, Liut G, Torri G (2006) Pullulans produced by strains of *Cryphonectria parasitica*—II. Nuclear magnetic resonance evidence. Carbohydr Polym 63:545–554

25. Duan X, Chi Z, Wang L, Wang X (2008) Influence of different sugars on pullulan production and activities of α- phosphoglucose mutase, UDPG pyrophosphorylase and glucosyltransferase involved in pullulan synthesis in *Aureobasidium pullulans* Y68. Carbohydr Polym 73:587–593
26. Finkelman MAJ, Vardanis A (1982) Simplified microassay for pullulan synthesis. Appl Environ Microbiol 43:483–485
27. Fogg SR, Patel D, Orr CL (2012) Naturally derived polymeric hair fixative systems with pullulan and mascara compositions comprising the same, US Patent No 36015410P
28. Forabosco A, Bruno G, Sparapano L, Liut G, Marino D, Delben F (2006) Pullulans produced by strains of *Cryphonectria parasitica*—I. Production and characterization of the exopolysaccharides. Carbohydr Polym 63:535–544
29. Fraser CG, Jennings HJ (1971) A glucan from *Tremella mesenterica* NRRL-Y6158. Can J Chem 49:1804–1807
30. Garg RK, Rennert RC, Duscher D, Sorkin M, Kosaraju R, Auerbach LJ, Lennon J, Chung MT, Paik K, Nimpf J (2014) Capillary force seeding of hydrogels for adipose-derived stem cell delivery in wounds. Stem Cells Transl Med 3:1079–1089
31. Gibbs PA, Seviour RJ (1996) Does the agitation rate and/or oxygen saturation influence exopolysaccharide production by *Aureobasidium pullulans* in batch? Appl Biochem Biotechnol 46:503–510
32. Glassey J, Ward AC (2015) Solid state fermentation. In: diversity, dynamics and functional role of actinomycetes on European smear ripened cheese. Springer, Cham, pp 217–225
33. Goksungur Y, Uzunogullari P, Dagbagli S (2011) Optimization of pullulan production from hydrolysed potato starch waste by response surface methodology. Carbohydr Polym 83:1330–1337
34. Gounga ME, Xu SY, Wang Z, Yang WG (2008) Effect of whey protein isolate-pullulan edible coatings on the quality and shelf-life of freshly roasted and freeze-dried Chinese chestnut. J Food Sci 73:151–161
35. Grigoras AG (2019) Drug delivery systems using pullulan, a biocompatible polysaccharide produced by fungal fermentation of starch. Environmental chemistry for a sustainable world, pp 99–141
36. Haghighatpanah N, Mirazee H, Khodaiyan F, Kennedy JF, Aghakhani A, Hosseini SS, Jahanbin K (2020) Optimization and characterization of pullulan produced by a newly identified strain of *Aureobasidium pullulans*. Int J Biol Macromol 152:305–313
37. Hamidi M, Kennedy JF, Khodaiyan F, Mousavi Z, Hossein SS (2019) Production optimization, characterization and gene expression of pullulan from a new strain of *Aureobasidium pullulan*. Int J Biol Macromol 138:725–735
38. Hansen GH, Lubeck M, Frisvad JC, Lubeck PS, Andersen B (2015) Production of cellulolytic enzymes from ascomycetes: Comparison of solid state and submerged fermentation. Process Biochem 50:1327–1341
39. Hasa Y, Tazaki H, Ohnishi M, Oda Y (2006) Preparation of antisticking substance for cooked noodles by fungal hydrolysis of potato pulp. Food Biotechnol 20:263–274
40. Heo S, Hwang HS, Jeong KN (2018) Skin protection efficacy from UV irradiation and skin penetration property of polysaccharide- benzophenone conjugates as a sunscreen agent. Carbohydr Polym 195:534–541
41. Hijiya H, Shiosaka M (1975) Process for the preparation of food containing pullulan and amylase. US Patent Office, Pat. No. 3872228
42. Huang L, Wang Y, Ling X, Chaurasiya B, Yang C, Du Y, Tu J, Xiong Y, Sun C (2017) Efficient delivery of paclitaxel into ASGPR over-expressed cancer cells using reversibly stabilized multifunctional pullulan nanoparticles. Carbohydr Polym 159:178–187
43. Ishikawa H, Jo J-I, Tabata Y (2012) Liver anti-fibrosis therapy with mesenchymal stem cells secreting hepatocyte growth factor. J Biomater Sci Polym Ed 23:2259–2272
44. Iyer A, Mody KH, Jha B (2005) Biosorption of heavy metals by a marine bacterium. Mar Poll Bull 50:340–343

45. Jiang H, Xue SJ, Li YF, Liu GL, Chi ZM, Hu Z (2018) Efficient transformation of sucrose into high pullulan concentrations by *Aureobasidium melanogenum* TN1-2 isolated from a natural honey. Food Chem 257:29–35
46. Kachhawa DK, Bhattacharjee P, Singhal RS (2003) Studies on downstream processing of pullulan. Carbohydr Polym 52:25–28
47. Kato T, Katsuki T, Takahashi A (1984) Static and dynamic solution properties of pullulan in a dilute-solution. Macromolecules 17:1726–1730
48. Kim JH, Kim MR, Lee JH, Lee JW, Kim SK (2000) Production of high molecular weight pullulan by *Aureobasidium pullulans* using glucosamine. Biotechnol Lett 22:987–990
49. Kumar AS, Mody K, Jha B (2007) Bacterial exopolysaccharides—a perception. J Basic Microbiol 47:103–117
50. Lavergne M, Derkaoui M, Delmau C, Letourneur D, Uzan G, Le Visage C (2012) Porous polysaccharide-based scaffolds for human endothelial progenitor cells. Macromol Biosci 12:901–910
51. Lazaridou A, Roukas T, Biliaderis CG, Vaikousi H (2002) Characterization of pullulan produced from beet molasses by *Aureobasidium pullulans* in stirred tank reactor under varying agitation. Enzyme Microbial Technol 31:122–132
52. Leathers TD (2003) Biotechnological production and applications of pullulan. Appl Microbiol Biotechnol 62:468–473
53. Leathers TD (1987) In: Kaplan DL (ed) First materials biotechnology symposium. Natick, US Army, pp 175–185
54. Leathers TD, Nofsinger GW, Kurtzman CP, Bothast RJ (1988) Pullulan production by color variant strains of *Aureobasidium pullulans*. J Ind Microbiol 3:231–239
55. Lee KY, Yoo YJ (1993) Optimization of pH for high molecular weight pullulan. Biotechnol Lett 15:1021–1024
56. Li H, Cui Y, Sui J, Liang J, Fan Y, Zhang X (2015) Efficient delivery of DOX to nuclei of hepatic carcinoma cells in the subcutaneous tumor model using pH-sensitive pullulan-DOX conjugates. ACS Appl Mater Interfaces 7:15855–15865
57. Li T, Song X, Weng C, Wang X, Wu J, Sun L, Gong X, Zeng W-N, Yang L, Chen C (2018) Enzymatically crosslinked and mechanically tunable silk fibroin/pullulan hydrogels for mesenchymal stem cells delivery. Int J Biol Macromol 115:300–307
58. Li BX, Zhang N, Peng Q, Yin T, Guan FF, Wang GL, Li Y (2009) Production of pigment-free pullulan by swollen cell in *Aureobasidium pullulans* NG which cell differentiation was affected by pH and nutrition. Appl Microbiol Biotechnol 84:293–300
59. Liu NN, Chi Z, Wang QQ, Hon J, Liu GL, Hu Z (2017) Simultaneous production of both high molecular weight pullulan and oligosaccharides by *Aureobasidium melanogenum* P16 isolated from a mangrove ecosystem. Int J Biol Macromol 102:1016–1024
60. Liu S, Liu YJ, Feng Y, Li B, Cui Q (2019) Construction of consolidated bio-saccharification biocatalyst and process optimization for highly efficient lignocellulose solubilization. Biotechnol Biofuels 12:35
61. Liu G, Zhao X, Chen C, Chi Z, Zhang Y, Cui Q, Chi Z (2020) Robust production of pigment- free pullulan from lignocellulosic hydrolysate by a new fungus co-utilizing glucose and xylose. Carbohydr Polym 241:2–6
62. Liu Y, Wang Y, Zhang C, Zhou P, Liu Y, An T, Sun D, Zhang N, Wang Y (2014) Core–shell nanoparticles based on Pullulan and Poly(β-amino) ester for hepatoma-targeted codelivery of gene and chemotherapy agent. ACS Appl Mater Interfaces 21:18712–18720
63. Liu S, Li B, Liu YJ, Feng Y, Cui Q (2020) Consolidated bio-saccharification: leading lignocellulose bioconversion into the real world. Biotechnol Adv 107535
64. Lynd LR, Liang X, Biddy MJ, Allee A, Cai H, Foust T (2017) Cellulosic ethanol: status and innovation. Curr Opin Biotech 45:202–211
65. Lynd LR, Van Zyl WH, McBride JE, Laser M (2005) Consolidated bioprocessing of cellulosic biomass: an update. Curr Opin Biotech 16:577–583

66. Ma ZC, Liu NN, Chi Z, Liu GL, Chi ZM (2015) Genetic modification of the marine-isolated yeast *Aureobasidium melanogenum* P16 for efficient pullulan production from inulin. Mar Biotechnol 17:511–522
67. Madi NS, Harvey LM, Mehlert A, McNeil B (1997) Synthesis of two distinct exopolysaccharide fractions by cultures of the polymorphic fungus *Aureobasidium pullulans*. Carbohydr Polym 32:307–314
68. Madi NS, McNeil B, Harvey LM (1996) Influence of culture pH and aeration on ethanol production and pullulan molecular weight by *Aureobasidium pullulans*. J Chem Technol Biotechnol 66:343–350
69. Manzoni M, Rollini M (2001) Isolation and characterization of the exopolysaccharide produced by *Daedalea quercina*. Biotechnol Lett 23:1491–1497
70. Matsunaga H, Tsuji K, Watanabe M (1978) Coated seed containing pullulan-based resin used as binder. US Patent Office, Pat. No. 4067141
71. Miller GL (1959) Use of DNS reagent for determination of reducing sugar. Anal Chem 31:426–428
72. Mirazee H, Khodaiyan F, Kennedy JF, Hosseini SS (2020) Production, optimization and characterization of pullulan from sesame seed oil cake as a new substrate by *Aureobasidium pullulans*. Carbohydr Polym Tech App 1:1–5
73. Mishra B, Suneetha V (2014) Biosynthesis and hyper production of pullulan by a newly isolated strain of Aspergillus japonicus-VIT-SB1. World J Microbiol Biotechnol 30:2045–2052
74. Mishra A, Jha B (2009) Microbial exopolysaccharides. In: Rosenberg E et al (eds) The prokaryotes-applied bacteriology and biotechnology, pp 180–192
75. Mishra B, Zamare D, Manikanta A (2018) Selection and utilization of agro-industrial waste for biosynthesis and hyper-production of pullulan: a review. In: Varjani SJ et al (eds) Biosynthetic technology and environmental challenges, energy, environment and sustainability, pp 80–103
76. Miyaka T (1979) Shaped matters of tobaccos and process for preparing the same. Canadian Patent Office, Pat. No. 1049245
77. NithyaBalaSundari S, Nivedita V, Chakravarthy M, Srisowmeya G, Usha Anthony, Dev GN (2007) Characterization of microbial polysaccharides and prebiotic enrichment of wheat bread with pullulan. LWT Food Sci Tech 122:2–5
78. NithyaBalaSundari S, Nivedita V, Chakravarthy M, Srisowmeya G, Antony U, Dev GN (2020) Characterization of microbial polysaccharides and prebiotic enrichment of wheat bread with pullulan. LWT—Food Sci Tech 122:2–5. https://doi.org/10.1016/j.lwt.2019.109002
79. Niu B, Shao P, Chen H, Sun P (2019) Structural and physiochemical characterization of novel hydrophobic packaging films based on pullulan derivatives for fruits preservation. Carbohydr Polym 208:76–284
80. Okada K, Yoneyama M, Mandai T, Aga H, Sakai S, Ichikawa T (1990) Digestion and fermentation of pullulan. J Jpn Soci Nutri Food Sci 43:23–29
81. Oku T, Yamada K, Hosoya N (1979) Effect of pullulan and cellulose on the gastrointestinal tract of rats. Nutr Food Sci 32:235–241
82. Orr D, Zheng W, Campbell BS, McDougall BM, Seviour RJ (2009) Culture conditions affecting the chemical composition of the exopolysaccharide synthesized by the fungus *Aureobasidium pullulans*. J Appl Microbiol 107:691–698
83. Parikh A, Madamwar D (2006) Partial characterization of extracellular polysaccharides from cyanobacteria. Bioresour Technol 97:1822–1827
84. De Philippis R, Sili C, Paperi R, Vincenzini M (2001) Exopolysaccharide producing cyanobacteria and their possible exploitation: a review. J Appl Phycol 13:293–299
85. Pollock TJ (1992) Isolation of new *Aureobasidium* strains that produce high-molecular-weight pullulan with reduced pigmentation. Soc Ind Microbiol News 42:147–156

86. Pometto AL III, Demirci A, Johnson KE (1997) Immobilization of microorganisms on a support made of synthetic polymer and plant material. US Patent No. 5595893
87. Popescu RA, Tăbăran FA, Bogdan S, Fărcăşanu A, Purdoiu R, Magyari K, Vulpoi A, Dreancă A, Sevastre B, Simon S (2019) Bone regeneration response in an experimental long bone defect orthotopically implanted with alginate-pullulanglass-ceramic composite scaffolds. J Biomed Mater Res B Appl Biomater 108:1129–1140
88. Radulovic MD, Cvetkovic OG, Nikolic SD, Dordevic DS, Makovljevic JD, Vrvic M (2008) Simultaneous production of pullulan and biosorption of metals by *Aureobasidium pullulans* strain CH-1 on peat hydrolysate. Bioresour Technol 99:6673–6677
89. Ray RC, Moorthy SN (2007) Exopolysaccharide (pullulan) production from cassava starch residue by *Aureobasidium pullulan* strain MTCC 1991. J Sci Ind Res 66:252–255
90. Raychaudhuri R, Naik S, Shreya AB, Kandpal N, Pandey A, Kalthur G, Mutalik S (2020) Pullulan based stimuli responsive and sub cellular targeted nanoplatforms for biomedical application: synthesis, nanoformulations and toxicological perspective. Int J of biol macromol 161:1189–1205
91. Reis RA, Tischer CA, Gorrin PA, Iacomini M (2002) A new pullulan and a branched (1→3)-, (1→6)-linked β-glucan from the lichenised ascomycete *Teloschistes flavicans*. FEMS Microbiol Lett 210:1–5
92. Rekha MR, Sharma CP (2007) Pullulan as a promising biomaterial for biomedical applications: a perspective. Trends Biomater Artif Organs 20:116–121
93. Ribot EJ, Tournier C, Aid-Launais R, Koonjoo N, Oliveira H, Trotier AJ, Rey S, Wecker D, Letourneur D, Vilamitjana JA (2017) 3D anatomical and perfusion MRI for longitudinal evaluation of biomaterials for bone regeneration of femoral bone defect in rats. Sci Rep 7:1–11
94. Roukas TC (1999) Pullulan production from brewery wastes by *Aureobasidium pullulans*. World J Microbiol Biotechnol 15:447–450
95. Roukas T, Mantzouridou F (2001) Effect of aeration rate on pullulan production and fermentation broth rheological properties in an airlift reactor. J Chem Technol Biotechnol 76:371–376
96. Sasaki Y, Asayama W, Niwa T, Sawada S, Ueda T, Taguchi H, Akiyoshi K (2011) Amphiphilic polysaccharide nanogels as artificial chaperones in cell-free protein synthesis. Macromol Biosci 11:814–820
97. Schuster R, Wenzig E, Mersmann A (1993) Production of the fungal exopolysaccharide pullulan by batch-wise and continuous fermentation. Appl Microbiol Biotechnol 39:155–158
98. Shin YC, Kim YH, Lee HS, Cho SJ, Byun SM (1987) Production of exopolysaccharide pullulan from inulin by a mixed culture of *Aureobasidium pullulans* and *Kluyveromyces fragilis*. Biotechnol Lett 9:621–624
99. Shingel KI (2004) Current knowledge on biosynthesis, biological activity, and chemical modification of the exopolysaccharide. Carbohydr Res 339:447–460
100. Shingel KI, Petrov PT (2002) Behavior of γ-ray-irradiated pullulan in aqueous solutions of cationic (cetyltrimethylammonium hydroxide) and anionic (sodium dodecyl sulfate) surfactants. Colloid Polym Sci 280:176–182
101. Simon L, Caye-Vaugien C, Bouchonneau M (1993) Relation between pullulan production, morphological state and growth conditions in *Aureobasidium pullulans*: new observations. J Gen Microbiol 139:979–985
102. Singh RS, Kaur N, Pandey A, Kennedy JF (2020) Hyper production of pullulan from de-oiled rice bran by *Aureobasidium pullulans* in a stirred tank reactor and its characterization. Biores Tech reports 11:1–5
103. Singh RS, Kaura N, Kennedy JF (2015) Pullulan and pullulan derivatives as promising biomolecules for drug and gene targeting. Carbohydr Polym 123:190–207
104. Singh RS, Saini GK (2007) Pullulan-hyperproducing color variant strain of *Aureobasidium pullulans* FB-1 newly isolated from phylloplane of *Ficus* sp. Bioresour Technol 99:3896–3899

105. Singh RS, Saini GK, Kennedy JF (2008) Pullulan: microbial sources, production and applications. Carbohydr Polym 73:515–531
106. Singh RS, Singh H, Saini GK (2009) Response surface optimization of the critical medium components for pullulan production by *Aureobasidium pullulans* FB-1. Appl Biochem Biotechnol 152:42–53
107. Srikanth S, Swathi M, Tejaswini M, Sharmila G, Muthukumaran C, Jaganathan M (2014) Statistical optimization of molasses based exopolysaccharide and biomass production by *Aureobasidium pullulans* MTCC 2195. Biocatal Agri Biotechnol 3:7–12
108. Sugumaran KR, Ponnusami V (2017) Review on production, downstream processing and characterization of microbial pullulan. Carbohydr Polym 173:573–591
109. Sugumaran KR, Shobana P, Mohan Balaji P, Ponnusami V, Gowdhaman D (2014) Statistical optimization of pullulan production from Asian palmkernel and evaluation of its properties. Int J Biol Macromol 66:229–235
110. Taha M, Foda M, Shahsavari E, Aburto-Medina A, Adetutu E, Ball A (2016) Commercial feasibility of lignocellulose biodegradation: possibilities and challenges. Curr Opin Biotechnol 38:190–197
111. Tamayo MZ, Choudat L, Aid-Launais R, Thibaudeau O, Louedec L, Letourneur D, Gueguen V, Meddahi-Pellé A, Couvelard A, Pavon-Djavid G (2019) Astaxanthin complexes to attenuate muscle damage after in vivo femoral ischemia-reperfusion. Marine Drugs 17:354
112. Tsujisaka Y, Mitsuhashi M (1993) Pullulan. In: Whistler RL, BeMiller JN (ed) Academic, San Diego, pp 447–460
113. Ueda S, Fujita K, Komatsu K, Nakashima Z (1963) Polysaccharide produced by the genus *Pullularia* I. Production of polysaccharide by growing cells. Appl Microbiol 11:211–215
114. Waksman N, De Lederkremer RM, Cerezo AS (1977) The structure of an α-D-glucan from *Cyttaria harioti* Fischer. Carbohydr Res 59:505–515
115. Wang D, Xiaomin J, Donghai Z, Gongyuan W (2014) Efficient production of pullulan using rice hull hydrolysate by adaptive laboratory evolution of Aureobasidium pullulans. Bioresource Technol 164:12–19
116. West TP (2000) Exopolysaccharide production by entrapped cells of the fungus *Aureobasidium pullulans* ATCC 201253. J Basic Microbiol 40:5–6
117. West TP, Strohfus B (2001) Polysaccharide production by immobilized *Aureobasidium pullulans* cells in batch bioreactors. Microbiol Res 156:285–288
118. Wu S, Chen H, Jin Z, Tong Q (2010) Effect of two-stage temperature on pullulan production by *Aureobasidium pullulans*. World J Microbiol Biotechnol 26:737–741
119. Wu S, Jin Z, Tong Q, Chen H (2009) Sweet potato: a novel substrate for pullulan production by *Aureobasidium pullulans*. Carbohydr Polym 76:645–649
120. Xia J, Zhang L, Qian M, Bao Y, Wang J, Li Y (2017) Specific light-up pullulan-based nanoparticles with reduction-triggered emission and activatable photoactivity for the imaging and photodynamic killing of cancer cells. J Colloid Interface Sci 498:170–181
121. Youssef F, Roukas T, Biliaderis CG (1999) Pullulan production by a non-pigmented strain of *Aureobasidium pullulans* using batch and fed-batch culture. Process Biochem 34:355–366
122. Yuen S (1974) Pullulan and its applications. Process Biochem 9:7–9
123. Zhang RL, Zhang AL, Zhang HZ, Liu J, Tang KY (2019) Preparation and application of dialdehyde pullulan for the construction of gelatin hydrogels. IOP Conf Ser Mater Sci Eng 504:1–5
124. Zou X, Sun M, Guo X (2006) Quantitative response of cell growth and polysaccharide biosynthesis by the medicinal mushroom *Phellinus linteus* to NaCl in the medium. World J Microbiol Biotechnol 22:1129–1133
125. Ürküt Z, Dağbaği S, Göksungur Y (2007) Optimization of pullulan production using Ca-alginate-immobilized *Aureobasidium pullulans* by response surface methodology. J Chem Technol Biotechnol 82:837–846

Exopolysaccharides in Drug Delivery Systems

Mozhgan Razzaghi, Azita Navvabi, Mozafar Bagherzadeh Homaee, Rajesh Sani, Philippe Michaud, and Ahmad Homaei

Abstract One of the most important challenges in drug delivery is the permeability of drugs as carriers to target cells. Depending on the medication interaction with target cell cause efficiency and high uptake of them. Hence drug loading happens in each cell of the body without the various side effects of the drugs. One example of various side effects of the drugs is enzyme interference in oral drug delivery. New drug delivery systems are designed to reduce the side effects of the drugs, optimize their therapeutic efficacy, and increase patient satisfaction. With modern drug delivery systems, we will be able to control and determine speed, time and place of the drug release. In recent years, coating a class of drugs with biodegradable polymeric compounds as controllers for drug delivery systems has become increasingly important in biomedical research, drug delivery, and pharmacy. Microbial exopolysaccharides (EPS), the most significant group of polymeric materials, are renewable and their structural and physicochemical diversity enables them to play diverse roles in various fields. Based on the recent findings in glycobiotechnologies, EPS will have wide applications in future notably in medicine and pharmacy as drug delivery systems.

M. Razzaghi · A. Navvabi · A. Homaei (✉)
Department of Marine Biology, Faculty of Marine Science and Technology, University of Hormozgan, P.O. Box 3995, Bandar Abbas, Iran
e-mail: a.homaei@hormozgan.ac.ir

M. B. Homaee
Department of Biology, Farhangian University, Tehran, Iran

R. Sani
Department of Chemical and Biological Engineering, South Dakota School of Mines and Technology, Rapid City, SD 57701, USA

P. Michaud
Université Clermont Auvergne, CNRS, SIGMA Clermont, Institut Pascal, 63000 Clermont-Ferrand, France

A. K. Nadda et al. (eds.), *Microbial Exopolysaccharides as Novel and Significant Biomaterials*, Springer Series on Polymer and Composite Materials,
https://doi.org/10.1007/978-3-030-75289-7_7

Keywords Drug delivery systems · Exopolysaccharides · Microorganisms · Biomedical

1 Introduction

Today, the pharmaceutical and biomedical industries are making great efforts to develop new therapies and materials that can benefit patients. Various drug delivery systems have been developed over the past decades. In drug delivery, we try to increase the bioavailability of drugs in certain tissues or organs of the body at certain times. According to the Ringsdorf model, the polymer drug delivery system consists of therapeutic molecules that are attached to a polymer column through a spacer. The spacer compound contains a breaking point ensuring the release of the drug after cell uptake from the conjugate. The system may also contain targeted ligands, such as tumor-specific antibodies, antibody fragments or saccharides [1]. Given the abundance and extent of protein and peptide drugs, the need to design new drug delivery systems seems to be essential. With traditional drug delivery systems, there is virtually no control over the time, place, and rate of drug release. In addition, the drug concentration fluctuates frequently in the blood and may exceed the therapeutic range, resulting in lower efficiency and more side effects. With modern drug delivery systems, also called controlled release drug delivery systems, the time, speed and release of drug in the body are under control [2–4]. The great advantage of these systems is the specificity of treatments and the opportunity to use highly toxic drugs as they are delivery only in a targeted tissue or organ.

First studies on drug delivery systems focused on slow-release ones. With them, it was possible to create a constant and uniform plasma concentration of the drug in the blood for a defined period of time. The fluctuations of traditional drug administrations were then eliminated, resulting in fewer side effects, greater efficiency and patient comfort. After a while, it was seen that this method does not work for all drugs as it was necessary to stop their releasing when they are not required. For example, insulin in the treatment of diabetes should be released in hyperglycemia context and not when the blood glucose returns to normal. On the other hand, the maximum amount of drug must be delivered to the target organ to reduce the side effects of the drug and increase its effectiveness. Therefore, with these interpretations, the need to design a drug delivery system that can release the drug in response to the needs of the body, at the required time and place, is vital [5–8].

All these systems have in common: the use of materials with special properties and carriers. In this method, the surface properties of materials such as surface energy and hydrophobicity play an important role. As a consequence of proper biocompatibility with living tissues, biodegradability, sufficiently eliminating substances from the human body, and preventing toxic reactions with blood or other tissues, biosourced molecules are preferred [9–11].

Among the various biopolymers described and proposed as biomaterials, polysaccharides possess numerous functional groups which can be grafted to obtain various physicochemical and biological characteristics [12–15]. Polysaccharides were also selected as part of drug delivery systems because they are, for a large majority of them, biocompatible biodegradable, non-toxic and edible [16]. Polysaccharides are carbohydrates, one of the main group of biomolecules [17, 18]. Depending on their origin, they are classified as plant, animal, and microbial polysaccharides [19]. Even if microbial resources facilitate high and rapid production yields of polysaccharides in a fully controlled fermentation environment, they are not always competitive with terrestrial plants and seaweeds one due to their operating costs [20]. Microbial polysaccharides have several cellular localizations and functions and can be divided into extracellular polysaccharides (exopolysaccharides (EPS)), structural and intracellular polysaccharides. Many species of organisms including gram-negative and gram-positive bacteria, fungi and also some algae have been described as EPS producers [21–24].

EPS are metabolic products produced by certain microorganisms. The biosynthesis of EPS in bacterial broths has been documented since the 1880s [25]. EPS are high molecular weight extracellular biopolymers composed of combinations of up to 40 different monosaccharides species linked by glycosidic bonds. They are mainly heteropolysaccharides with or without repeating units in their structures and sometimes associated with non-sugars components such as sulfates. These macromolecules are linear or branched and soluble in water for the major part of them [26]. Many EPS have found widespread industrial applications in regard to their good water solubility, temperature stability, unique rheologic properties, good dispersibility, compatibility in solutions with different amounts of solutes, and emulsifying properties [27]. These compounds are used in the food processing industry, in the field of pharmacy, drug delivery, and biomedicine. The use of EPS and their derivatives in medical fields increases rapidly. They have found applications for numerous medical purposes such as ophthalmology [28], orthopedic surgery [29], tissue engineering, bone repair, and many other medical fields [27]. In addition, they have therapeutic applications that can control the slow release of drugs into the body. They also make it possible to place drugs in target areas such as areas of inflammation or tumors to treat the disease, and they can also be used to rejuvenate the skin and heal wounds (Fig. 1) [30, 31]. The health effects of EPS include lowering blood cholesterol, lowering blood sugar, and stimulating the immune system and antiinflammatory activities. Also, Some EPS are very stable in the gastrointestinal tract where they reinforce a healthy lifestyle. Recently, the publications on biological functions of these compounds, such as anti-tumor and antioxidative activities, have increased. Several of these microbial polysaccharides are commercial products, but the majority of them are still under development and not yet consolidated [29, 32, 33]. The results of some recent research on the role of microbial EPS in drug delivery systems are discussed in the present study.

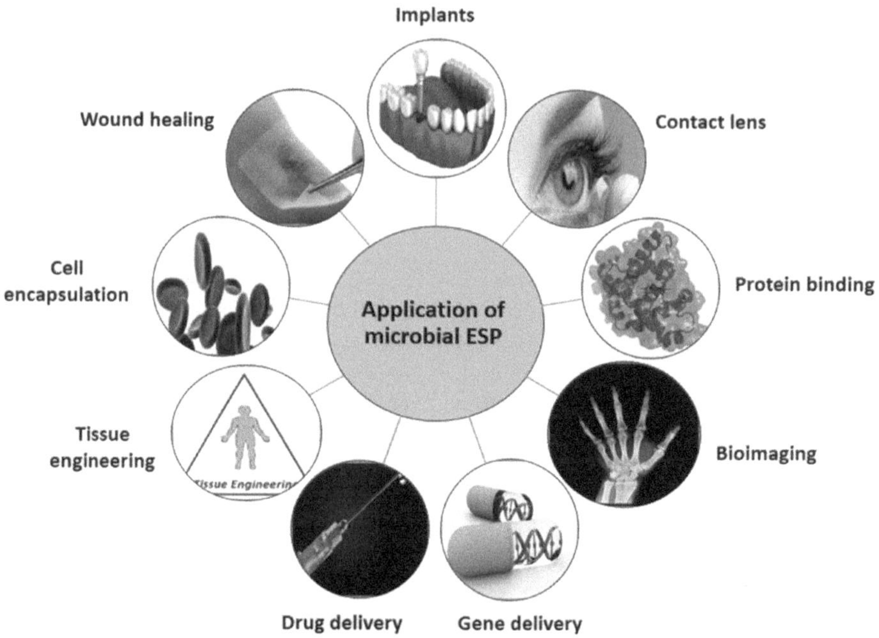

Fig. 1 The different pharmaceutical applications of exopolysaccharides (EPSs)

2 Chemical Structure of Exopolysaccharides

In the preceding years, EPS from microorganisms have received much attention in industry as an alternative to other hydrocolloids. The extraction process of EPS is relatively simple. Crude EPS can be obtained by adding ethanol to the fermentation broth. The crude polysaccharide that has been deproteinized and decolorized is further purified by chromatography columns such as ion exchange chromatography, and then the dialysis concentrate is collected, and finally the lyophilized polysaccharide is obtained [34]. Microbial EPS can be classified into homoexopolysaccharides and heteroexopolysaccharides groups (Fig. 2). Non-sugar groups are usual in heteropolysaccharides structures and are often organic acids such as acetate, succinate, pyruvate or other. Xanthan is a classic example of an anionic heteroexopolysaccharides including pyruvate and glucuronic acid in its structure. The unique physical properties of EPS are due to the molecular arrangements of monosaccharides and notably angles between the glycosidic linkages which determine their basic structures and intra- and intermolecular relationships in solution [35, 36].

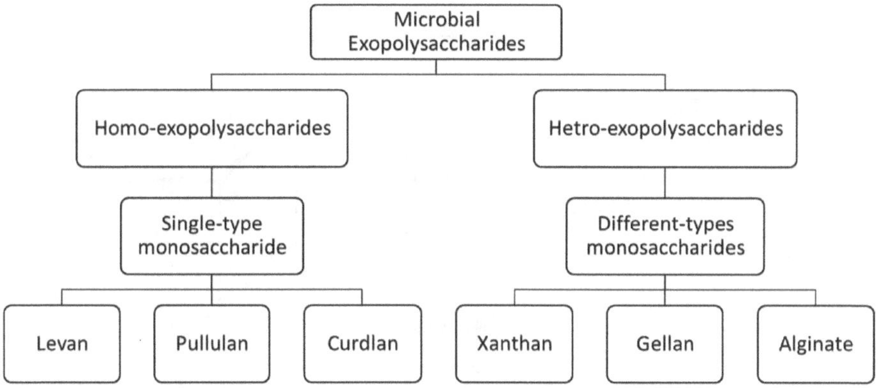

Fig. 2 Classification of microbial exopolysaccharides (EPSs)

3 Homoexopolysaccharides

Homohexopolysaccharides are composed of a unique species of monosaccharides linked together by a simple enzymatic system. Homoexopolysaccharides are generally β-D-or α-D-glucans, galactans and fructans, and but some polyglucuronic acids called glucuronans have been also described. They include levan, pullulan and curdlan. The most common examples of homopolysaccharides are produced by lactic acid-producing bacteria after action of extracellular transglycosidases. They belong to glucan and fructan families [34, 37, 38]. The structures of some of them are given in Fig. 3.

3.1 Levan as Drug Delivery System

Levan is a β-(2,6)-D-fructan extracellularly synthesized using sucrose by levansucrase also called β-(2,6)-fructosyltransferase (EC 2.4.1.10) from *Zymomonas mobilis* or *Bacillus subtilis* [39–41]. A number of gram-positive and gram-negative bacteria, respectively *Bacillus* sp. and *Zaimomonas mobilis* species, produce this enzyme. Levansucrase activity can be controlled by glucose at temperatures above 45 °C [42, 43]. The pure levan obtain toughly hence. The applications of levan are less than other carbohydrate polymers but despite this limitation, it has many advantages such as high biodegradability, biocompatibility, amphiphilic characteristic, medefication usage for imaging and drug delivery agent to tumor cells [44]. In addition, levan exhibited great potential as encapsulating agent to produce nanocapsules. For example, Sarilmiser et al. (2015) used levan as efficient microparticles for drug delivery system in encapsulating VANCO. Particle sizes and potentials were respectively between 400 nm and 1.3 μm and between 4 and 6.5 mV [45]. Bovine serum albumin (BSA) was also encapsulated using levan and

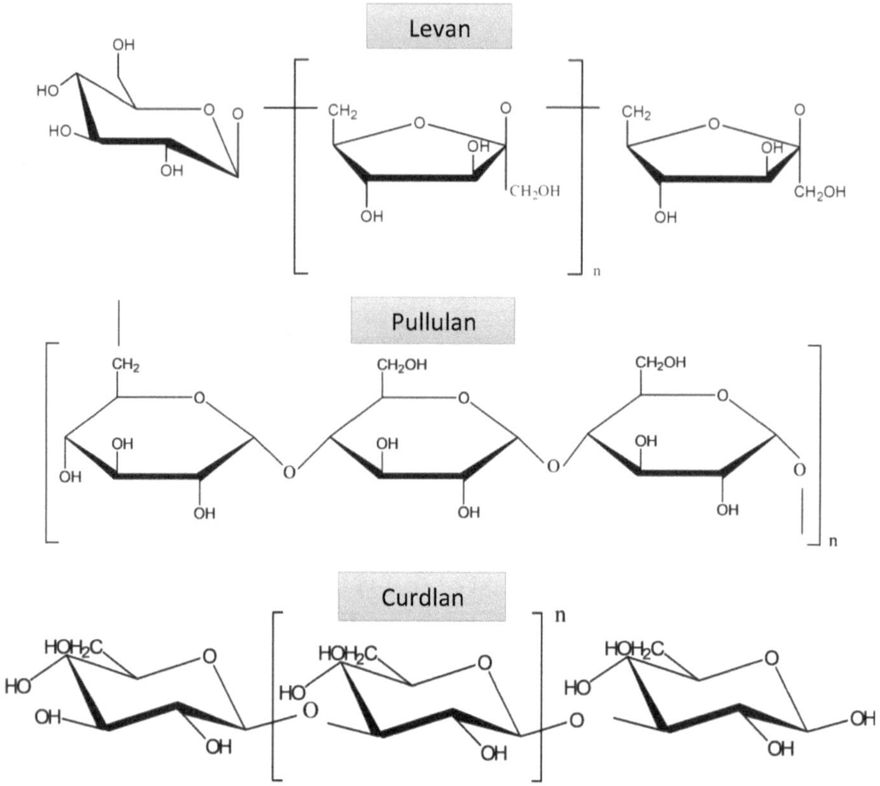

Fig. 3 Chemical structures of some of industrial homoexopolysaccharides

particles between 200 and 540 nm with an output efficiency of 50–70% were obtained. Again, the potential value of ζ was about 4–7 mV. Finally, Kim et al. [44] used the above affinity between fructose and a fructose transporter called GLUT5 to test the efficiency of levan nanoparticles and green indocyanine to diagnose breast cancer. The particles were obtained in the range of 150–200 nm, but with a weak encapsulation efficiency of 14–30% potential was determined. During in vivo experiments (mice model) these nanoparticles targeted breast cancer cells. Low potential values of ζ (suitable values of about 15–30 in absolute value) indicated the tendency of particles to form aggregates, which can prevent their use for various applications. This weakness may be overcome by changing the polymer surface or anchoring a negatively charged drug to the hydroxyl groups of levan by electrostatic interactions. In this regard, a drug delivery system composed of levan or its carboxymethyl form 5-fluorouracil has been produced [46]. Because of the orientation of the fluoride atom, Tabernero et al. [46] obtained a nanoparticle system of 300–400 nm with a potential of ζ of −20 mV. Stability studies revealed a decrease of aggregation of these particles at basic pH [46]. Aldehyde groups on levan were obtained after periodic oxidation [47]. They were useful for inducing cell proliferation [47].

Moreover, carboxymethylation of levans may be performed to produce carboxymethyl levans. These carboxymethylated levans can be used as anionic biopolymers for ionic binding processes [46]. In addition, levan shows great potential as restorative drugs. Its ability to promote cell proliferation and its healing properties make it a promising polymer for creating scaffolds with cells incorporated [46]. Levan has been used also as a coating agent for various metals. Cerium oxide nanoparticles were coated with levan, resulting in an improvement of their antioxidant activity. These nanoparticles are good candidates for the treatment of diseases caused by elevated reactive oxygen species (ROS) [48]. Coating particles of selenium, iron and cobalt using levan has been also suggested [49]. Nanoparticles with sizes between 150 and 200 nm were always obtained from different systems. These nanoparticles had a polydispersity index of about 0.2–0.3, indicating they are spherical. Appropriate values of zeta potential were measured for only half of the experiments. Numerous experiments highlighted the improvement of cellular metabolism by levan nanoparticles. Nevertheless, as reported by authors, more studies are required regarding the effect of products from nanoparticle degradation on the colon [49]. Levan has been employed as a reducing agent to obtain different nanoparticles. Indeed, silver and gold nanoparticles were obtained reducing the ions of the corresponding compounds (gold chloride and silver nitrate) to levan coated nanoparticles [50]. The catalytic activity of nanoparticles was tested with success to reduce various compounds. Some studies reported the production of levan magnetic nanoparticles by coprecipitation. In this case, the particle size was from 100 to 200 μm. These particles were used to stabilize trypsin which was reused 10 times and the reused enzyme lost only 16% of its original specific activity. Thus, these magnetic levan nanoparticles can be presented as a matrix for enzyme stabilization and drug delivery systems [51].

3.2 Pullulan as Drug Delivery System

Some fungal EPS have been identified for their potential in numerous industrial applications. Pullulan, a water-soluble biopolymer from *A. pullulans* is one of such fungal EPS commercially produced not only for medicinal applications but also in foods and cosmetics areas. It is generally recognized as safe (GRAS). Pullulan was discovered in 1938, but was further studied after its description in 1959 [52]. The first studies on the production of pullulan were performed by Bauer in 1938. The polysaccharide was then isolated and characterized by Bernier in 1958 [53]. It is a linear, extracellular, neutral, and water-soluble glucan composed of maltotriose units composed of three glucose units α-(1,4) linked, associated by α-(1,6) glycosidic linkages. The presence of alternating α-(1,4) and α-(1,6) bonds increased its solubility in hot and cold water compared to amylose or amylopectin [54, 55]. Pullulan is mainly produced by the yeast-like *Aureobasidium pullulans* widely found in humid places, especially on the leaf surface of various plants, tree trunks, soil, rocks, water and animal tissues [56, 57]. Then, Bender et al. [58] studied the structure of this biopolymer and named it pullulan. The molecular mass of pullulan varies from several to 2000 kDa depending on microbial strain, culture medium conditions and type of substrate used [53]. Pullulan and its derivatives are widely used in food, pharmaceutical, fiber, and electronics industries since their introduction in the Japanese food industry as food additives. Its current applications as a drug are based on its bonding properties and film formation, as well as its low oxygen permeability [16]. Pullulan and its derivatives have found many applications in the food, pharmaceutical, and medical industries. Pullulan and its derivatives can be conjugated with hydrophobic drugs to create systems with controlled loading and targeted release of drugs. Pullulan can form films, hydrogels microparticles, nanoparticles and micelles [59–62]. Chemical changes in polysaccharides can lead to derivatives with improved medicinal properties. Due to the growing demand for anticancer drugs in the last decade, various non-toxic, biodegradable and biocompatible systems based on pullulan derivatives for the loading and unloading of doxorubicin, epirubicin, paclitaxel, mitoxantrone or 10-hydroxy-amphetamine have been designed and tested. pH-sensitive pullulan nanoparticles, bioconjugates, or self-assembling hydrophobized pullulan are pullulan-based release systems for anticancer drugs. Acid-sensitive hydrazone bond, stable at physiological pH but hydrolyzable under acidic conditions, has been used to conjugate the drugs to pullulan backbone [63]. Many efforts have also been made to find the best ligand to selectively target drug delivery systems to tumor tissues. In addition to other factors such as pH of the medium, size of nanoparticles, microparticles, micelles, drug-loaded hydrogels, performance of antineoplastic drug delivery systems in pullulan derivatives, drug content, encapsulation efficiency, and drug loading efficiency have been determined. Typically, new antineoplastic drug delivery systems based on pullulan derivatives in the range of 50–300 nm have been recorded (Table 1). According to Bae and Park [64], this size criterion will ensure the continuity of new drug delivery systems in the blood circulation for a long time. So they can fulfill their therapeutic role. Due to its antioxidant properties and stability, pullulan has been used as a reducing agent in the preparation and stabilization of various metal nanoparticles including silver [65]. For example, non-toxic acetylated pullulan coated magnetic particles. Depending on power and frequency of the magnetic field (in a hyperthermia study) temperature change was obtained for the nanoparticles [66]. A histone antibody was conjugated to pullulan nanoparticles to release this hepatic drug specifically in liver considering the affinity of this organ for this biopolymer. Nanoparticles of about 20–30 nm were obtained and in vivo experiments with mice showed a decrease in inflammatory response and mortality [67]. A summary of the function and role of pullulan and its derivatives in drug delivery systems is provided in Table 1.

Table 1 Performances of drug delivery systems (DDSs) based on pullulan derivatives

Polymeric component	Drug component	Mean sizes (nm)	Performances	References
Poly(D,L-lactide-co-glycolide)-graft pullulan	Adriamycin	75–150	~20–30% w/w drug content	[68]
Nanospheres based on pullulan poly (caprolactone) (PULL-PCL)	Ciprofloxacin	142	Less than 2 μg/mL of nanosphere/cipro effectively inhibited the proliferation of cultures inoculated with 107 or 108 bacteria/mL of *S. aureus* and *P. aeruginosa*, respectively	[69]
Pullulan–riboflavin nanohydrogel	Levofloxacin	210	15.0 ± 2.0% w/w drug encapsulation efficiency	[70]
Novel pemulen/pullulan blended hydrogel	Clotrimazole	–	Drug release 20.14 μg/cm^2 in 8 h	[71]
Pullulan nanobased nail formulation	Tioconazole	155–162	Drug content close to theoretical values (1 mg/mL) Association efficiency close to 100% (HPLC)	[72]
Liposome-loaded pullulan films	Terbinafine	–	31% accumulated drug in the nail	[73]
Pullulan	Doxorubicin	50–170	Drug release in 2 h at pH = 5	[74, 75]
Pullulan	Co-delivery of doxorubicin and pyrrolidinedithiocarbamate	130–180	–	[74, 75]
Ligand combination of pullulan with arabinogalactan	Doxorubicin	220	~20% drug loading	[76]
Folate-decorated maleilated pullulan	Co-delivery of doxorubicin and pyrrolidinedithiocarbamate	150	–	[77]
Nanogels from poly(L-lactide)-g-pullulan copolymers	Doxorubicin	121–163	~4% w/w drug content	[78]
Nanoparticles based on folic acid-conjugated pullulan and poly (D,L-lactide-coglycolide) graft copolymer	Doxorubicin	<200	~7% w/w drug content	[79]

(continued)

Table 1 (continued)

Polymeric component	Drug component	Mean sizes (nm)	Performances	References
Folate–biotin–pullulan nanoparticles	Doxorubicin	~170	1.72% drug content 69% drug loading efficiency	[80]
Pullulan–retinoic acid–biotin conjugate	Doxorubicin	~192	92% entrapment efficiency	[81]
All-trans-retinoic acid–pullulan conjugate	Doxorubicin	230–260	1.1–1.4% drug content 38–47% drug loading efficiency	[82]
Pullulan–disulfide–stearic acid conjugate	Doxorubicin	~190	65.53% encapsulation efficiency	[83–85]
Pullulan derivative containing stearic acid and low molecular weight branched poly(ethylenimine)	Co-delivery of doxorubicin and p53 protein	~189	~5.10% doxorubicin loading content 56.07% encapsulation efficiency for doxorubicin	[86]
Reducible cholesterol-modified pullulan	Doxorubicin	80–160	–	[87]
Pullulan–cholesterol succinate–urocanic acid conjugate	Doxorubicin	150–300	–	[83–85]
Pullulan acetate nanoparticles	Epirubicin	340	–	[88]
Pullulan acetate nanoparticles Folic acid–pullulan acetate nanoparticles	Epirubicin	–	52% drug loading efficiency over 72 h 92% drug loading efficiency over 72 h	[89, 90]
Micelles based on pullulan, tocopherol succinate and folic acid	Epirubicin	150	–	[91]
Cholesterol-modified pullulan	Epirubicin	160	–	[92]
Pullulan acetate nanoparticles	Paclitaxel	130–250	7–13% w/w drug content	[93, 94]

(continued)

Table 1 (continued)

Polymeric component	Drug component	Mean sizes (nm)	Performances	References
Acetylated pullulan particles coated with hyaluronic acid	Paclitaxel	200–250	–	[95]
Paclitaxel-incorporated pullulan acetate nanoparticles	Paclitaxel	160	10% w/w drug content 98% drug loading efficiency	[96]
Cholesterol-modified pullulan Cholesterol-modified carboxyethyl pullulan	Mitoxantrone	168 192	58% encapsulation efficiency 7.12% loading capacity 50% encapsulation efficiency 6.14% loading capacity	[97]
Cholesterol-modified pullulan nanoparticles	Mitoxantrone	153–174	4.35–14.29% drug loading capacity	[98]
Biotin-modified cholesteryl pullulan nanoparticles	Mitoxantrone	146, 170 or 205	53, 85 or 80% drug loading efficiency	[99]
Amphiphilic nanomicelles consisted of α-tocopheryl succinate-modified pullulan	10-Hydroxycamptothecin	170–250	56–95% entrapment efficiency 1.65–16.42% loading capacity	[85]
Pullulan (PUL) nanofibers containing rutin-pluronic solid dispersions (SDs)	Rutin	155 ± 25	4% w/v pluronic concentration provided the optimal drug loading efficiency	[100]
Pullulan acetate nanoparticles	Curcumin	123.4 ± 2	22.72 ± 1.02 to 85.87 ± 1.09% curcumin loading efficiency	[101]
Glycyrrhetic acid (GA)-modified pullulan nanoparticles (GAP NPs)	Curcumin	109.3 ± 8.6	72.7 ± 1.45% encapsulation efficiency	[102]
Smart nanoparticles based on pullulan-g-poly(*N*-isopropylacrylamide)	Indometacin	145	80% drug entrapment efficiency	[103]

(continued)

Table 1 (continued)

Polymeric component	Drug component	Mean sizes (nm)	Performances	References
Triacetate-pullulan thin films	Sodium diclofenac	–	Delayed delivery of the drug for up to 30 min	[104]
Tertiary mixture of pullulan, polyvinylpyrrolidone and hypromellose	Sodium diclofenac	–	Dissolution of the optimized film almost immediately with 50% of the drug released within 1 min	[105]
Anionic pullulan nanoparticles	Diclofenac	200	At 40 °C, 70% drug released in 1 h	[106]
pH-sensitive pullulan acetate microsphere	Naproxen	–	80% drug loading efficiency	[107]
Hydroxypropyl cyclosophoraose-pullulan (HPCys-Pul) microspheres	Naproxen	–	The naproxen levels in the plasma after oral administration of naproxen-loaded microsphere were maintained for 72 h	[108]
Pullulan–riboflavin nanohydrogel	Piroxicam	210	11% w/w entrapping efficiency	[70]
Succinylated pullulan-g-oligo(L-lactide); SPL	Etanercept	~250 nm	Long-term stability of etanercept in an aqueous environment was improved by temperature-induced noncovalent interaction controllable complex	[109]
Pullulan acetate nanoparticles	Silymarin	720	In the first hour these released only 19.05% of a drug; after the seventh hour, the drug release was 61.14%	[110]
Polyethylene sebacate silymarin pullulan nanoparticles	Silymarin	283.4 ± 1.2	42.77 ± 0.68% entrapment efficiency	[111]
Poly etyleneimine-pullulan (PEI-PUL) modified apolipoprotein B siRNA	Apolipoproteins B	<260	High MW pullulan complexes (>20,000) resulted in more efficient gene expression than low MW pullulan complexes (<20,000)	[112]

(continued)

Table 1 (continued)

Polymeric component	Drug component	Mean sizes (nm)	Performances	References
Mixtures of pullulan–tween 80, pullulan–sodium dodecyl sulfate (SDS)	Fenofibrate	–	Relative order of effectiveness: pullulan–tween 80 > pullulan–SDS	[113]
Hydroxypropyl methylcellulose and pullulan	Atenolol	–	98.83% drug release in 150 s from the wafer	[114]
Films of chitosan-pullulan	Metoprolol	–	5.1199 ± 0.28 mg drug content	[115]
Hydroxypropyl methylcellulose, pullulan, and poly (vinylpyrrolidone)	Nebivolol	–	High procent of drug release from the film in simulated saliva and simulated gastric fluid	[116]

3.3 Curdlan as Drug Delivery System

Curdlan, which is a bacterial polysaccharide, is a water-insoluble extracellular glucan consisting of 400–500 units of D-glucose linked by β-(1,3) glycosidic linkages [117]. It is the third microbial EPS approved by the FDA for use in USA food and pharmaceutical industries [118], the two others being dextran and xanthan. Only a few species of bacteria such as *Alcaligenes faecalis* and *Agrobacterium* spp. produce this linear polysaccharide. Curdlan can also be produced by some species of *Rhizobium* and *Pseudomonas*. The optimum pH range for *Agrobacterium* culture medium is about 7, while the optimum pH and temperature for curdlan production are about 4.5 and 32 °C, respectively. By regulating the optimal concentration of mineral elements such as phosphate, sulfate and with nitrogen starvation, the growth of microorganisms in the constant phase is intensified and curdlan is produced [119]. The growth rate of curdlan-producing microorganisms decreases with increasing ammonium concentration [43]. Use of this EPS in drug delivery systems has increased due to the ability to control the structure of curdlan-based carriers. Curdlan can be used as gels, nano- and microparticles or complexes. Curdlan is insoluble in water, alcohol, and most organic solvents, but still soluble in alkaline solutions such as sodium hydroxide (NaOH) and trisodium phosphate (Na_3PO_4). In aqueous suspension, it formed a gel after heating. Depending on the temperature applied to it, two types of gels are obtained: low-set (50–60 °C) and high-set (above 80 °C) gels. The low-set thermally reversible gel has a similar behavior to agar and gelatin whereas high-set thermally irreversible gel has a thermal irreversible state and is stable even at low temperatures such as freezing temperatures as well as temperatures above 90 °C [120, 121]. The molecular mechanisms leading to high-set and low-set gels formation are different. In fast-setting gels, hydrophobic interactions between the curdlan chains give rise to multi-helical and three-stranded structures whereas the cross-links between the macromolecules in the slow-setting gels are hydrogen bonding interactions which lead to single helical structures. Curdlan-based gels can also be obtained after neutralization (with an acid) of curdlan alkali solutions. In this case, intermolecular and intramolecular hydrogen bonds of native curdlan are broken by the alkali and the helix gel structure is open leading to the solubilization of curdlan macromolecules. When the solution is neutralized, a series of new hydrogen bonds are formed, leading to gel formation. These "alkali gels" are of the slow-set type and have a heat reversible state. Changes of temperatures applied to the curdlan gum suspension to produce reversible to thermally irreversible gels depend also on its concentration. Increasing the concentration of curdlan suspension causes a thermal irreversible gel at temperature of 80 °C as the amount of heat reversible gel is reduced. The average molecular weight of this gum has not yet been determined precisely because it is insoluble. Three types of curdlan conformational structures in alkaline solutions are considered, including single helix, triple helix, and random coil [122, 123]. Under alkaline conditions (more than 0.2 M), this gum dissolves completely and becomes a random coil. In this case, its molecular weight is reported to be about 2.0×10^6—5.3×10^4 Da [124]. Drugs such prednisolone, indomethacin and salbutamol sulfate have been included in curdlan hydrogels. The gels were prepared by mixing curdlan at a concentration of 5–10% (w/v) in drug solutions in a glass test tube and heating the mixture in a water bath for 10 min at the required temperature. After return to room temperature, the gel is unmolded. Experiments performed by Kanke et al. [121] showed that a high-set gel can lower the rate of drug release. Curdlan-based gels are able to deliver drugs consistently compared to existing commercial formulations. These gels can be used as drug delivery suppositories for rectal administration [121].

3.4 *Other (1,3)-β-D-Glucans as Drug Delivery Systems*

(1,3)-β-D-glucans have been used to encapsulate insoluble drugs [125]. The release of curcumin is enhanced by encapsulating it inside the curdlan Curdlan nanoparticle with a size of 50 nm. These nanoparticles were obtained by adding curcumin solubilized in ethanol to an aqueous suspension of curdlan at room temperature. After solvents evaporation, the nanoparticles were purified by centrifugation to remove the excess of curdlan and larger aggregates. These curcumin-loaded curdlan nanoparticles inhibited tumor growth in the Hep-G2 cell line [126].

Amphiphilic derivatives of (1,3)-β-D-glucans have also been investigated for the development of drug-encapsulated micellar nanoparticles as cholesterol carboxymethyl curdlan one. Curdlan hydrophilicity increases with carboxymethyl groups facilitating the loading of hydrophobic drugs such as epirubcin [127]. Curdlan can also play the role of a hydrophobic component in an amphiphilic drug delivery system. Indeed, a copolymer of curdlan and polyethylene glycol (PEG) graft has been produced [128]. Doxorubicin has been combined with this graft copolymer nanoparticles by nanoprecipitation. Mixing led to the self-assembly of polymers with hydrophobic components at the core of the nanoparticles and hydrophilic components at the surface. In the case of doxorubicin and curdlan graft-PEG, dimethyl sulfoxide was used as solvent [128]. Thus, curdlan-based nano- and microparticles provide promising opportunities for drug release as structural units and as targeted ligands. The combination of these properties can be used to make drug delivery vectors with increased potential.

4 Heteroexopolysaccharides

Heteroexopolysaccharides consist of oligosaccharides with 2–8 monosaccharides different species including notably D-glucose, L-rhamnose, D-mannose, D-galactose, D-guluronic acid, L-rhamnose and their derivatives. Examples of bacterial heteroexopolysaccharides are gellan, xanthan, succinoglycan and others. Bacteria producing heteropolysaccharides include without being exhaustive,

Enterobacteriaceae family, *Klebsiella* genus, and lactic acid bacteria, including *Streptococcus* and *Lactobacillus* genus [129]. The chemical structures of some useful and industrial heteroexopolysaccharides are shown in Fig. 4.

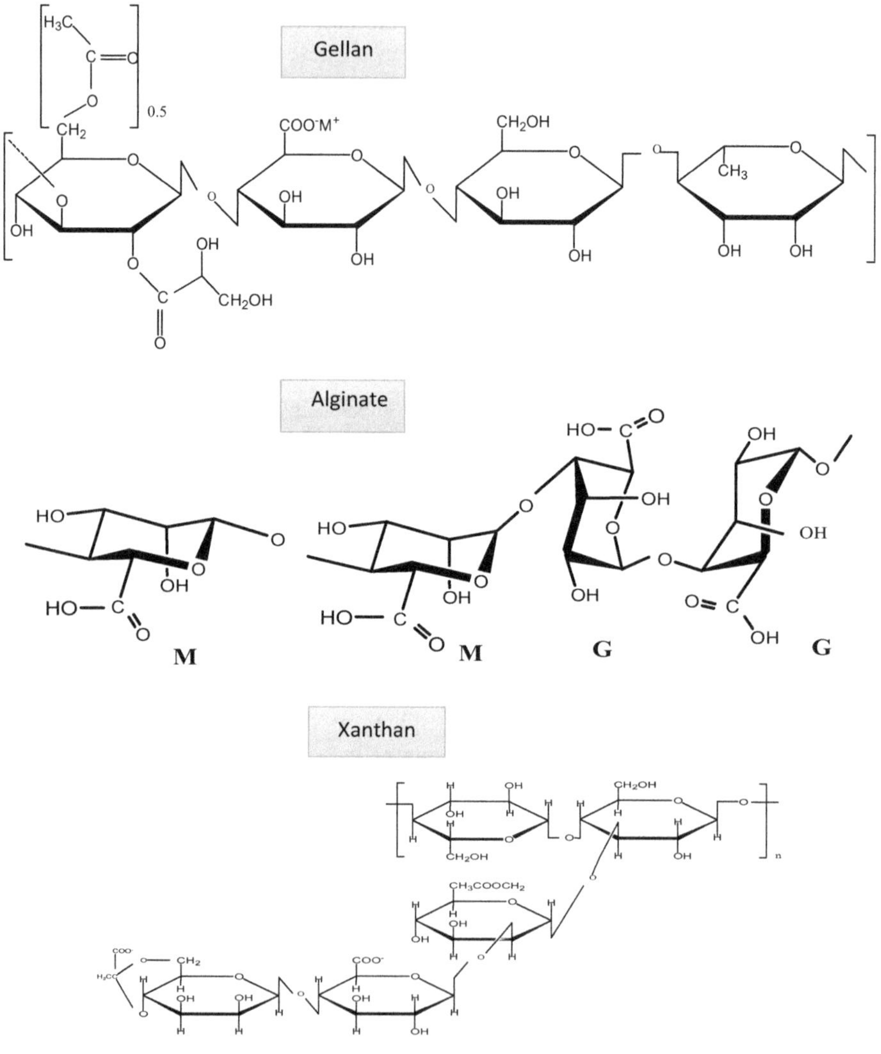

Fig. 4 Chemical structures of some of industrial heteroexopolysaccharides

4.1 Xanthan as Drug Delivery System

Xanthan is the most important commercial microbial exopolysaccharide, produced by *Xanthomonas campestris* in 1967. This EPS production efficiency depends on the conditions of the fermentation environment. The production of this polysaccharide is controlled under the steady-state fermentation, in which nutrients are supplied optimally and cell growth is controlled by nutrient limitation. The highest expected yield for xanthan is due to the use of nitrogen as a limiting agent. The main xanthan chain is a β-(1,4)-D-glucan ramified at C3 of every alternate glucose of the main chain by a trisaccharide side chain. This compound is composed of two D-mannoses with a D-glucuronic acid between them [130]. This aqueous soluble polysaccharide has many rheological and stability properties in a wide range of pHs and temperatures. Indeed, xanthan solutions are highly viscous even at low concentration. Xanthan is also used as a bulking and stabilizing agent. It can give solutions with viscosity 100 times higher than gelatin for the same concentrations. The World Pharmaceutical Organization has classified this substance as a safe substance for humans. The amount of this substance consumed by humans is very high as it is a widely used food additive [131, 132].

Xanthan is a non-toxic, non-irritating, non-inflammatory and readily available hydrocolloid. As it has exceptional rheological properties, it is abundantly used in pharmaceutical industries to manufacture a variety of solid, semisolid and liquid dosage forms. In solid dosage form, it has been studied and traded in capsule formulations and controlled release tablets. In the liquid oral dosage form, it is used to enhance viscosity of formulations and to stabilize them. More recently, advanced drug delivery systems including microparticles, nanoparticles, liposomes, buccal mucosal patches, hydrogels, microspheres and matrix coating have been developed with xanthan. It also acts as a neutral diluent in nasal gels with strong adhesion and slowing release properties. In addition, it increases bioavailability in many drug formulations [133]. Xanthan forms hydrogels in which polymer chains are cross-linked. So, they do not dissolve in solvents and can absorb water up to several times their own mass [134]. Rapid absorption of water in the macromolecular structure occurs through penetration and capillary effect. Hydrogels contract or expand when exposed to stimuli such as electric field, temperature and pH [134].

Jackson and Ofoefule [135] reported that when xanthan and ethylcellulose-based tablets were used for colonic drug release [135]. Shinde and Kanojiya [136] observed that using xanthan in the preparation of serratiopeptidase niosomal gel led to good dispersion, changes in particle size compared to formulations without xanthan. Moreover, niosomal formulation with xanthan showed pseudoplastic behavior. It was also found that physical stability of xanthan-containing formulations increased, and, the enzyme leakage from the gels is less likely when the niosomal formulation is converted to gel using xanthan. Therefore, xanthan was successfully used as a gelling agent to prepare serratiopeptidase niosomal hydrogel [136]. Polysaccharide nanoparticles can be synthesized using covalent crosslinking, ion crosslinking, polyelectrolyte complexation, etc. Pooja et al. [137] investigated

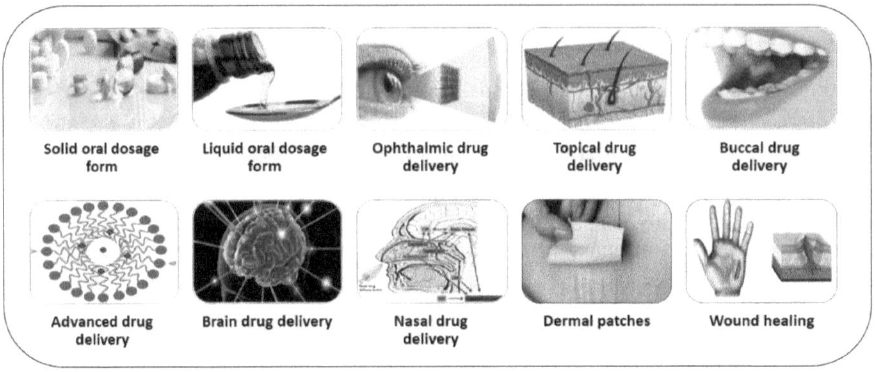

Fig. 5 Applications of xanthan gum in pharmaceutical

the use of xanthan as a reducing agent in the production of gold nanoparticles. These nanoparticles are involved in drug release due to the size and efficient and targeted release of drug. Gold nanoparticles synthesized using xanthan were found to be biocompatible and non-toxic. They had also stability, high drug loading and increased cytotoxicity toward lung cancer cells [137]. Chitosan is a natural polymer used to increase vesicle stability. In a specific study, a xanthan polyanionic compound undergone macromolecular complexation with the polycationic chitosan. Results showed that liposomal formulation has a positive effect on pulmonary release when the liposome is coated with a polyelectrolyte complex created by chitosan and xanthan complexation [138]. Therefore, xanthan is a promising material for further development, modifications, and wider applications in clinical practice and advanced drug delivery. A summary of the various applications of xanthan is shown in Fig. 5 and Table 2.

4.2 *Gellan as Drug Delivery System*

Gellan is a water-soluble anionic heteroexopolysaccharide with a high molecular weight. This heteropolysaccharide, in the presence of heat, forms a reversible gel with high strength and stability as well as a thin layer with high resolution. Gellan-producing organisms are aerobic, gram-negative, and non-pathogenic bacteria belonging to *Sphingomonas* genus. It consists of a tetrasaccharidic repeating unit composed of one β-(1,4)-D-glucuronic acid, two β-(1,4)-D-glucoses and one α-(1,3)-L-rhamnose. It has been currently used as an alternative to agar and carrageenan. It is recommended due to its relative advantages over agar, namely the transparency and thickness of the gel. Gellan has been studied for different solid, liquid and semisolid dosages [165]. To improve bioavailability and modify drug pharmacokinetics many gellan-based formulations have been developed for specific

Table 2 Different applications of xanthan gum (XG) and its derivatives in drug delivery

Polymeric component	Drug	Dosage form	Performances	References
XG	Aceclofenac	Solid oral dosage form	Drug release was approximately 80% in 8 h.	[139]
XG, hydoxypropyl Methylcellulose and carbopol	Verapamil hydrochloride	Solid oral dosage form	Formulation containing XG demonstrated prolonged drug release up to 24 h and tablet remained buoyant for more than 24 h	[140]
Xanthan and guar gum	Rosiglitazone	Solid oral dosage form	Drug release studies showed sustained release (98%) for 12 h	[141]
XG and hypromellos	Venlafaxine	Solid oral dosage form	Drug release studies showed that it can provide sustained action for 24 h	[142]
XG and acrylic polymer	Ibuprofen	Liquid oral dosage form	XG at a concentration of more than 0.6% showed stability in the suspension even after standing for 10 long days	[143]
XG	Herbal drugs (herbal extracts of *Glycyrrhiza glabra*, *Terminalia chebula*, *Terminalia belerica*, *Emblica Officinalis*, and *Turbinella rapa*)	Liquid oral dosage form	Suspension containing 0.3% of XG along with herbal powders showed better consistency and redispersibility	[144]
XG	–	Ophthalmic drug delivery	Effect of sonification studies revealed that 0.2% dispersion of XG does not show any interaction whereas 1% dispersion of XG showed clear interaction	[145]
XG	Sodium hyaluronate	Ophthalmic drug delivery	XG and sodium hyaluronate both were highly effective in healing corneal wound in 3 days and complete healing of eye was done in 9 days	[146]
XG	Netilmicin	Ophthalmic drug delivery	Administration of eye gel containing XG was able to decrease the length of occlusive patching	[147]

(continued)

Table 2 (continued)

Polymeric component	Drug	Dosage form	Performances	References
XG	Zolmitriptan	Buccal drug delivery	Drug release studies revealed that 43.15% of the drug was released within 15 min and then it showed a sustained release for 5 h	[148]
XG	–	Buccal drug delivery	Modified XG for treating sialorrhea showed 1.5 times more water uptake capacity compared to simple XG Reduction of 2.61% of saliva when modified XG was used in contrast to unmodified XG which showed only 1.54% reduction	[149]
XG gel	Flurbiprofen	Topical drug delivery	Gel formulations showed sustained released for 24 h	[150]
XG hydrogel	Liranaftate	Topical drug delivery	Skin retention study of designed formulation showed six times higher drug retention capacity than the saturated liranaftate solution The skin sensitivity studies indicated no signs of irritation or erythema	[151]
XG	Serratiopeptidase	Topical drug delivery	XG can be successfully used as a gelling agent with the help of dimethyl sulfoxide, that is, permeation enhancer in the formation of serratiopeptidase niosomal gel	[152]
XG and guar gum	Lamivudine	Advance drug delivery	Formulation containing a higher concentration of polymer showed decreased release rate and sustained release up to 24 h	[153]

(continued)

Table 2 (continued)

Polymeric component	Drug	Dosage form	Performances	References
XG and guar gum	Metformin	Advance drug delivery	Designed microsphere showed drug release up to 94.96 and 92.98% at the end of 10 h Formulation containing 93.72–95.94% was stable up to 3 months during storage without any physical changes	[154]
XG	Carbamazepine	Brain drug delivery	The findings of in vitro release study depicted that XG can be a better choice as natural mucoadhesive and rate controlling polymer for brain targeting delivery systems via olfactory mucosa	[155]
XG	Curcumin	Brain drug delivery	Higher drug distribution from liposomes was seen in brain, that is, approximately 1,240 ng in contrast to drug solution which showed 65 ng	[156]
XG and 4 aminophenylmercuric acetate	Metalloproteinase-9	Wound healing	Till 7 days 12% larger wounds were found in the MMP-9 injected group as compared to control	[157]
Xanthan-based biofilm containing silver nanoparticles	Methicillin	Wound healing	When nanoparticles of silver were incorporated into xanthan-based film, it showed larger inhibition zone as compared to film without silver nanoparticles	[158]
Xanthan-chitosan complex	Neomycin	Wound healing	No significant weight gain was found in rats treated with neomycin-xanthan-chitosan complex whereas when rats were treated with neomycin alone decrease in weight was seen	[159]

(continued)

Table 2 (continued)

Polymeric component	Drug	Dosage form	Performances	References
XG	Paracetamol	Dermal patches	With an increase in the concentration of XG release rate decreases XG-based patches showed the extended drug release (98.65%) at the end of 12 h	[160]
XG, and carbopol-934 and ethyl cellulose	Nicotine	Dermal patches	XG-based patches demonstrated initial fast release followed by extended release up to 10 h	[161]
XG and carbopol 934 polymer	Loratadine	Nasal drug delivery	Formulation containing 0.2% XG showed sustained drug release for 8 h	[162]
XG and gellan gum	Lamotrigine	Nasal drug delivery	Use of XG in formulation increases the mucoadhesion and residence time, and also enhances the gelling property of the formulation	[163]
XG—chitosan	Ciprofloxacin	Hydrogel	The gel with high drug loading efficiency (3.52 ± 0.07%) displayed faster and higher release rate than that of gel containing a smaller amount of drug (0.44 ± 0.01%)	[164]

site release, delayed release, stable release, and receptor targeting [166]. Delayed drug release is most used in oral dosage forms to avoid drug-induced mucosal irritation or to protect drugs from gastric enzymes and acidic pH and as well as to deliver drugs in a specific site [167]. The mechanism of drug release is based on time and changes in the physicochemical environment (pH, temperature, presence of enzymes and others). Once initiated, further drug releases may be sustained or immediate [166]. Sustainable drug delivery systems are used in the preparation of optimal drug concentrations over a long period of time to maintain the therapeutic level of the drug in the desired location, or to prevent side effects and the degree of drug poisoning. Gellan has a wide variety of interesting properties including its easy bio-production, cell adhesion ability, biodegradability, biocompatibility, drug release after rapid gelation, water holding capacity (hydrogel), and non-toxicity. Gellan is particularly interesting because of its relatively high mucosal properties at acidic pH and its ability to sol-gel transition in physiological fluids. These properties, especially for development of different physiologically responsive types of drug delivery formulations. Another very interesting formulation is the so-called liquid gels or shear gels, which are produced by applying shear force during the gelling process. They are especially suitable for oral formulations in patients with difficulty swallowing, as the gel particles are microscopic in size. However, in designing new dosage forms, the correct choice of formulation materials is important [168]. Gellan is gaining more attention today due to its desirable properties as a formulation material. It is prescribed in different routes and used in many different dosage forms. Gellan-based nanohydrogel systems for multi-functional drug delivery applications have been recently studied. For example, paclitaxel and prednisolone were covalently bound to gellan and their anticancer and antiinflammatory effects on malignant cells were investigated [169]. The multi-particulate drug delivery system with numerous particles of about 0.05–2 nm in size offers numerous advantages over a single system because of their smaller size. They are weakly dependent on gastric emptiness, and they have higher bioavailability limiting irritation and systemic toxicity. They also have better pharmacokinetic behavior and degradability compared with traditional formulations [170]. Several antibiotics and other therapeutic agents were encapsulated in floating gellan beads to increase their stomachal retention time [171, 172]. Gellan has more beneficial properties than other available materials, such as being able to contact cations in physiological fluids, stability and high capacity of water retention, mucosal adhesion, non-toxicity, temperature and enzyme resistances, biodegradability and persistence at low pH in the gastrointestinal tract. Due to these properties, it can be easily adjusted to various forms with stable and controlled drug release such as particles, films, hydrogels, fibers, in situ gel, and others [173]. Effects of gellan-based gold nanoparticles (AuNPs) were investigated with mouse embryonic fibroblasts (LN-229) and human glioma cell lines (NIH 3T3) [174]. The cytotoxic and antibacterial activities of silver nanoparticles (AgNPs) stabilized with gellan were also studied using NIH 3T3 cells model [175]. In another study, gellan-coated gold nanorods (AuNRs) were experimented for imaging and intercellular drug delivery [176]. Recent studies revealed that gellan can be used in nasal, gastric or ocular drug delivery systems [177, 178]. Gellan has also been used for protein delivery systems (implants) to release insulin in mice having diabetes. The levels of glucose in blood of diabetic rats with this kind of implant were halved for one week compared with control [179]. Levofloxacin hemihydrate with gellan was used as an ophthalmic gel. In this study gelling time, drug release, stability, and gel uptake time were investigated [180]. Vashisth et al. [181] used ofloxacin-loaded polyvinyl alcohol/GG nanofibers for gastroretentive/mucoadhesive drug release applications, and their results showed a two-phase drug release pattern with mucosal adhesion and significant gastric retention in the mucous membranes of mice. Resveratrol-loaded chitosan/gellan nanofibers used as gastrointestinal release system showed a resveratrol encapsulation efficiency of 86.6%. The antioxidant activity of resveratrol-loaded nanofibers was higher than that of control showing the great potential of resveratrol-loaded chitosan/gellan nanofibers as drug release carriers [182] (Table 3).

4.3 Alginate as Drug Delivery Systems

Alginates or alginic acids are one of the biomaterials for mucoadhesive drug delivery systems, due to their cell compatibility, biodegradability, Sol-Gel transfer properties and chemical versatility. Even if this substance is industrially extracted from different types of seaweeds, it can also be excreted as EPS by some bacteria. Alginate appears as an interesting candidate platform for the promotion of new nanodrug delivery [312]. This heteropolysaccharide is a linear polymer consisting of repeats of β-manuronic (M) and α-L-gulucuronic (G) acids (1,4) linked, the two uronic acids being in pyranosic conformation. The two uronic acids are arranged in homogeneous (MM or GG) and heterogeneous (MG or GM) blocks leading to a large diversity of structures and physicochemical properties [313]. The word alginate is derived from the word alga meaning algae. Commercial alginates are isolated from the cell wall of brown algae such as *Macrocystis pyrifera*, *Laminaria hyperborean*, and *Ascophyllum nodosum* (Fig. 6). Alginate can also be extracted from the polysaccharide capsule of some bacteria, such as *Azotobacter vinelandii*. Bacterial alginates have very weak gel-forming properties due to their mannuronic acid-rich building blocks. In addition, the production of alginates from bacteria is not cost-effective for commercial use. So, alginates from marine algae are often used for mass production [314–316]. Bacterial alginate can be produced in aerobically continuous culture system at a temperature of 30 °C and a pH of 7. Excessive oxygen concentration does not stimulate alginate synthesis. The source of carbon required for the production of extracellular alginate is sucrose at a rate of 20 g/l. Limiting phosphorus and molybdenum in the culture medium or slowing down the dilution rate induces an increase of production rate of this biopolymer [317]. The extraction process of seaweed alginates is a simple but multi-step process that usually begins with the application of dilute mineral acid to the dried algae. In the next steps, the alginic acid obtained from the previous step is converted

Table 3 Drug delivery formulation types based on GG

Polymeric component	Drug	Dosage form	Formulation type	Applications	References
GG/gum cordia	Metformin HCl	Oral drug delivery	Beads	Antidiabetic	[183]
GG	Amoxicillin	Oral drug delivery	Macrobeads	Antibiotic	[184]
GG (disintegrant)	Metronidazole	Oral drug delivery	Tablets	Antibacterial	[185]
GG/poly(*N*-isopropylacrylamide)	Atenolol	Oral drug delivery	Thermoresponsie microspheres	Antihypertensive	[186]
GG/chitosan	Amoxicillin	Oral drug delivery	Coated beads	Antibiotic	[187]
Microwave-modified GG (disintregant)	Diclofenac sodium	Oral drug delivery	Tablets	Antiinflammatory	[188]
GG/hydroxypropyl methyl cellulose/ carbopol 934	Clarithromycin	Oral drug delivery	Floating oil-entrapped beads	Antibiotic	[189]
GG	Ambroxol HCl	Oral drug delivery	Immediate release soft gel	Oral expectorant and mucolytic agent	[190]
GG/chitosan/gelatin	Metronidazole and metronidazole benzoate	Oral drug delivery	Beads	Antibiotic	[191]
GG/low methoxy pectin	Ofloxacin	Oral drug delivery	Floating beads	Antibiotic	[192]
GG/egg albumin	Diltiazem-resin complex	Oral drug delivery	Microcapsules	Antihypertensive	[193]
GG	Glipizide	Oral drug delivery	Beads	Type 2 diabetes	[194]

(continued)

Table 3 (continued)

Polymeric component	Drug	Dosage form	Formulation type	Applications	References
GG; GG/chitosan	Stavudine	Oral drug delivery	Hydrogel beads	Antiretroviral	[195]
GG/tamarind xyloglucan	Diltiazem	Oral drug delivery	Matrix tablets	Antihypertensive	[196]
GG	Rifabutin	Oral drug delivery	Floating beads	Antibiotic	[197]
GG	Prednisolone	Oral drug delivery	Nanohydrogel	Antiinflammatory	[198]
GG/chitosan	Metronidazole	Oral drug delivery	Floating beads	Antibacterial	[199]
GG	Verapamil HCl	Oral drug delivery	Floating in situ gelling raft system	Antihypertensive	[200]
Partially hydrolyzed poly (acrylamide)-grafted GG	Salbutamol sulfate	Oral drug delivery	Gel	Bronchodilator	[201]
GG/hydroxypropyl methyl cellulose; GG/carbopol 934	Amoxicillin	Oral drug delivery	Floating-coated pH-sensitive oil eentrapped beads	Antibiotic	[172]
GG/chitosan	Metronidazole, metronidazole benzoate	Oral drug delivery	Beads	Antibiotic	[202]
Acrylamide-grafted GG (swelling agent)	Metformin HCl	Oral drug delivery	Tablets	Antidiabetic	[203]
Carboxymethylated GG	Metformin	Oral drug delivery	Beads	Antidiabetic	[204]
GG	Tranexamic acid	Oral drug delivery	Microbeads	Antifibrinolytic	[205]

(continued)

Table 3 (continued)

Polymeric component	Drug	Dosage form	Formulation type	Applications	References
GG	Paclitaxel, prednisolone	Oral drug delivery	Nanohydrogel	Anticancer, Antiinflammatory	[169]
GG/chitosan	–	Oral drug delivery	Self-destructing macrobeads loaded with microbeads	–	[206]
GG/alginate	Aceclofenac	Oral drug delivery	Microspheres for prolonged release	Antiinflammatory	[207]
GG	Ornidazole	Oral drug delivery	Floating in situ gelling system	Antibiotic	[208]
GG; GG/ethyl cellulose	5-Fluorouracil	Oral drug delivery	Microbeads	Anticancer	[209]
GG/chitosan	Bovine serum albumin	Oral drug delivery	Beads	–	[210]
GG (swelling agent)	Levofloxacin hemihydrate	Oral drug delivery	Swelling tables	Antibiotic	[211]
GG	Ibuprofen	Oral drug delivery	Oral liquid gel	Antiinflammatory	[212]
GG/fenugreek seed mucilage	Metformin HCl	Oral drug delivery	Mucoadhesive beads	Antidiabetic	[213]
GG/jackfruit seed starch	Metformin HCl	Oral drug delivery	Mucoadhesive beads	Antidiabetic	[214]
GG/ispaghula husk mucilage	Metformin HCl	Oral drug delivery	Mucoadhesive beads	Antidiabetic	[215]
GG/tamarind seed polysaccharide	Metformin HCl	Oral drug delivery	Mucoadhesive beads	Antidiabetic	[216]
GG/pectin	Ketoprofen	Oral drug delivery	Beads	Antiinflammatory	[217]

(continued)

Table 3 (continued)

Polymeric component	Drug	Dosage form	Formulation type	Applications	References
Cetyl GG	Simvastatin	Oral drug delivery	Micelle-loaded beads	Lipid lowering	[218]
GG	Glipizide	Oral drug delivery	Microparticles	Type 2 diabetes	[219]
GG; polymethacrylamide- grafted GG	Diclofenac sodium	Oral drug delivery	Sustained release tablets	Antiinflammatory	[220]
GG	Itopride HCl	Oral drug delivery	Floating in situ gelling raft system	*Gastroprokinetic*	[221]
GG	Metformin	Oral drug delivery	Microbeads for sustained release	Antidiabetic	[222]
GG	Ketoprofen	Oral drug delivery	Microspheres	Antiinflammatory	[223]
GG	Baclofen	Oral drug delivery	Superporous hydrogel	CNS depressant and SM relaxant	[224]
Polymethacrylamide-g-GG/tamarind seed gum	Metformin	Oral drug delivery	Extended release tablet	Type 2 diabetes	[225]
GG/retrograded starch	Ketoprofen	Oral drug delivery	Gel	Antiinflammatory	[226]
GG/polyethylene oxide	Sulpiride	Oral drug delivery	Xerogel	Antipsychotic	[227]
GG	Meloxicam	Oral drug delivery	Macrobeads	Antiinflammatory	[228]
GG/polyvinyl alcohol	Ofloxacin	Oral drug delivery	Nanofibers	Antibiotic	[181]

(continued)

Table 3 (continued)

Polymeric component	Drug	Dosage form	Formulation type	Applications	References
GG/LM-pectin 101	Gabapentin	Oral drug delivery	Floating raft system	Antiepileptic, antineuropathic	[229]
GG/pectin	Metformin HCl	Oral drug delivery	Bionanofiller composite	Antidiabetic	[230]
GG/starch/pectin	Insulin	Oral drug delivery	Coated microparticles	Antidiabetic	[231]
GG	Naproxen	Oral drug delivery	Macrobeads	Antiinflammatory	[232]
GG; GG/carrageenan; GG/guar gum; GG/cellulose sulfate; GG/dextran sulfates	Naproxen	Oral drug delivery	Macrobeads	Antiinflammatory	[233]
GG/pectin	Resveratrol	Oral drug delivery	Beads	Antiapoptotic	[234]
GG (binder)	Theophylline	Oral drug delivery	Pellets	Phosphodiesterase inhibitor	[235]
gellan gum/pectin	Curcumin	Oral drug delivery	Mucoadhesive film	–	[236]
GG	Apigenin	Oral drug delivery	Hydrogel	pH-sensitive microemulsion	[237]
gellan gum/pectin	Resveratrol	Oral drug delivery	Oral nanoparticles	Colon targeted delivery	[238]
GG	Vildagliptin	Oral drug delivery	mucoadhesive beads	Antidiabetic	[239]
GG/chitosan	Timolol maleate	Ophthalmic drug delivery	In situ gelling solution	Antiglaucoma	[240]

(continued)

Table 3 (continued)

Polymeric component	Drug	Dosage form	Formulation type	Applications	References
GG/alginate; GG/carboxymethyl cellulose	Gatifloxacin	Ophthalmic drug delivery	In situ gelling solution	Antibiotic	[241]
GG; GG/alginate	Matrine	Ophthalmic drug delivery	In situ gelling solution	Antibiotic/ antiinflammatory	[242]
GG	Flurbiprofen axetil	Ophthalmic drug delivery	In situ gelling nanoemulsion	Antiinflammatory	[243]
GG	Moxifloxacin	Ophthalmic drug delivery	In situ gelling solution	Antibiotic	[244]
GG	Pilocarpine HCl	Ophthalmic drug delivery	In situ gelling solution	Antiglaucoma	[245, 246]
GG	Cx43 antisense oligodeoxynucleotide	Ophthalmic drug delivery	In situ gelling solution	Wound closure	[247]
GG	Aesculin	Ophthalmic drug delivery	In situ gelling solution	Antiinflammatory/ vasoprotective	[56, 248]
GG	Brimonidine tartrate	Ophthalmic drug delivery	In situ gelling solution	Antiglaucoma	[249]

(continued)

Table 3 (continued)

Polymeric component	Drug	Dosage form	Formulation type	Applications	References
GG/alginate GG/carbopol 934	Ketotifen fumarate	Ophthalmic drug delivery	In situ gelling solution	Antihistamine	[110, 250]
GG	Terbinafine HCl	Ophthalmic drug delivery	In situ gelling nanoemulsion	Fungal keratitis	[251]
GG/cyclodextrin	Fluconazole	Ophthalmic drug delivery	In situ gelling solution for low water-soluble drugs	Antifungal	[252]
GG/pluronic 123/TPGS	Curcumin	Ophthalmic drug delivery	In situ gelling nanomicellar solution	Antiinflammatory, antimicrobic	[253]
GG/chitosan	Levofloxacin	Ophthalmic drug delivery	In situ gelling solution	Antibiotic	[254]
GG/chitosan	Sparfloxacin	Ophthalmic drug delivery	In situ gelling solution	Antibiotic	[255]
GG	Doxycycline HCl	Ophthalmic drug delivery	In situ gelling nanoparticle solution	Antibiotic	[256]
GG	Timolol	Ophthalmic drug delivery	In situ gelling liposome solution	Antiglaucoma	[257]

(continued)

Table 3 (continued)

Polymeric component	Drug	Dosage form	Formulation type	Applications	References
GG	Ketotifen fumarate	Ophthalmic drug delivery	In situ gelling solution	Allergic conjunctivitis	[258]
GG	Moxifloxacin	Ophthalmic drug delivery	In situ gelling nanoparticle solution	Antibiotic	[259]
GG/carboxymethyl cellulose	Gatifloxacin	Ophthalmic drug delivery	In situ gelling solution	Antibiotic	[260]
GG/calcium gluconate/polyvinylpyrrolidone	Timolol	Ophthalmic drug delivery	In situ gelling solution	Antiglaucoma	[261]
GG/poloxamer 407	Pilocarpine HCl	Ophthalmic drug delivery	In situ gelling solution	Antiglaucoma	[262]
GG	Estradiol	Ophthalmic drug delivery	In situ gelling solution	Anticataract	[263]
GG/carrageenan	Cysteamine	Ophthalmic drug delivery	In situ gelling solution	Cystinosis treatment	[264]
GG/xanthan gum; GG/hydroxypropyl methyl cellulose; GG/Carbopol	Acetazolamide	Ophthalmic drug delivery	In situ gelling nanoemulsions	Antiglaucoma	[265]

(continued)

Table 3 (continued)

Polymeric component	Drug	Dosage form	Formulation type	Applications	References
GG/carrageenan	Econazole	Ophthalmic drug delivery	In situ gelling solution	Antifungal	[266]
GG/chitosan/polyvinylalcohol	Besifloxacin	Ophthalmic drug delivery	In situ gelling solution	Antimicrobic	[267]
GG	Natamycin	Ophthalmic drug delivery	In situ gelling solution	Antifungal	[268]
GG/carbopol 934P	Benzododecinium bromide	Ophthalmic drug delivery	In situ gelling solution	Antimicrobic	[269]
GG	Brinzolamide	Ophthalmic drug delivery	In situ gelling solution	Antiglaucoma	[270]
GG	Natamycin	Ophthalmic drug delivery	In situ hydrogel	–	[271]
GG and its methacrylated derivatives	Pilocarpine	Ophthalmic drug delivery	In situ gelling mucoadhesive	Antiglaucoma	[272]
GG and hydroxyethylcellulose	Phenylephrine and tropicamide	Ophthalmic drug delivery	In situ gelling	–	[273]
GG/Carbopol	Metoclopramide HCl	Nasal drug delivery	In situ nasal gel	Antiemetic	[274]

(continued)

Table 3 (continued)

Polymeric component	Drug	Dosage form	Formulation type	Applications	References
GG	Ondansetron	Nasal drug delivery	Dry intranasal microspheres	Antiemetic	[275]
GG	Sildenafil citrate	Nasal drug delivery	Dry intranasal microspheres	Erectile dysfunction	[276]
GG	Gastrodin	Nasal drug delivery	In situ nasal gel	CNS sedation	[277]
GG	Curcumin	Nasal drug delivery	In situ nasal microemulsion gel	Antidepressant	[278]
GG	Carvedilol	Nasal drug delivery	In situ nasal nanosuspensin gel	Beta-blocker	[279]
GG	Almotriptan malate	Nasal drug delivery	Dry intranasal microspheres	Antimigraine	[280]
GG	Sumatriptan succinate	Nasal drug delivery	In situ nasal gel	Antimigraine	[281]
GG	Saquinavir mesylate	Nasal drug delivery	In situ nasal nanosized	Protease inhibitor	[282]
C18-grafted GG	Budesonide	Nasal drug delivery	Dry intranasal nanomicelles	Antiallergic	[283]
GG	Granisetron HCl	Nasal drug delivery	In situ nasal droppable gel	Antiemetic	[284]
GG/poloxamer 407	Ketorolac tromethamine	Nasal drug delivery	Thermo- and ionsensitive in situ nasal gel	Analgesic	[285]
HA GG/LA GG	–	Nasal drug delivery	In situ nasal sprayable gels	–	[286]

(continued)

Table 3 (continued)

Polymeric component	Drug	Dosage form	Formulation type	Applications	References
GG/Lutrol F 127	Rivastigmine	Nasal drug delivery	In situ nasal nanosuspension gel	Cholinesterase inhibitor	[287]
GG	Resveratrol	Nasal drug delivery	In situ nasal nanosuspension gel	Antiapoptotic	[288]
GG/hydroxypropyl methyl cellulose	Salbutamol sulfate	Nasal drug delivery	In situ nasal gel	Bronchodilator	[289]
GG/chitosan	Ondansetron HCl	Nasal drug delivery	Lyophilized nasal inserts	Antiemetic	[290]
GG/xanthan gum	Lamotrigine	Nasal drug delivery	In situ nasal gel	Antiepileptic	[163]
GG/carbopol 934	Lorazepam	Nasal drug delivery	In situ nasal microemulsion gel	Anxiolytic	[291]
GG/konjac gum	Donepezil HCl	Nasal drug delivery	In situ nasal cubosome gel	Cholinesterase inhibitor	[292]
GG/polyethylenimine	Plasmid DNA	Topical drug delivery	Nanocomposites	Nonviral gene vector	[293]
GG/hyaluronic acid ester/polyvinyl alcohol	Silver	Topical drug delivery	Nonwoven dressings	Wound healing	[294]
GG/chitosan	Levofloxacin and TiO_2	Topical drug delivery	Bilayer film	Wound healing	[295]
GG/palm kernel oil esters	Ibuprofen	Topical drug delivery	Modified nanoemulsion	Antiinflammatory	[296]

(continued)

Table 3 (continued)

Polymeric component	Drug	Dosage form	Formulation type	Applications	References
Quaternized GG/chitosan	Ciprofloxacin	Topical drug delivery	Particles	Antibiotic	[297]
GG/chitosan	Ibuprofen	Topical drug delivery	Nanogel	Antiinflammatory	[298]
GG	Diclofenac sodium	Topical drug delivery	Thermoresponsive nanogels: semisolid gel and solid hydrogel film	Antiinflammatory	[299]
GG	Diclofenac	Topical drug delivery	Fluid gels	Antiinflammatory	[300]
GG/chitosan/PEG	Apigenin	Topical drug delivery	Gel	Wound healing	[301]
GG/pomegranate oil	Silibinin	Topical drug delivery	Semisolid gel with nanocapsules	Antiinflammatory	[302]
GG	Coenzyme Q10 and vitamin E acetate	Topical drug delivery	Nanocapsule suspensions	Antiinflammatory	[303]
GG	Eumelanin	Topical drug delivery	Hydrogel subcutaneous application	Wound healing	[323]
Cholesterol derivatized GG	Baicalin	Topical drug delivery	Nanohydrogels	Wound healing	[305]

(continued)

Table 3 (continued)

Polymeric component	Drug	Dosage form	Formulation type	Applications	References
GG	Baicalin	Topical drug delivery	Core-shell transfersomes	Wound healing	[306]
GG	Piroxicam	Topical drug delivery	Nanohydrogels for cutaneous administration	Antiinflammatory	[307]
GG	Vancomycin	Topical drug delivery	Hydrogel; hydrogel with nanoparticles	Wound healing	[308]
GG/glucosamine	Clioquinol	Topical drug delivery	Patch	Oral cancer treatment	[328]
GG	Nebivolol	Topical drug delivery	Hydrogel	–	[310]
GG	Diindolylmethane	–	Hydrogel containing nanocapsules	Anti-trichomonas vaginalis	[311]

to water-soluble sodium salt in the presence of sodium carbonate. Finally, alginates are purified [318]. The molecular weights of alginates range between 32 to 400 kDa and their mannuronic/guluronic ratios (M/G) between 0.2 and 6 [319]. Chain arrays based on extraction source and algae age have led to the commercialization of more than 200 types of alginates. In addition, the efficiency of drug release by alginate is modulated by M/G ratio, molecular weight, concentration and pH of the environment [315]. Alginates in aqueous solutions have ability to crosslink by a mechanism through the carboxylic acid moiety of G units with Ca^{2+} ions and other divalent cations (such as Zn^{2+}, Sr^{2+}, Ba^{2+}) to form a three-dimensional network. This gelling mechanism is explained by the egg-box model, in which a dual-valent cation with four COOH groups reacts to encapsulate a wide range of drugs, proteins, genes, and cells. Another cation used for crosslinking alginate is Fe^{3+} [312, 320]. The gelifying ability of alginates and their high viscosity in aqueous solutions are widely used in pharmaceutical, food and cosmetic industries [321].

Properties of alginate such as biocompatibility and non-immunogenicity can explain its exclusive biological activities [86]. In a purified form, it prevents immune responses after transplantation. Because cells are not damaged during the process of gel formation and ionic crosslinking, this substance is used for drug delivery, cell encapsulation and tissue regeneration [322]. The characteristics of alginate hydrogels can be controlled by the molecular weight of the polymer, its M/G ratio, the types of cross-linkage and the concentration of multivalent cations [323]. G-rich alginate gels have more pores and less shrinkage, while M-containing gels are softer and have less elasticity and pores [324]. In addition, alginate gels at neutral and basic pH are more prone to degradation than under acidic conditions. These properties encourage their use in drug stabilization and oral administration of biological substances poorly stable in gastrointestinal fluids [324]. On the other hand, the production of high purified alginates can pave the way for applications aimed at greater compatibility with living systems. To improve the properties of alginates, their surface can be modified with other inorganic materials. Due to their hydrophilic nature, the release of drugs in alginate hydrogels can follow diverse mechanisms. Polar drugs are mostly released through diffusion, while hydrophobic ones are discharged through erosion of the alginate matrix. The release of small molecules is also faster due to the fact that a hole about 5 nm in diameter is created in the swollen alginate matrices. However, to prolong the release time of drugs,

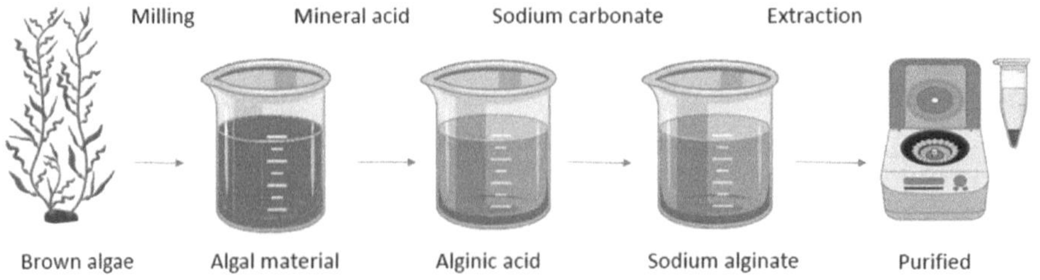

Fig. 6 Production of sodium alginate isolated from brown algae

various changes can be made in the physical and chemical connections of drugs encapsulated in the polymer network [93]. It is then possible to prolong its shelf life and release time in various mucosal tissues, including the intestine, lung, nose, and eye [312, 314, 316, 320].

One of the challenges associated with the production of alginate nanoparticles is the difficulty in production methods to achieve small and nanometer sizes. To overcome this problem, nanoparticles, chitosan, and liposomes as model, which sustain more surface changes due to the nature of the alginate mucosa adhesive [325]. Also, Hong et al. [326] used liposome nuclei with high bilayer melting point as reaction channels. In this study, alginate encapsulated in a liposomal nucleus was exposed to $CaCl_2$ solution at a temperature above the melting point of the two layers. This approach made it possible to transfer calcium ions into the nucleus and to gradually gelify alginate. Then, liposomes were removed by surfactants and nanoparticles of 120–200 nm were obtained [326]. Polyelectrolyte layers can also be assembled with pure drug crystals. Based on this approach, artemisinin nanocapsules (766 nm) were coated with chitosan, gelatin and alginate [327]. Moreover, Saraei et al. [328] produced alginate nanoparticles containing protein (BSA) with a diameter of 50 nm [328].

Clotrimazole and econazole, two antifungal compounds, and pyrazinamide, rifampicin, ethambutol and isoniazid, four anti-tuberculosis drugs were encapsulated in alginate nanoparticles which were prepared with modified cation induced gelation in a controlled manner [329]. After oral administration, free drugs were detectable during 6–24 h, while those encapsulated samples were stable for 8–15 days. In addition, 8 doses of nanoparticles loaded with econazole had an antibacterial effect similar to that of 112 free drug doses taken twice daily. These findings confirmed the ability of alginate nanoparticles to bind to the intestinal barrier and reach the bloodstream, in contrast to alginate microparticles that remained mainly in the intestinal mucosa. This beneficial effect has also been investigated for the fabrication of mucosal alginate/chitosan nanoparticles to release amoxicillin against *Helicobacter pylori*, a pathogen that implants in the deep mucosal layers of the stomach [329–331]. Alginate nanoparticles containing 5-aminolaevulinic acid (5-ALA) were used to detect cancer cells. Alginate nanoparticles containing 5-ALA enable photodynamic detection when taken up by cancer cells. Inside the cancer cell, 5-ALA is converted to fluorescent protoporphyrin IX. Folic acid is conjugated to target cancer cells with high folic acid receptor [312, 332].

These results confirmed the efficacy of alginate nanoparticles for drug encapsulation and release and showed that alginate nanoparticles could be used as an effective tool for designing more sophisticated vaccination and treatment systems.

5 Conclusion

Microbial exopolysaccharides are macromolecules produced by different microorganisms and have various structures depending on their monomeric composition. The chemical diversity is linked to their physicochemical properties and consequently attribute to their various functions and applications including new drug delivery systems. Production of microbial exopolysaccharides is influenced by various conditions and factors controlling microbial growth, such as concentration, carbon source, temperature and pH. However, a better and deeper understanding of the mechanism involved in the synthesis and extraction of microbial exopolysaccharides is needed. Controlling the expression of a particular gene will increase the production of microbial extracellular polysaccharides and ultimately lead to controlling the structure and properties of microbial extracellular polysaccharides in the coming years. Therefore, exopolysaccharides are still limited in drug delivery systems as drug delivery methods encountered some difficulties for drug loading, prolonging drug influence time, interference, side effects, the amount of drug absorption, adverse breakdown, and even convenience of administration of an ideal drug delivery system should deliver the right amount of drug to a target site in the human or animal body to maintain the drug concentration for an optimal period of time, without side effects. Therefore, in vivo and in vitro cytotoxicity tests for new systems and knowledge of drug specifications are necessary to optimize drug formulation.

References

1. Kratz F, Warnecke A, Riebeseel K, Rodrigues P (2001) Anticancer drug conjugates with macromolecular carriers. In: Polymeric biomaterials, revised and expanded. CRC Press
2. Bhowmik D, Gopinath H, Duraivel S, Kumar KS (2013) Silk-based drug delivery systems. Pharma Innov 1:42
3. Bhowmik D, Gopinath H, Kumar BP, Duraivel S, Kumar KS (2012) Microchip drug delivery: new era of drug delivery system. Pharma Innov 1
4. Patel T (2007) Recent trends in novel drug delivery system. Latest Rev
5. Ligade PC, Jadhav KR, Kadam VJ (2010) Brain drug delivery system: an overview. Curr Drug Ther 5:105–110
6. Dhoble T (2016) Drug delivery systems. Int J Ind Biotechnol Biomater 2:6–10
7. Jain KK (2008) Drug delivery systems. Springer
8. Khan MG (2017) Topic—the novel drug delivery system
9. Falde EJ, Yohe ST, Colson YL, Grinstaff MW (2016) Superhydrophobic materials for biomedical applications. Biomaterials 104:87–103
10. Pillai O, Panchagnula R (2001) Polymers in drug delivery. Curr Opin Chem Biol 5:447–451
11. Yao Q, Zheng Y-W, Lan Q-H, Kou L, Xu H-L, Zhao Y-Z (2019) Recent development and biomedical applications of decellularized extracellular matrix biomaterials. Mater Sci Eng C 104:109942
12. Ngwuluka NC (2018) Responsive polysaccharides and polysaccharides-based nanoparticles for drug delivery. In: Stimuli responsive polymeric nanocarriers for drug delivery applications, vol 1. Elsevier

13. Rahmati M, Alipanahi Z, Mozafari M (2019) Emerging biomedical applications of algal polysaccharides. Curr Pharm Des 25:1335–1344
14. Seidi F, Jenjob R, Phakkeeree T, Crespy D (2018) Saccharides, oligosaccharides, and polysaccharides nanoparticles for biomedical applications. J Controlled Release 284:188–212
15. Zhu Y, Liu Y, Jin K, Pang Z (2019) Polysaccharide nanoparticles for cancer drug targeting. Polysaccharide carriers for drug delivery. Elsevier
16. Grigoras AG (2019) Drug delivery systems using pullulan, a biocompatible polysaccharide produced by fungal fermentation of starch. Environ Chem Lett 1–15
17. de Melo Barbosa A, da Cunha PDT, Pigatto MM, da Silva MDLC (2004) Produção e aplicações de exopolissacarídeos fúngicos. Semina: Ciências Exatas e Tecnológicas 25:29–42
18. de Souza DM, Garcia-Cruz CH (2004) Produção fermentativa de polissacarídeos extracelulares por bactérias. Semina Ciências Agrárias Londrina 25:253–262
19. Nayak A, Pal D (2012) Natural polysaccharides for drug delivery in tissue engineering. Everyman's Sci XLVI
20. Bhatia S (2016) Microbial polysaccharides as advance nanomaterials. In: Systems for drug delivery. Springer
21. Bergmaier D (2002) Production d'exopolysaccharides par fermentation avec des cellules immobilisées de LB. Rhamnosus RW-9595 M d'un milieu à base de perméat de lactosérum
22. Boels IC, van Kranenburg R, Hugenholtz J, Kleerebezem M, de Vos WM (2001) Sugar catabolism and its impact on the biosynthesis and engineering of exopolysaccharide production in lactic acid bacteria. Int Dairy J 11:723–732
23. Lahaye É (2006) Rôle structurant des exopolysaccharides dans un biofilm bactérien. Lorient
24. Suresh Kumar A, Mody K, Jha B (2007) Bacterial exopolysaccharides—a perception. J Basic Microbiol 47:103–117
25. Whitfield C (1988) Bacterial extracellular polysaccharides. Can J Microbiol 34:415–420
26. Pichavant L (2009) Design, synthèse et réactivité de monomères issus de ressources renouvelables pour la polymérisation radicalaire. Reims
27. Yildiz H, Karatas N (2018) Microbial exopolysaccharides: resources and bioactive properties. Process Biochem 72:41–46
28. Dubashynskaya N, Poshina D, Raik S, Urtti A, Skorik YA (2020) Polysaccharides in ocular drug delivery. Pharmaceutics 12:22
29. Shih IL (2010) Microbial exo-polysaccharides for biomedical applications. Mini Rev Med Chem 10:1345–1355
30. Cordeiro AS, Alonso MJ, de la Fuente M (2015) Nanoengineering of vaccines using natural polysaccharides. Biotechnol Adv 33:1279–1293
31. Shariatinia Z (2019) Pharmaceutical applications of natural polysaccharides. In: Natural polysaccharides in drug delivery and biomedical applications. Elsevier
32. Sutherland IW (1990) Biotechnology of microbial exopolysaccharides. Cambridge University Press
33. Vasiliu S, Racovita S, Lungan MA, Desbrieres J, Popa M (2016) Microbial exopolysaccharides for biomedical applications. In: Pillay V, Choonara YE, Kumar P (eds) Frontiers in biomaterials: unfolding the biopolymer landscape. Bentham Sci. Publ., Sharjah, UAE, pp 180–238
34. Sutherland IW (1997) Microbial exopolysaccharides-structural subtleties and their consequences. Pure Appl Chem 69:1911–1918
35. Nichols CM, Guezennec J, Bowman J (2005) Bacterial exopolysaccharides from extreme marine environments with special consideration of the southern ocean, sea ice, and deep-sea hydrothermal vents: a review. Mar Biotechnol 7:253–271
36. Nwodo UU, Green E, Okoh AI (2012) Bacterial exopolysaccharides: functionality and prospects. Int J Mol Sci 13:14002–14015

37. Bajpai VK, Rather IA, Majumder R, Shukla S, Aeron A, Kim K, Kang SC, Dubey R, Maheshwari D, Lim J (2016) Exopolysaccharide and lactic acid bacteria: perception, functionality and prospects. Bangladesh J Pharmacol 11:1–23
38. Harutoshi T (2013) Exopolysaccharides of lactic acid bacteria for food and colon health applications. Lactic acid bacteria-R&D for food, health and livestock purposes. IntechOpen
39. Meng G, Fütterer K (2003) Structural framework of fructosyl transfer in Bacillus subtilis levansucrase. Nat Struct Mol Biol 10:935–941
40. Notararigo S, Nácher-Vázquez M, Ibarburu I, Werning ML, de Palencia PF, Dueñas MT, Aznar R, López P, Prieto A (2013) Comparative analysis of production and purification of homo-and hetero-polysaccharides produced by lactic acid bacteria. Carbohyd Polym 93:57–64
41. Shih L, Chen L-D, Wu J-Y (2010) Levan production using Bacillus subtilis natto cells immobilized on alginate. Carbohyd Polym 82:111–117
42. Hernández L, Arrieta J, Betancourt L, Falcón V, Madrazo J, Coego A, Menéndez C (1999) Levansucrase from *Acetobacter diazotrophicus* SRT4 is secreted via periplasm by a signal-peptide-dependent pathway. Curr Microbiol 39:146–152
43. Vandamme E, De Baets S, Steinbuchel A (2002) Polysaccharides I: polysaccharides and prokaryotes (biopolymers series). In: Vandamme EJ, De Baets S, Steinbüchel A (eds) Biopolymers vol. 5: polysaccharides I, vol 5, 10 vols. Wiley, p 532
44. Kim S-J, Bae PK, Chung BH (2015) Self-assembled levan nanoparticles for targeted breast cancer imaging. Chem Commun 51:107–110
45. Kazak Sarilmişer H, Öner Toksoy E, Sezer AD, Rayaman E, Çevikbaş A, Akbuğa J (2015) Development and characterization of vancomycin-loaded levan-based microparticular system for drug delivery
46. Tabernero A, González-Garcinuño Á, Sánchez-Álvarez JM, Galán MA, del Valle EMM (2017) Development of a nanoparticle system based on a fructose polymer: stability and drug release studies. Carbohyd Polym 160:26–33
47. Sarilmiser HK, Oner ET (2014) Investigation of anti-cancer activity of linear and aldehyde-activated levan from Halomonas smyrnensis AAD6T. Biochem Eng J 92:28–34
48. Kim S-J, Chung BH (2016) Antioxidant activity of levan coated cerium oxide nanoparticles. Carbohyd Polym 150:400–407
49. Bondarenko OM, Ivask A, Kahru A, Vija H, Titma T, Visnapuu M, Joost U, Pudova K, Adamberg S, Visnapuu T (2016) Bacterial polysaccharide levan as stabilizing, non-toxic and functional coating material for microelement-nanoparticles. Carbohyd Polym 136:710–720
50. Ahmed KBA, Kalla D, Uppuluri KB, Anbazhagan V (2014) Green synthesis of silver and gold nanoparticles employing levan, a biopolymer from Acetobacter xylinum NCIM 2526, as a reducing agent and capping agent. Carbohyd Polym 112:539–545
51. Maciel J, Andrad P, Neri D, Carvalho Jr L, Cardoso C, Calazans G, Aguiar JA, Silva M (2012) Preparation and characterization of magnetic levan particles as matrix for trypsin immobilization. J Magn Magn Mater 324:1312–1316
52. Moscovici M (2015) Present and future medical applications of microbial exopolysaccharides. Front Microbiol 6:1012
53. Rekha M, Sharma CP (2007) Pullulan as a promising biomaterial for biomedical applications: a perspective. Trends Biomater Artif Organs 20:116–121
54. Leathers TD (2003) Biotechnological production and applications of pullulan. Appl Microbiol Biotechnol 62:468–473
55. Madi N, Harvey L, Mehlert A, McNeil B (1997) Synthesis of two distinct exopolysaccharide fractions by cultures of the polymorphic fungus Aureobasidium pullulans. Carbohyd Polym 32:307–314
56. Chen J, Wu S, Pan S (2012) Optimization of medium for pullulan production using a novel strain of *Auerobasidium pullulans* isolated from sea mud through response surface methodology. Carbohyd Polym 87:771–774
57. Sutherland IW (1998) Novel and established applications of microbial polysaccharides. Trends Biotechnol 16:41–46

58. Bender H, Lehmann J, Wallenfels K (1959) Pullulan, ein extracelluläres Glucan von Pullularia pullulans. Biochimica et biophysica acta 36(2):309–316
59. Cristescu R, Dorcioman G, Ristoscu C, Axente E, Grigorescu S, Moldovan A, Mihailescu I, Kocourek T, Jelinek M, Albulescu M (2006) Matrix assisted pulsed laser evaporation processing of triacetate-pullulan polysaccharide thin films for drug delivery systems. Appl Surf Sci 252:4647–4651
60. Liang Y, Zhao X, Ma PX, Guo B, Du Y, Han X (2019) pH-responsive injectable hydrogels with mucosal adhesiveness based on chitosan-grafted-dihydrocaffeic acid and oxidized pullulan for localized drug delivery. J Colloid Interface Sci 536:224–234
61. Scomparin A, Salmaso S, Bersani S, Satchi-Fainaro R, Caliceti P (2011) Novel folated and non-folated pullulan bioconjugates for anticancer drug delivery. Eur J Pharm Sci 42:547–558
62. Zhang H, Li F, Yi J, Gu C, Fan L, Qiao Y, Tao Y, Cheng C, Wu H (2011) Folate-decorated maleilated pullulan–doxorubicin conjugate for active tumor-targeted drug delivery. Eur J Pharm Sci 42:517–526
63. Scomparin A, Salmaso S, Eldar-Boock A, Ben-Shushan D, Ferber S, Tiram G, Shmeeda H, Landa-Rouben N, Leor J, Caliceti P (2015) A comparative study of folate receptor-targeted doxorubicin delivery systems: dosing regimens and therapeutic index. J Controlled Release 208:106–120
64. Bae YH, Park K (2011) Targeted drug delivery to tumors: myths, reality and possibility. J Controlled Release 153(3):198
65. Saranya D, Rajan R, Suganthan V, Murugeswari A, Raj NAN (2015) Synthesis and characterization of pullulan acetate coated magnetic nanoparticle for hyperthermic therapy. Procedia Mater Sci 10:2–9
66. Kanmani P, Lim ST (2013) Synthesis and characterization of pullulan-mediated silver nanoparticles and its antimicrobial activities. Carbohyd Polym 97:421–428
67. Rekha M, Pal K, Bala P, Shetty M, Mittra I, Bhuvaneshwar G, Sharma CP (2013) Pullulan-histone antibody nanoconjugates for the removal of chromatin fragments from systemic circulation. Biomaterials 34:6328–6338
68. Jeong Y-I, Na H-S, Oh J-S, Choi K-C, Song C-E, Lee H-C (2006) Adriamycin release from self-assembling nanospheres of poly(DL-lactide-co-glycolide)-grafted pullulan. Int J Pharm 322:154–160
69. Garhwal R, Shady SF, Ellis EJ, Ellis JY, Leahy CD, McCarthy SP, Crawford KS, Gaines P (2012) Sustained ocular delivery of ciprofloxacin using nanospheres and conventional contact lens materials. Invest Ophthalmol Vis Sci 53:1341–1352
70. di Meo C, Montanari E, Manzi L, Villani C, Coviello T, Matricardi P (2015) Highly versatile nanohydrogel platform based on riboflavin-polysaccharide derivatives useful in the development of intrinsically fluorescent and cytocompatible drug carriers. Carbohyd Polym 115:502–509
71. de Lima JA, Paines TC, Motta MH, Weber WB, dos Santos SS, Cruz L, da Silva CDB (2017) Novel pemulen/pullulan blended hydrogel containing clotrimazole-loaded cationic nanocapsules: evaluation of mucoadhesion and vaginal permeation. Mater Sci Eng C 79:886–893
72. Flores FC, Rosso RS, Cruz L, Beck RC, Silva CB (2017) An innovative polysaccharide nanobased nail formulation for improvement of onychomycosis treatment. Eur J Pharm Sci 100:56–63
73. Tuncay Tanriverdi S, Hilmioğlu Polat S, Yeşim Metin D, Kandiloğlu G, Özer Ö (2016) Terbinafine hydrochloride loaded liposome film formulation for treatment of onychomycosis: in vitro and in vivo evaluation. J Liposome Res 26:163–173
74. Li H, Cui Y, Sui J, Bian S, Sun Y, Liang J, Fan Y, Zhang X (2015) Efficient delivery of DOX to nuclei of hepatic carcinoma cells in the subcutaneous tumor model using pH-sensitive pullulan–DOX conjugates. ACS Appl Mater Interfaces 7:15855–15865

75. Li H, Sun Y, Liang J, Fan Y, Zhang X (2015) pH-Sensitive pullulan–DOX conjugate nanoparticles for co-loading PDTC to suppress growth and chemoresistance of hepatocellular carcinoma. J Mater Chem B 3:8070–8078
76. Pranatharthiharan S, Patel MD, Malshe VC, Pujari V, Gorakshakar A, Madkaikar M, Ghosh K, Devarajan PV (2017) Asialoglycoprotein receptor targeted delivery of doxorubicin nanoparticles for hepatocellular carcinoma. Drug Deliv 24:20–29
77. Li F, Zhang H, Gu C, Fan L, Qiao Y, Tao Y, Cheng C, Wu H, Yi J (2013) Self-assembled nanoparticles from folate-decorated maleilated pullulan–doxorubicin conjugate for improved drug delivery to cancer cells. Polym Int 62:165–171
78. Seo S, Lee C-S, Jung Y-S, Na K (2012) Thermo-sensitivity and triggered drug release of polysaccharide nanogels derived from pullulan-g-poly (L-lactide) copolymers. Carbohyd Polym 87:1105–1111
79. Lee SJ, Shim Y-H, Oh J-S, Jeong Y-I, Park I-K, Lee HC (2015) Folic-acid-conjugated pullulan/poly(DL-lactide-co-glycolide) graft copolymer nanoparticles for folate-receptor-mediated drug delivery. Nanoscale Res Lett 10:1–11
80. Wang M, Huang M, Wang J, Ye M, Deng Y, Li H, Qian W, Zhu B, Zhang Y, Gong R (2016) Facile one-pot synthesis of self-assembled folate-biotin-pullulan nanoparticles for targeted intracellular anticancer drug delivery. J Nanomaterials
81. Hassanzadeh F, Varshosaz J, Khodarahmi G, Rostami M, Hassanzadeh F (2016) Biotin-encoded pullulan-retinoic acid engineered nanomicelles: preparation, optimization and in vitro cytotoxicity assessment in MCF-7 cells. Indian J Pharm Sci 78:557–565
82. Lee YC, Chen CT, Chiu YT, Wu KCW (2013) An effective cellulose-to-glucose-to-fructose conversion sequence by using enzyme immobilized Fe3O4-loaded mesoporous silica nanoparticles as recyclable biocatalysts. Chem Cat Chem 5:2153–2157
83. Wang J, Cui S, Bao Y, Xing J, Hao W (2014) Tocopheryl pullulan-based self assembling nanomicelles for anti-cancer drug delivery. Mater Sci Eng C 43:614–621
84. Wang X, Wang J, Bao Y, Wang B, Wang X, Chen L (2014) Novel reduction-sensitive pullulan-based micelles with good hemocompatibility for efficient intracellular doxorubicin delivery. RSC Adv 4:60064–60074
85. Wang Y, Liu Y, Liu Y, Wang Y, Wu J, Li R, Yang J, Zhang N (2014) pH-sensitive pullulan-based nanoparticles for intracellular drug delivery. Polym Chem 5:423–432
86. Chen C-Y, Ke C-J, Yen K-C, Hsieh H-C, Sun J-S, Lin F-H (2015) 3D porous calcium-alginate scaffolds cell culture system improved human osteoblast cell clusters for cell therapy. Theranostics 5:643
87. Li H, Cui Y, Liu J, Bian S, Liang J, Fan Y, Zhang X (2014) Reduction breakable cholesteryl pullulan nanoparticles for targeted hepatocellular carcinoma chemotherapy. J Mater Chem B 2:3500–3510
88. Tang H-B, Li L, Chen H, Zhou Z-M, Chen H-L, Li X-M, Liu L-R, Wang Y-S, Zhang Q-Q (2010) Stability and in vivo evaluation of pullulan acetate as a drug nanocarrier. Drug Deliv 17:552–558
89. Zhang C, Kim S-K (2010) Research and application of marine microbial enzymes: status and prospects. Mar Drugs 8:1920–1934
90. Zhang H-Z, Li X-M, Gao F-P, Liu L-R, Zhou Z-M, Zhang Q-Q (2010) Preparation of folate-modified pullulan acetate nanoparticles for tumor-targeted drug delivery. Drug Deliv 17:48–57
91. Hassanzadeh F, Mehdifar M, Varshosaz J, Khodarahmi GA, Rostami M (2018) Folic acid targeted polymeric micelles based on tocopherol succinate-pulluan as an effective carrier for Epirubicin: preparation, characterization and in-vitro cytotoxicity assessment. Curr Drug Deliv 15:235–246
92. Shen S, Li H, Yang W (2014) The preliminary evaluation on cholesterol-modified pullulan as a drug nanocarrier. Drug Deliv 21:501–508
93. Lee KY, Mooney DJ (2012) Alginate: properties and biomedical applications. Prog Polym Sci 37:106–126

94. Lee SJ, Hong G-Y, Jeong Y-I, Kang M-S, Oh J-S, Song C-E, Lee HC (2012) Paclitaxel-incorporated nanoparticles of hydrophobized polysaccharide and their antitumor activity. Int J Pharm 433:121–128
95. Yim H, Park S-J, Bae YH, Na K (2013) Biodegradable cationic nanoparticles loaded with an anticancer drug for deep penetration of heterogeneous tumours. Biomaterials 34:7674–7682
96. Hong G-Y, Jeong Y-I, Lee SJ, Lee E, Oh JS, Lee HC (2011) Combination of paclitaxel-and retinoic acid-incorporated nanoparticles for the treatment of CT-26 colon carcinoma. Arch Pharmacal Res 34:407–417
97. Tao X, Zhang Q, Ling K, Chen Y, Yang W, Gao F, Shi G (2012) Effect of pullulan nanoparticle surface charges on HSA complexation and drug release behavior of HSA-bound nanoparticles. PLoS ONE 7:e49304
98. Yang W-Z, Chen H-L, Gao F-P, Chen M-M, Li X-M, Zhang M-M, Zhang Q-Q, Liu L-R, Jiang Q, Wang Y-S (2010) Self-aggregated nanoparticles of cholesterol-modified pullulan conjugate as a novel carrier of mitoxantronep. Curr Nanosci 6:298–306
99. Yang W, Wang M, Ma L, Li H, Huang L (2014) Synthesis and characterization of biotin modified cholesteryl pullulan as a novel anticancer drug carrier. Carbohyd Polym 99:720–727
100. Lee IW, Li J, Chen X, Park HJ (2017) Fabrication of electrospun antioxidant nanofibers by rutin-pluronic solid dispersions for enhanced solubility. J Appl Polym Sci 134
101. Ganeshkumar M, Ponrasu T, Subamekala MK, Janani M, Suguna L (2016) Curcumin loaded on pullulan acetate nanoparticles protects the liver from damage induced by DEN. RSC Adv 6:5599–5610
102. Yuan R, Zheng F, Zhong S, Tao X, Zhang Y, Gao F, Yao F, Chen J, Chen Y, Shi G (2014) Self-assembled nanoparticles of glycyrrhetic acid-modified pullulan as a novel carrier of curcumin. Molecules 19:13305–13318
103. Constantin M, Bucătariu S, Stoica I, Fundueanu G (2017) Smart nanoparticles based on pullulan-g-poly (N-isopropylacrylamide) for controlled delivery of indomethacin. Int J Biol Macromol 94:698–708
104. Cristescu R, Popescu C, Popescu A, Socol G, Mihailescu I, Caraene G, Albulescu R, Buruiana T, Chrisey D (2012) Pulsed laser processing of functionalized polysaccharides for controlled release drug delivery systems. In: Technological innovations in sensing and detection of chemical, biological, radiological, nuclear threats and ecological terrorism. Springer
105. El-Malah Y, Nazzal S (2013) "Real-time" disintegration analysis and D-optimal experimental design for the optimization of diclofenac sodium fast-dissolving films. Pharm Dev Technol 18:1355–1360
106. Mocanu G, Nichifor M, Picton L, About-Jaudet E, le Cerf D (2014) Preparation and characterization of anionic pullulan thermoassociative nanoparticles for drug delivery. Carbohyd Polym 111:892–900
107. Bishwambhar M, Suneetha V (2012) Release study of naproxen, a modern drug from pH sensitive pullulan acetate microsphere. Int J Drug Dev Res 4:184–191
108. Choi JM, Lee B, Jeong D, Park KH, Choi E-J, Jeon Y-J, Dindulkar SD, Cho E, Do SH, Lee K (2017) Characterization and regulated naproxen release of hydroxypropyl cyclosophoraose-pullulan microspheres. J Ind Eng Chem 48:108–118
109. Jung Y-S, Park W, Na K (2013) Temperature-modulated noncovalent interaction controllable complex for the long-term delivery of etanercept to treat rheumatoid arthritis. J Controlled Release 171:143–151
110. Kumar BS, Kumar MG, Suguna L, Sastry T, Mandal A (2012) Pullulan acetate nanoparticles based delivery system for hydrophobic drug. Int J Pharma Biol Sci 3:24–32
111. Guhagarkar SA, Shah D, Patel MD, Sathaye SS, Devarajan PV (2015) Polyethylene sebacate-silymarin nanoparticles with enhanced hepatoprotective activity. J Nanosci Nanotechnol 15:4090–4093

112. Kang J-H, Tachibana Y, Obika S, Harada-Shiba M, Yamaoka T (2012) Efficient reduction of serum cholesterol by combining a liver-targeted gene delivery system with chemically modified apolipoprotein B siRNA. J Controlled Release 163:119–124
113. Zhu W, Romanski FS, Meng X, Mitra S, Tomassone MS (2011) Atomistic simulation study of surfactant and polymer interactions on the surface of a fenofibrate crystal. Eur J Pharm Sci 42:452–461
114. Sushmitha S, Priyanka SR, Krishna LM, Murthy MS (2014) Formulation and evaluation of mucoadhesive fast melt-away wafers using selected polymers. Res J Pharm Technol 7:176–180
115. Bhat V, Shivakumar H, Sheshappa R, Sanjeev G, Bhavya B (2012) Influence of blending of chitosan and pullulan on their drug release behavior: an in-vitro study. Int J Pharm Pharm Sci 4:313–317
116. Parejiya PB, Patel RC, Mehta DM, Shelat PK, Barot BS (2013) Quick dissolving films of nebivolol hydrochloride: formulation and optimization by a simplex lattice design. J Pharm Invest 43:343–351
117. Shih L, Yu J-Y, Hsieh C, Wu J-Y (2009) Production and characterization of curdlan by Agrobacterium sp. Biochem Eng J 43:33–40
118. Pszczola DE (1997) Curdlan differs from other gelling agents. Food Technology
119. Jung SB, Choi GG, Kim YB, Rhee YH (2001) Biosynthesis of polyhydroxyalkanoate copolyester containing cyclohexyl groups by *Pseudomonas oleovorans*. Int J Biol Macromol 29:145–150
120. Funami T, Unami M, Yada H, Nakao Y (1999) Gelation mechanism of curdlan by dynamic viscoelasticity measurements. J Food Sci 64:129–132
121. Kanke M, Tanabe E, Katayama H, Koda Y, Yoshitomi H (1995) Application of curdlan to controlled drug delivery. III. Drug release from sustained release suppositories in vitro. Biol Pharm Bull 18:1154–1158
122. Funami T, Yada H, Nakao Y (1998) Curdlan properties for application in fat mimetics for meat products. J Food Sci 63:283–287
123. Funami T, Yotsuzuka F, Yada H, Nakao Y (1998) Thermoirreversible characteristics of curdlan gels in a model reduced fat pork sausage. J Food Sci 63:575–579
124. Freimund S, Sauter M, Käppeli O, Dutler H (2003) A new non-degrading isolation process for 1,3-β-D-glucan of high purity from baker's yeast *Saccharomyces cerevisiae*. Carbohyd Polym 54:159–171
125. Sa G, Das T (2008) Anti cancer effects of curcumin: cycle of life and death. Cell Div 3:14
126. Huong LM et al (2011) Preparation and antitumor-promoting activity of curcumin encapsulated by 1, 3-β-glucan isolated from Vietnam medicinal mushroom *Hericium erinaceum*. Chem Lett 40:846–848
127. Li L, Gao F-P, Tang H-B, Bai Y-G, Li R-F, Li X-M, Liu L-R, Wang Y-S, Zhang Q-Q (2010) Self-assembled nanoparticles of cholesterol-conjugated carboxymethyl curdlan as a novel carrier of epirubicin. Nanotechnology 21:265601
128. Lehtovaara BC, Verma MS, Gu FX (2012) Synthesis of curdlan-graft-poly(ethylene glycol) and formulation of doxorubicin-loaded core–shell nanoparticles. J Bioact Compatible Polym 27:3–17
129. Ruffing A, Mao Z, Chen RR (2006) Metabolic engineering of Agrobacterium sp. for UDP-galactose regeneration and oligosaccharide synthesis. Metab Eng 8:465–473
130. Sutherland IW (2001) Microbial polysaccharides from gram-negative bacteria. Int Dairy J 11:663–674
131. Deligkaris K, Tadele TS, Olthuis W, van den Berg A (2010) Hydrogel-based devices for biomedical applications. Sens Actuators B Chem 147:765–774
132. Gulrez SK, AL-Assaf S, Phillips GO (2011) Hydrogels: methods of preparation, characterisation and applications. In: Progress in molecular and environmental bioengineering-from analysis and modeling to technology applications, pp 117–150
133. Singhvi G, Hans N, Shiva N, Dubey SK (2019) Xanthan gum in drug delivery applications. In: Natural polysaccharides in drug delivery and biomedical applications. Elsevier

134. Rosiak J, Janik I, Kadlubowski S, Kozicki M, Kujawa P, Stasica P, Ulanski P (2003) Nano-, micro-and macroscopic hydrogels synthesized by radiation technique. Nucl Instrum Methods Phys Res Sect B 208:325–330
135. Jackson C, Ofoefule S (2011) Use of xanthan gum and ethylcellulose in formulation of metronidazole for colon delivery. J Chem Pharm Res 3:11–20
136. Shinde UA, Kanojiya SS (2014) Serratiopeptidase niosomal gel with potential in topical delivery. J Pharm
137. Pooja D, Panyaram S, Kulhari H, Rachamalla SS, Sistla R (2014) Xanthan gum stabilized gold nanoparticles: characterization, biocompatibility, stability and cytotoxicity. Carbohyd Polym 110:1–9
138. Sandolo C, Coviello T, Matricardi P, Alhaique F (2007) Characterization of polysaccharide hydrogels for modified drug delivery. Eur Biophys J 36:693–700
139. Ramasamy T, Kandhasami UDS, Ruttala H, Shanmugam S (2011) Formulation and evaluation of xanthan gum based aceclofenac tablets for colon targeted drug delivery. Braz J Pharm Sci 47:299–311
140. Patel VM, Prajapati BG, Patel AK (2009) Controlled release gastroretentive dosage form of verapamil hydrochloride. Int J PharmTech Res 2:215–221
141. Kavitha K, Puneeth KP, Mani TT (2010) Development and evaluation of Rosiglitazone maleate floating tablets using natural gums. Int J PharmTech Res 2(3):1662–1669
142. Butani SB (2013) Development and optimization of venlafaxine hydrochloride sustained release triple layer tablets adopting quality by design approach
143. Devrim B, Bozkir A, Canefe K (2011) Formulation and evaluation of reconstitutable suspensions containing ibuprofen-loaded Eudragit microspheres. Acta Pol Pharm 68(4):593–599
144. Roopa G, Bhat RS, Dakshina MS (2015) Formulation and evaluation of an antacid and anti-ulcer suspension containing herbal drugs. Biomed Pharmacol J 3(1):01–06
145. Ceulemans J, Vinckier I, Ludwig A (2002) The use of xanthan gum in an ophthalmic liquid dosage form: rheological characterization of the interaction with mucin. J Pharm Sci 91 (4):1117–1127
146. Milazzo G, Russo S, Rasà D, Perri P, Ferri M, Monari P, Sebastiani A, et al (2006) Effect of an ophthalmic gel containing xanthan gum and sodium hyaluronate on corneal wound healing after photorefractive keratectomy. Invest Ophthalmol Vis Sci 47(13):2737
147. Faraldi F, Papa V, Santoro D, Rasà D, Mazza AL, Rabbione MM, Russo S (2012) A new eye gel containing sodium hyaluronate and xanthan gum for the management of post-traumatic corneal abrasions. Clin Ophthalmol (Auckland, NZ) 6:727
148. Shiledar RR, Tagalpallewar AA, Kokare CR (2014) Formulation and in vitro evaluation of xanthan gum-based bilayered mucoadhesive buccal patches of zolmitriptan. Carbohydr Polym 101:1234–1242
149. Laffleur F, Michalek M (2017) Modified xanthan gum for buccal delivery—a promising approach in treating sialorrhea. Int J Biol Macromol 102:1250–1256
150. Bhaskar K, Anbu J, Ravichandiran V, Venkateswarlu V, Rao YM (2009) Lipid nanoparticles for transdermal delivery of flurbiprofen: formulation, in vitro, ex vivo and in vivo studies. Lipids Health Dis 8(1):1–15
151. Mishra B, Sahoo SK, Sahoo S (2018) Liranaftate loaded Xanthan gum based hydrogel for topical delivery: Physical properties and ex-vivo permeability. Int J Biol Macromol 107:1717–1723
152. Shinde UA, Kanojiya SS (2014) Serratiopeptidase niosomal gel with potential in topical delivery. J Pharm 2014
153. Chaturvedi M, Kumar M, Sinhal A, Saifi A (2011) Recent development in novel drug delivery systems of herbal drugs. Int J Green Pharm (IJGP) 5(2)
154. Nethaji R, Narayanan A, Palanivelu M, Surendiran NS, Ganesan B (2016) Formulation and evaluation of metformin hydrochloride loaded mucoadhesive microspheres. Int J Pharm Chem Biol Sci 6(2)

155. Samia O, Hanan R, Kamal ET (2012) Carbamazepine mucoadhesive nanoemulgel (MNEG) as brain targeting delivery system via the olfactory mucosa. Drug Deliv 19(1):58–67
156. Samudre S, Tekade A, Thorve K, Jamodkar A, Parashar G, Chaudhari N (2015) Xanthan gum coated mucoadhesive liposomes for efficient nose to brain delivery of curcumin. Drug Deliv Lett 5(3):201–207
157. Reiss MJ, Han YP, Garcia E, Goldberg M, Yu H, Garner WL (2010). Matrix metalloproteinase-9 delays wound healing in a murine wound model. Surgery 147 (2):295–302
158. Huang J, Ren J, Chen G, Deng Y, Wang G, Wu X (2017) Evaluation of the xanthan-based film incorporated with silver nanoparticles for potential application in the nonhealing infectious wound. J Nanomaterials 2017
159. Merluşcă IP, Plămădeală P, Gîrbea C, Popa IM (2016) Xanthan-chitosan complex as a potential protector against injurious effects of neomycin
160. Gorle A, Pawara I, Achaliya A (2017) Design development and evaluation of transdermal drug delivery system of antipyretic agent. Int J Pharma Res Health Sci 5(4):1743e9
161. Abu-Huwaij R, Obaidat RM, Sweidan K, Al-Hiari Y (2011) Formulation and in vitro evaluation of xanthan gum or carbopol 934-based mucoadhesive patches, loaded with nicotine. Aaps Pharmscitech 12(1):21–27
162. Singhvi G, Hans N, Shiva N, Dubey SK (2019) Xanthan gum in drug delivery applications. In: Natural polysaccharides in drug delivery and biomedical applications. Academic Press, pp. 121–144
163. Paul A, Fathima KM, Nair SC (2017) Intra nasal in situ gelling system of lamotrigine using ion activated mucoadhesive polymer. Open Med Chem J 11:222
164. Hanna DH, Saad GR (2019) Encapsulation of ciprofloxacin within modified xanthan gum-chitosan based hydrogel for drug delivery. Bioorg Chem 84:115–124
165. Becerra FYG, Acosta EJ, Allen DG (2010) Alkaline extraction of wastewater activated sludge biosolids. Biores Technol 101:6972–6980
166. Oh JK, Lee DI, Park JM (2009) Biopolymer-based microgels/nanogels for drug delivery applications. Prog Polym Sci 34:1261–1282
167. Liu L, Yao W, Rao Y, Lu X, Gao J (2017) pH-Responsive carriers for oral drug delivery: challenges and opportunities of current platforms. Drug Deliv 24:569–581
168. Zhang Y, Chan HF, Leong KW (2013) Advanced materials and processing for drug delivery: the past and the future. Adv Drug Deliv Rev 65:104–120
169. D'Arrigo G, Navarro G, di Meo C, Matricardi P, Torchilin V (2014) Gellan gum nanohydrogel containing anti-inflammatory and anti-cancer drugs: a multi-drug delivery system for a combination therapy in cancer treatment. Eur J Pharm Biopharm 87(1):208–216
170. Monica R, Mayuri K (2019) Multiparticulate drug delivery system for gastrointestinal tuberculosis. Int J Pharm Sci Drug Res 11:210–220
171. Srinatha A, Pandit JK (2008) Multi-unit floating alginate system: effect of additives on ciprofloxacin release. Drug Deliv 15:471–476
172. Tripathi GK, Singh S, Nath G (2012) Formulation and In-vitro evaluation of pH-sensitive oil entrapped polymeric blend amoxicillin beads for the eradication of Helicobacter pylori. Iranian J Pharm Res IJPR 11(2):447
173. Milivojevic M, Pajic-Lijakovic I, Bugarski B, Nayak AK, Hasnain MS (2019) Gellan gum in drug delivery applications. In: Natural polysaccharides in drug delivery and biomedical applications, pp 145–186
174. Dhar S, Mali V, Bodhankar S, Shiras A, Prasad B, Pokharkar V (2011) Biocompatible gellan gum-reduced gold nanoparticles: cellular uptake and subacute oral toxicity studies. J Appl Toxicol 31:411–420
175. Dhar S, Murawala P, Shiras A, Pokharkar V, Prasad B (2012) Gellan gum capped silver nanoparticle dispersions and hydrogels: cytotoxicity and in vitro diffusion studies. Nanoscale 4:563–567

176. Vieira S, Vial S, Maia FR, Carvalho M, Reis RL, Granja PL, Oliveira JM (2015) Gellan gum-coated gold nanorods: an intracellular nanosystem for bone tissue engineering. RSC Adv 5:77996–78005
177. Fialho AM, Moreira LM, Granja AT, Popescu AO, Hoffmann K, Sá-Correia I (2008) Occurrence, production, and applications of gellan: current state and perspectives. Appl Microbiol Biotechnol 79:889
178. Rozier A, Mazuel C, Grove J, Plazonnet B (1989) Gelrite®: A novel, ion-activated, in-situ gelling polymer for ophthalmic vehicles. Effect on bioavailability of timolol. Int J Pharm 57:163–168
179. Li J, Kamath K, Dwivedi C (2001) Gellan film as an implant for insulin delivery. J Biomater Appl 15:321–343
180. Bhalerao H, Koteshwara K, Chandran S (2019) Levofloxacin hemihydrate in situ gelling ophthalmic solution: formulation optimization and in vitro and in vivo evaluation. AAPS PharmSciTech 20:272
181. Vashisth P, Raghuwanshi N, Srivastava AK, Singh H, Nagar H, Pruthi V (2017) Ofloxacin loaded gellan/PVA nanofibers-Synthesis, characterization and evaluation of their gastroretentive/mucoadhesive drug delivery potential. Mater Sci Eng: C 71:611–619
182. Rostami M, Ghorbani M, Delavar M, Tabibiazar M, Ramezani S (2019) Development of resveratrol loaded chitosan-gellan nanofiber as a novel gastrointestinal delivery system. Int J Biol Macromol 135:698–705
183. Ahuja M, Yadav M, Kumar S (2010) Application of response surface methodology to formulation of ionotropically gelled gum cordia/gellan beads. Carbohydr Polym 80(1):161–167
184. Babu B, Wu JT (2010) Production of phthalate esters by nuisance freshwater algae and cyanobacteria. Sci Total Environ 408(21):4969–4975
185. Emeje MO, Franklin-Ude PI, Ofoefule SI (2010) Evaluation of the fluid uptake kinetics and drug release from gellan gum tablets containing metronidazole. Int J Biol Macromol 47 (2):158–163
186. Mundargi RC, Shelke NB, Babu VR, Patel P, Rangaswamy V, Aminabhavi TM (2010) Novel thermo-responsive semi-interpenetrating network microspheres of gellan gum-poly (N-isopropylacrylamide) for controlled release of atenolol. J Appl Polym Sci 116(3):1832–1841
187. Narkar M, Sher P, Pawar A (2010) Stomach-specific controlled release gellan beads of acid-soluble drug prepared by ionotropic gelation method. Aaps Pharmscitech 11(1):267–277
188. Shah DP, Jani GK (2010) A newer application of physically modified gellan gum in tablet formulation using factorial design. Ars Pharmaceutica 51(1):28–40
189. Tripathi G, Singh S (2010) Formulation and in vitro evaluation of pH sensitive oil entrapped polymeric blended gellan gum buoyant beads of clarithromycin. DARU: J Fac Pharm Tehran Univ Med Sci 18(4):247
190. Dabhi M, Gohel M, Parikh R, Sheth N, Nagori S (2011) Formulation development of ambroxol hydrochloride soft gel with application of statistical experimental design and response surface methodology. PDA J Pharm Sci Technol 65(1): 20–31
191. Dixit R, Verma A, Singh UP, Soni S, Mishra AK, Bansal AK, Pandit JK (2011) Preparation and characterization of gellan-chitosan polyelectrolyte complex beads. Lat Am J Pharm 30 (6):1186–1195
192. Kabbur N, Rajendra A, Sridhar BK (2011) Design and evaluation of intragastric floating drug delivery system for ofloxacin. Int J Pharm Pharm Sci 3:93–98
193. Kulkarni RV, Mangond BS, Mutalik S, Sa B (2011) Interpenetrating polymer network microcapsules of gellan gum and egg albumin entrapped with diltiazem–resin complex for controlled release application. Carbohydr Polym 83(2):1001–1007
194. Maiti S, Ranjit S, Mondol R, Ray S, Sa B (2011) Al+ 3 ion cross-linked and acetalated gellan hydrogel network beads for prolonged release of glipizide. Carbohydr Polym 85 (1):164–172

195. Patil JS, Kamalapur MV, Marapur SC, Shiralshetti SS (2011) Ionotropically gelled novel hydrogel beads: Preparation, characterization and in vitro evaluation. Indian J Pharm Sci 73 (5):504
196. Prashant P, Ashwini R, Shivakumar S, Sridhar BK (2011) Preparation and evaluation of extended release matrix tablets of diltiazem using blends of tamarind xyloglucan with gellan gum and sodium carboxymethyl cellulose. Der Pharmacia Lettre 3(4):380–392
197. Verma A, Pandit JK (2011) Rifabutin-loaded floating gellan gum beads: effect of calcium and polymer concentration on incorporation efficiency and drug release. Trop J Pharm Res 10(1)
198. D'Arrigo G, Di Meo C, Gaucci E, Chichiarelli S, Coviello T, Capitani D, Matricardi P, et al (2012) Self-assembled gellan-based nanohydrogels as a tool for prednisolone delivery. Soft Matter 8(45):11557–11564
199. Dixit R, Verma A, Mishra AK, Verma N, Pandit JK (2012) Floating gellan-chitosan polyelectrolyte complex beads: effect of gelucires incorporation on encapsulation efficiency and drug release. Lat Am J Pharm 31(1):37–42
200. Gulecha BS, Shahi S, Lahoti SR (2012) Floating in situ gelling drug delivery system of verapamil hydrochloride. Am J PharmTech Res 2(4):954–969
201. Maiti S, Ghosh S, Mondol R, Ray S, Sa B (2012) Smart reticulated hydrogel of functionally decorated gellan copolymer for prolonged delivery of salbutamol sulphate to the gastro-luminal milieu. J Microencapsul 29(8):747–758
202. Verma A, Ramesh CN, Sharma SD, Pandit JK (2012) Preparation and characterization of floating gellan-chitosan polyelectrolyte complex beads. Lat Am J Pharm31(1):138–146
203. Vijan V, Kaity S, Biswas S, Isaac J, Ghosh A (2012) Microwave assisted synthesis and characterization of acrylamide grafted gellan, application in drug delivery. Carbohydr Polym 90(1):496–506
204. Ahuja M, Singh S, Kumar A (2013) Evaluation of carboxymethyl gellan gum as a mucoadhesive polymer. Int J Biol Macromol 53:114–121
205. Bhattacharya SS, Banerjee S, Chowdhury P, Ghosh A, Hegde RR, Mondal R (2013) Tranexamic acid loaded gellan gum-based polymeric microbeads for controlled release: In vitro and in vivo assessment. Colloids Surf B: Biointerfaces 112:483–491
206. Dowling MB, Bagal AS, Raghavan SR (2013) Self-destructing "mothership" capsules for timed release of encapsulated contents. Langmuir 29(25):7993–7998
207. Jana S, Das A, Nayak AK, Sen KK, Basu SK (2013) Aceclofenac-loaded unsaturated esterified alginate/gellan gum microspheres: in vitro and in vivo assessment. Int J Biol Macromol 57:129–137
208. Parthiban S, Shivaraju SG, Vikneswari A (2013) Formulation and evaluation of gastroretentive drug delivery of ornidazole in situ gellingsystem using gellan gum. Int J Res Pharm Nano Sci 2:747–756
209. Sahoo SK, Sahoo SK, Behera A, Patil SV, Panda SK (2013) Formulation, in vitro drug release study and anticancer activity of 5-fluorouracil loaded gellan gum microbeads. Acta Pol Pharm 70(1):123–127
210. Yang F, Xia S, Tan C, Zhang X (2013). Preparation and evaluation of chitosan-calcium-gellan gum beads for controlled release of protein. Eur Food Res Technol 237(4):467–479
211. El-Zahaby SA, Kassem AA, El-Kamel AH (2014) Formulation and in vitro evaluation of size expanding gastro-retentive systems of levofloxacin hemihydrate. Int J Pharm 464(1–2):10–18
212. Mahdi MH, Conway BR, Smith AM (2014) Evaluation of gellan gum fluid gels as modified release oral liquids. Int J Pharm 475(1–2):335–343
213. Nayak AK, Pal D (2014) Trigonella foenum-graecum L. seed mucilage-gellan mucoadhesive beads for controlled release of metformin HCl. Carbohydr Polym 107:31–40
214. Nayak AK, Pal D, Santra K (2014) Artocarpus heterophyllus L. seed starch-blended gellan gum mucoadhesive beads of metformin HCl. Int J Biol Macromol 65:329–339

215. Nayak AK, Pal D, Santra K (2014) Ispaghula mucilage-gellan mucoadhesive beads of metformin HCl: Development by response surface methodology. Carbohydr Polym 107:41–50
216. Nayak AK, Pal D, Santra K (2014) Tamarind seed polysaccharide–gellan mucoadhesive beads for controlled release of metformin HCl. Carbohydr Polym 103:154–163
217. Prezotti FG, Cury BSF, Evangelista RC (2014) Mucoadhesive beads of gellan gum/pectin intended to controlled delivery of drugs. Carbohydr Polym 113:286–295
218. Kundu P, Maiti S (2015) Cetyl gellan copolymer micelles and hydrogels: in vitro and pharmacodynamic assessment for drug delivery. Int J Biol Macromol 72:1027–1033
219. Maiti S, Laha B, Kumari L (2015) Gellan micro-carriers for pH-responsive sustained oral delivery of glipizide. Farmacia 63(6):913e21
220. Nandi G, Patra P, Priyadarshini R, Kaity S, Ghosh LK (2015) Synthesis, characterization and evaluation of methacrylamide grafted gellan as sustained release tablet matrix. Int J Biol Macromol 72:965–974
221. Rao MRP, Shelar SU (2015) Controlled release ion sensitive floating oral in situ gel of a prokinetic drug using gellan gum. Indian J Pharm Educ Res 49:158–167
222. Allam AN, Mehanna MM (2016) Formulation, physicochemical characterization and in-vivo evaluation of ion-sensitive metformin loaded-biopolymeric beads. Drug Dev Ind Pharm 42 (3):497–505
223. Boni FI, Prezotti FG, Cury BSF (2016) Gellan gum microspheres crosslinked with trivalent ion: Effect of polymer and crosslinker concentrations on drug release and mucoadhesive properties. Drug Dev Ind Pharm 42(8):1283–1290
224. El-Said IA, Aboelwafa AA, Khalil RM, ElGazayerly ON (2016) Baclofen novel gastroretentive extended release gellan gum superporous hydrogel hybrid system: In vitro and in vivo evaluation. Drug Deliv 23(1):101–112
225. Priyadarshini R, Nandi G, Changder A, Chowdhury S, Chakraborty S, Ghosh LK (2016) Gastroretentive extended release of metformin from methacrylamide-g-gellan and tamarind seed gum composite matrix. Carbohydr Polym 137:100–110
226. Cardoso VM, Cury BSF, Evangelista RC, Gremião MPD (2017) Development and characterization of cross-linked gellan gum and retrograded starch blend hydrogels for drug delivery applications. J Mech Behav Biomed Mater 65:317–333
227. Hoosain FG, Choonara YE, Kumar P, Tomar LK, Tyagi C, du Toit LC, Pillay V (2017) In vivo evaluation of a PEO-gellan gum semi-interpenetrating polymer network for the oral delivery of sulpiride. AAPS PharmSciTech 18(3):654–670
228. Osmałek T, Milanowski B, Froelich A, Szybowicz M, Białowąs W, Kapela M, Ancukiewicz K, et al (2017) Design and characteristics of gellan gum beads for modified release of meloxicam. Drug Dev Ind Pharm 43(8):1314–1329
229. Abouelatta SM, Aboelwafa AA, El-Gazayerly ON (2018) Gastroretentive raft liquid delivery system as a new approach to release extension for carrier-mediated drug. Drug Deliv 25 (1):1161–1174
230. Bera H, Kumar S, Maiti S (2018) Facile synthesis and characterization of tailor-made pectin-gellan gum-bionanofiller composites as intragastric drug delivery shuttles. Int J Biol Macromol 118:149–159
231. Meneguin AB, Beyssac E, Garrait G, Hsein H, Cury BS (2018) Retrograded starch/pectin coated gellan gum-microparticles for oral administration of insulin: A technological platform for protection against enzymatic degradation and improvement of intestinal permeability. Eur J Pharm Biopharm 123:84–94
232. Osmalek T, Froelich A, Milanowski B, Bialas M, Hyla K, Szybowicz M (2018). pH-dependent behavior of novel gellan beads loaded with naproxen. Curr Drug Deliv 15 (1):52–63
233. Osmałek TZ, Froelich A, Soból M, Milanowski B, Skotnicki M, Kunstman P, Szybowicz M, et al (2018) Gellan gum macrobeads loaded with naproxen: The impact of various naturally derived polymers on pH-dependent behavior. J Biomater Appl 33(1):140–155

234. Prezotti FG, Boni FI, Ferreira NN, Campana-Filho SP, Almeida A, Vasconcelos T, Sarmento B, et al (2018) Gellan gum/pectin beads are safe and efficient for the targeted colonic delivery of resveratrol. Polymers 10(1):50
235. Barbosa EJ, Ferraz HG (2019) Gellan gum and polyvinylpyrrolidone (PVP) as binding agents in extrusion/spheronization pellet formulations. Acta Pharmaceutica 69(1):99–109
236. Prezotti FG, Siedle I, Boni FI, Chorilli M, Müller I, Cury BSF (2020) Mucoadhesive films based on gellan gum/pectin blends as potential platform for buccal drug delivery. Pharm Dev Technol 25(2):159–167
237. Zhao X, Wang Z (2019) A pH-sensitive microemulsion-filled gellan gum hydrogel encapsulated apigenin: Characterization and in vitro release kinetics. Colloids Surf B: Biointerfaces 178:245–252
238. Prezotti FG, Boni FI, Ferreira NN, Silva DS, Almeida A, Vasconcelos T, Cury BSF, et al (2020) Oral nanoparticles based on gellan gum/pectin for colon-targeted delivery of resveratrol. Drug Dev Ind Pharm 46(2):236–245
239. Shirsath NR, Goswami AK (2020) Vildagliptin-loaded gellan gum mucoadhesive beads for sustained drug delivery: design, optimisation and evaluation. Mater Technol, 1–13
240. Gupta H, Velpandian T, Jain S (2010) Ion-and pH-activated novel in-situ gel system for sustained ocular drug delivery. J Drug Target 18(7):499–505
241. Kesavan K, Nath G, Pandit JK (2010) Preparation and in vitro antibacterial evaluation of gatifloxacin mucoadhesive gellan system. Daru: J Fac Pharm Tehran Univ Med Sci 18 (4):237
242. Liu Y, Liu J, Zhang X, Zhang R, Huang Y, Wu C (2010) In situ gelling gelrite/alginate formulations as vehicles for ophthalmic drug delivery. Aaps PharmSciTech 11(2):610–620
243. Shen JQ, Gan Y, Gan L, Zhu CL, Zhu JB (2010) Ion-sensitive nanoemulsion-in situ gel system for ophthalmic delivery of flurbiprofen axetil. Yao xue xue bao—Acta pharmaceutica Sinica 45(1):120–125
244. El-Laithy HM, Nesseem DI, El-Adly AA, Shoukry M (2011) Moxifloxacin-Gelrite in situ ophthalmic gelling system against photodynamic therapy for treatment of bacterial corneal inflammation. Arch Pharm Res 34(10):1663–1678
245. Rupenthal ID, Green CR, Alany RG (2011) Comparison of ion-activated in situ gelling systems for ocular drug delivery. Part 1: physicochemical characterisation and in vitro release. Int J Pharm 411(1–2):69–77
246. Rupenthal ID, Green CR, Alany RG (2011) Comparison of ion-activated in situ gelling systems for ocular drug delivery. Part 2: Precorneal retention and in vivo pharmacodynamic study. Int J Pharm 411(1–2):78–85
247. Rupenthal ID, Alany RG, Green CR (2011) Ion-activated in situ gelling systems for antisense oligodeoxynucleotide delivery to the ocular surface. Mol Pharm 8(6):2282–2290
248. Chen Q, Zheng Y, Li Y, Zeng Y, Kuang J, Hou S, Li X (2012) The effect of deacetylated gellan gum on aesculin distribution in the posterior segment of the eye after topical administration. Drug Deliv 19(4):194–201
249. Geethalakshmi A, Karki R, Kumar Jha S, P Venkatesh D, Nikunj B (2012) Sustained ocular delivery of brimonidine tartrate using ion activated in situ gelling system. Curr Drug Deliv 9 (2):197–204
250. Kumar JRK, Muralidharan S (2012) Formulation and in vitro evalution of gellan gum/ carbopol and sodium alginate based solution to gel depot of ketotifen fumarate system. J Pharm Sci Res 4(11):1973
251. Tayel SA, El-Nabarawi MA, Tadros MI, Abd-Elsalam WH (2013) Promising ion-sensitive in situ ocular nanoemulsion gels of terbinafine hydrochloride: design, in vitro characterization and in vivo estimation of the ocular irritation and drug pharmacokinetics in the aqueous humor of rabbits. Int J Pharm 443(1–2):293–305
252. Fernández-Ferreiro A, Bargiela NF, Varela MS, Martínez MG, Pardo M, Ces AP, Otero-Espinar FJ, et al (2014) Cyclodextrin–polysaccharide-based, in situ-gelled system for ocular antifungal delivery. Beilstein J Org Chem 10(1):2903–2911

253. Duan Y, Cai X, Du H, Zhai G (2015) Novel in situ gel systems based on P123/TPGS mixed micelles and gellan gum for ophthalmic delivery of curcumin. Colloids Surf B: Biointerfaces 128:322–330
254. Gupta H, Aqil M, Khar RK, Ali A, Bhatnagar A, Mittal G (2015) An alternative in situ gel-formulation of levofloxacin eye drops for prolong ocular retention. J Pharm Bioallied Sci 7(1):9
255. Gupta H, Malik A, Khar RK, Ali A, Bhatnagar A, Mittal G (2015) Physiologically active hydrogel (in situ gel) of sparfloxacin and its evaluation for ocular retention using gamma scintigraphy. J Pharm Bioallied Sci 7(3):195
256. Pokharkar V, Patil V, Mandpe L (2015) Engineering of polymer–surfactant nanoparticles of doxycycline hydrochloride for ocular drug delivery. Drug Deliv 22(7):955–968
257. Yu S, Wang QM, Wang X, Liu D, Zhang W, Ye T, Pan W et al (2015) Liposome incorporated ion sensitive in situ gels for opthalmic delivery of timolol maleate. Int J Pharm 480(1–2), 128–136
258. Zhu L, Ao J, Li P (2015) A novel in situ gel base of deacetylase gellan gum for sustained ophthalmic drug delivery of ketotifen: in vitro and in vivo evaluation. Drug Des Dev Therapy 9:3943
259. Kesarla R, Tank T, Vora PA, Shah T, Parmar S, Omri A (2016) Preparation and evaluation of nanoparticles loaded ophthalmic in situ gel. Drug Deliv 23(7):2363–2370
260. Kesavan K, Kant S, Pandit JK (2016) Therapeutic effectiveness in the treatment of experimental bacterial keratitis with ion-activated mucoadhesive hydrogel. Ocular Immunol Inflamm 24(5):489–492
261. Reed K, Li A, Wilson B, Assamoi T (2016) Enhancement of ocular in situ gelling properties of low acyl gellan gum by use of ion exchange. J Ocular Pharm Therap 32(9):574–582
262. Dewan M, Sarkar G, Bhowmik M, Das B, Chattoapadhyay AK, Rana D, Chattopadhyay D (2017) Effect of gellan gum on the thermogelation property and drug release profile of Poloxamer 407 based ophthalmic formulation. Int J Biol Macromol 102:258–265
263. Kotreka UK, Davis VL, Adeyeye MC (2017) Development of topical ophthalmic in situ gel-forming estradiol delivery system intended for the prevention of age-related cataracts. PloS One 12(2), e0172306
264. Luaces-Rodríguez A, Díaz-Tomé V, González-Barcia M, Silva-Rodríguez J, Herranz M, Gil-Martínez M, Fernández-Ferreiro A, et al (2017) Cysteamine polysaccharide hydrogels: study of extended ocular delivery and biopermanence time by PET imaging. Int J Pharm 528 (1–2), 714–722
265. Morsi N, Ibrahim M, Refai H, El Sorogy H (2017) Nanoemulsion-based electrolyte triggered in situ gel for ocular delivery of acetazolamide. Eur J Pharm Sci 104:302–314
266. Díaz-Tomé V, Luaces-Rodríguez A, Silva-Rodríguez J, Blanco-Dorado S, García-Quintanilla L, Llovo-Taboada J, Fernández-Ferreiro A, et al (2018) Ophthalmic econazole hydrogels for the treatment of fungal keratitis. J Pharm Sci 107(5):1342–1351
267. Imam SS, Bukhari SNA, Ali A (2018) Preparation and evaluation of novel chitosan: gelrite ocular system containing besifloxacin for topical treatment of bacterial conjunctivitis: scintigraphy, ocular irritation and retention assessment. Artif Cells Nanomed Biotechnol 46 (5):959–967
268. Janga KY, Tatke A, Balguri SP, Lamichanne SP, Ibrahim MM, Maria DN, Majumdar S et al (2018) Ion-sensitive in situ hydrogels of natamycin bilosomes for enhanced and prolonged ocular pharmacotherapy: in vitro permeability, cytotoxicity and in vivo evaluation. Artif Cells Nanomed Biotechnol 46(sup1):1039–1050
269. Ranch KM, Maulvi FA, Naik MJ, Koli AR, Parikh RK, Shah DO (2019) Optimization of a novel in situ gel for sustained ocular drug delivery using Box-Behnken design: in vitro, ex vivo, in vivo and human studies. Int J Pharm 554:264–275
270. Sun J, Zhou Z (2018) A novel ocular delivery of brinzolamide based on gellan gum: in vitro and in vivo evaluation. Drug Des Dev Therapy 12:383

271. Janga KY, Tatke A, Dudhipala N, Balguri SP, Ibrahim MM, Maria DN, Majumdar S, et al (2019) Gellan gum based sol-to-gel transforming system of natamycin transfersomes improves topical ocular delivery. J Pharmacol Exp Therap 370(3):814–822
272. Agibayeva LE, Kaldybekov DB, Porfiryeva NN, Garipova VR, Mangazbayeva RA, Moustafine RI, Khutoryanskiy VV, et al (2020). Gellan gum and its methacrylated derivatives as in situ gelling mucoadhesive formulations of pilocarpine: In vitro and in vivo studies. Int J Pharm 577:119093
273. Destruel PL, Zeng N, Seguin J, Douat S, Rosa F, Brignole-Baudouin F, Boudy V, et al (2020) Novel in situ gelling ophthalmic drug delivery system based on gellan gum and hydroxyethylcellulose: Innovative rheological characterization, in vitro and in vivo evidence of a sustained precorneal retention time. Int J Pharm 574:118734
274. Mahajan HS, Gattani S (2010) In situ gels of metoclopramide hydrochloride for intranasal delivery: in vitro evaluation and in vivo pharmacokinetic study in rabbits. Drug Deliv 17 (1):19–27
275. Mahajan HS, Gattani SG (2010) Nasal administration of ondansetron using a novel microspheres delivery system Part II: ex vivo and in vivo studies. Pharm Dev Technol 15 (6):653–657
276. Shah V, Sharma M, Parmar V, Upadhyay U (2010) Formulation of sildenafil citrate loaded nasal microsphers: an in vitro, ex vivo characterization. Int J Drug Deliv 2(3)
277. Cai Z, Song X, Sun F, Yang Z, Hou S, Liu Z (2011) Formulation and evaluation of in situ gelling systems for intranasal administration of gastrodin. Aaps Pharmscitech 12(4):1102–1109
278. Wang S, Chen P, Zhang L, Yang C, Zhai G (2012) Formulation and evaluation of microemulsion-based in situ ion-sensitive gelling systems for intranasal administration of curcumin. J Drug Target 20(10):831–840
279. Saindane NS, Pagar KP, Vavia PR (2013) Nanosuspension based in situ gelling nasal spray of carvedilol: development, in vitro and in vivo characterization. Aaps Pharmscitech 14 (1):189–199
280. Abbas Z, Marihal S (2014) Gellan gum-based mucoadhesive microspheres of almotriptan for nasal administration: Formulation optimization using factorial design, characterization, and in vitro evaluation. J Pharm Bioallied Sci 6(4):267
281. Galgatte UC, Kumbhar AB, Chaudhari PD (2014) Development of in situ gel for nasal delivery: design, optimization, in vitro and in vivo evaluation. Drug Deliv 21(1):62–73
282. Hosny KM, Hassan AH (2014) Intranasal in situ gel loaded with saquinavir mesylate nanosized microemulsion: preparation, characterization, and in vivo evaluation. Int J Pharm 475(1–2), 191–197
283. Maiti S, Chakravorty A, Chowdhury M (2014) Gellan co-polysaccharide micellar solution of budesonide for allergic anti-rhinitis: an in vitro appraisal. Int J Biol Macromol 68:241–246
284. Ibrahim HK, Abdel Malak NS, Abdel Halim SA (2015) Formulation of convenient, easily scalable, and efficient granisetron HCl intranasal droppable gels. Mol Pharm 12(6):2019–2025
285. Li X, Du L, Chen X, Ge P, Wang Y, Fu Y, Jin Y, et al (2015) Nasal delivery of analgesic ketorolac tromethamine thermo-and ion-sensitive in situ hydrogels. Int J Pharm 489(1–2):252–260
286. Mahdi MH, Conway BR, Smith AM (2015) Development of mucoadhesive sprayable gellan gum fluid gels. Int J Pharm 488(1–2), 12–19
287. Wavikar PR, Vavia PR (2015) Rivastigmine-loaded in situ gelling nanostructured lipid carriers for nose to brain delivery. J Liposome Res 25(2):141–149
288. Hao J, Zhao J, Zhang S, Tong T, Zhuang Q, Jin K, Tang H, et al (2016) Fabrication of an ionic-sensitive in situ gel loaded with resveratrol nanosuspensions intended for direct nose-to-brain delivery. Colloids Surf B: Biointerfaces 147:376–386
289. Salunke SR, Patil SB (2016) Ion activated in situ gel of gellan gum containing salbutamol sulphate for nasal administration. Int J Biol Macromol 87:41–47

290. Sonje AG, Mahajan HS (2016) Nasal inserts containing ondansetron hydrochloride based on Chitosan–gellan gum polyelectrolyte complex: In vitro–in vivo studies. Mater Sci Eng: C 64:329–335
291. Shah V, Sharma M, Pandya R, Parikh RK, Bharatiya B, Shukla A, Tsai HC (2017) Quality by Design approach for an in situ gelling microemulsion of Lorazepam via intranasal route. Mater Sci Eng: C 75:1231–1241
292. Patil RP, Pawara DD, Gudewar CS, Tekade AR (2019) Nanostructured cubosomes in an in situ nasal gel system: an alternative approach for the controlled delivery of donepezil HCl to brain. J Liposome Res 29(3):264–273
293. Goyal R, Tripathi SK, Tyagi S, Ram KR, Ansari KM, Shukla Y, Gupta KC, et al (2011) Gellan gum blended PEI nanocomposites as gene delivery agents: evidences from in vitro and in vivo studies. Eur J Pharm Biopharm 79(1):3–14
294. Cencetti C, Bellini D, Pavesio A, Senigaglia D, Passariello C, Virga A, Matricardi P (2012) Preparation and characterization of antimicrobial wound dressings based on silver, gellan, PVA and borax. Carbohydr Polym 90(3):1362–1370
295. Mat Amin KA, Gilmore KJ, Matic J, Poon S, Walker MJ, Wilson MR, in het Panhuis M (2012) Polyelectrolyte complex materials consisting of antibacterial and cell-supporting layers. Macromol Biosci 12(3):374–382
296. Salim N, Basri M, Rahman MB, Abdullah DK, Basri H (2012) Modification of palm kernel oil esters nanoemulsions with hydrocolloid gum for enhanced topical delivery of ibuprofen. Int J Nanomed 7:4739
297. Novac O, Lisa G, Profire L, Tuchilus C, Popa MI (2014) Antibacterial quaternized gellan gum based particles for controlled release of ciprofloxacin with potential dermal applications. Mater Sci Eng: C 35:291–299
298. Abioye AO, Issah S, Kola-Mustapha AT (2015) Ex vivo skin permeation and retention studies on chitosan–ibuprofen–gellan ternary nanogel prepared by in situ ionic gelation technique—a tool for controlled transdermal delivery of ibuprofen. Int J Pharm 490(1–2), 112–130
299. Carmona-Moran CA, Zavgorodnya O, Penman AD, Kharlampieva E, Bridges Jr SL, Hergenrother RW, Wick TM, et al (2016) Development of gellan gum containing formulations for transdermal drug delivery: Component evaluation and controlled drug release using temperature responsive nanogels. Int J Pharm 509(1–2):465–476
300. Mahdi MH, Conway BR, Mills T, Smith AM (2016) Gellan gum fluid gels for topical administration of diclofenac. Int J Pharm 515(1–2):535–542
301. Shukla R, Kashaw SK, Jain AP, Lodhi S (2016) Fabrication of Apigenin loaded gellan gum–chitosan hydrogels (GGCH-HGs) for effective diabetic wound healing. Int J Biol Macromol 91:1110–1119
302. Marchiori MCL, Rigon C, Camponogara C, Oliveira SM, Cruz L (2017) Hydrogel containing silibinin-loaded pomegranate oil based nanocapsules exhibits anti-inflammatory effects on skin damage UVB radiation-induced in mice. J Photochem Photobiol B: Biol 170:25–32
303. Pegoraro NS, Barbieri AV, Camponogara C, Mattiazzi J, Brum ES, Marchiori MC, Cruz L, et al (2017) Nanoencapsulation of coenzyme Q10 and vitamin E acetate protects against UVB radiation-induced skin injury in mice. Colloids Surf B: Biointerfaces 150:32–40
304. Da Silva LP, Oliveira S, Pirraco RP, Santos TC, Reis RL, Marques AP, Correlo VM (2017) Eumelanin-releasing spongy-like hydrogels for skin re-epithelialization purposes. Biomed Mater 12(2):025010
305. Manconi M, Manca ML, Caddeo C, Cencetti C, di Meo C, Zoratto N, Matricardi P, et al (2018) Preparation of gellan-cholesterol nanohydrogels embedding baicalin and evaluation of their wound healing activity. Eur J Pharm Biopharm 127:244–249
306. Manconi M, Manca ML, Caddeo C, Valenti D, Cencetti C, Diez-Sales O, Matricardi P et al (2018) Nanodesign of new self-assembling core-shell gellan-transfersomes loading baicalin and in vivo evaluation of repair response in skin. Nanomed: Nanotechnol Biol Med 14 (2):569–579

307. Musazzi UM, Cencetti C, Franzé S, Zoratto N, Di Meo C, Procacci P, Cilurzo F et al (2018) Gellan nanohydrogels: novel nanodelivery systems for cutaneous administration of piroxicam. Mol Pharm 15(3):1028–1036
308. Shukla S, Shukla A (2018) Tunable antibiotic delivery from gellan hydrogels. J Mater Chem B 6(40):6444–6458
309. Tsai W, Tsai H, Wong Y, Hong J, Chang S, Lee M (2018) Preparation and characterization of gellan gum/glucosamine/clioquinol film as oral cancer treatment patch. Mater Sci Eng: C 82:317–322
310. Nair AB, Shah J, Aljaeid BM, Al-Dhubiab BE, Jacob S (2019) Gellan gum-based hydrogel for the transdermal delivery of nebivolol: Optimization and evaluation. Polymers 11 (10):1699
311. Osmari BF, Giuliani LM, Reolon JB, Rigo GV, Tasca T, Cruz L (2020) Gellan gum-based hydrogel containing nanocapsules for vaginal indole-3-carbinol delivery in trichomoniasis treatment. Eur J Pharm Sci 151:105379
312. Sosnik A (2014) Alginate particles as platform for drug delivery by the oral route: state-of-the-art. Int Sch Res Not
313. Liu F, Chen Q, Liu C, Ao Q, Tian X, Fan J, Tong H, Wang X (2018) Natural polymers for organ 3D bioprinting. Polymers 10:1278
314. Draget K, Bræk GS, Smidsrød O (1994) Alginic acid gels: the effect of alginate chemical composition and molecular weight. Carbohyd Polym 25:31–38
315. Laurienzo P (2010) Marine polysaccharides in pharmaceutical applications: an overview. Mar Drugs 8:2435–2465
316. Sun J, Tan H (2013) Alginate-based biomaterials for regenerative medicine applications. Materials 6:1285–1309
317. Grasdalen H (1983) High-field, 1H-nmr spectroscopy of alginate: sequential structure and linkage conformations. Carbohydr Res 118:255–260
318. Fertah M, Belfkira A, Taourirte M, Brouillette F (2017) Extraction and characterization of sodium alginate from *Moroccan Laminaria* digitata brown seaweed. Arab J Chem 10: S3707–S3714
319. Rinaudo K, Bleris L, Maddamsetti R, Subramanian S, Weiss R, Benenson Y (2007) A universal RNAi-based logic evaluator that operates in mammalian cells. Nat Biotechnol 25 (7):795–801
320. Steinbüchel A, Rhee SK (2005) Polysaccharides and polyamides in the food industry: properties, production, and patents. Wiley-VCH Verlag GmbH & CO, KGaA
321. Li X, Feng J, Zhang R, Wang J, Su T, Tian Z, Han D, Zhao C, Fan M, Li C (2016) Quaternized chitosan/alginate-Fe3O4 magnetic nanoparticles enhance the chemosensitization of multidrug-resistant gastric carcinoma by regulating cell autophagy activity in mice. J Biomed Nanotechnol 12:948–961
322. Haseeb MT, Hussain MA, Yuk SH, Bashir S, Nauman M (2016) Polysaccharides based superabsorbent hydrogel from linseed: dynamic swelling, stimuli responsive on–off switching and drug release. Carbohyd Polym 136:750–756
323. Mallardi A, Angarano V, Magliulo M, Torsi L, Palazzo G (2015) General approach to the immobilization of glycoenzyme chains inside calcium alginate beads for bioassay. Anal Chem 87:11337–11344
324. Kabu S, Gao Y, Kwon BK, Labhasetwar V (2015) Drug delivery, cell-based therapies, and tissue engineering approaches for spinal cord injury. J Controlled Release 219:141–154
325. Borges O, Silva M, de Sousa A, Borchard G, Junginger HE, Cordeiro-da-Silva A (2008) Alginate coated chitosan nanoparticles are an effective subcutaneous adjuvant for hepatitis B surface antigen. Int Immunopharmacol 8:1773–1780
326. Hong JS, Vreeland WN, Depaoli Lacerda SH, Locascio LE, Gaitan M, Raghavan SR (2008) Liposome-templated supramolecular assembly of responsive alginate nanogels. Langmuir 24:4092–4096
327. Chen Y, Lin X, Park H, Greever R (2009) Study of artemisinin nanocapsules as anticancer drug delivery systems. Nanom Nanotechnol Biol Med 5:316–322

328. Saraei F, Mohammadpour DN, Zolfagharian H, Moradi BS, Khaki P, Inanlou F (2013) Design and evaluate alginate nanoparticles as a protein delivery system
329. Ahmad Z, Sharma S, Khuller GK (2007) Chemotherapeutic evaluation of alginate nanoparticle-encapsulated azole antifungal and antitubercular drugs against murine tuberculosis. Nanomed Nanotechnol Biol Med 3:239–243
330. Arora S, Gupta S, Narang RK, Budhiraja RD (2011) Amoxicillin loaded chitosan–alginate polyelectrolyte complex nanoparticles as mucopenetrating delivery system for *H. pylori*. Sci Pharm 79:673–694
331. Pandey R, Ahmad Z, Sharma S, Khuller G (2005) Nano-encapsulation of azole antifungals: potential applications to improve oral drug delivery. Int J Pharm 301:268–276
332. Yang S-J, Lin F-H, Tsai H-M, Lin C-F, Chin H-C, Wong J-M, Shieh M-J (2011) Alginate-folic acid-modified chitosan nanoparticles for photodynamic detection of intestinal neoplasms. Biomaterials 32:2174–2182

Exopolysaccharides in Food Processing Industrials

Dilhun Keriman Arserim Ucar, Dilara Konuk Takma, and Figen Korel

Abstract Microbial exopolysaccharides are a class of extracellular carbohydrates based on biopolymeric materials produced and secreted by bacteria, yeast, molds, and microalgae. Cellulose, pullulan, xanthan gum, dextran, kefiran, curdlan, emulsan, alginate, gellan, carrageenans, hyaluronic acid, levan, colanic acid, welan, glucuronides, succinoglycans, and mutan are the exopolysaccharides (EPSs) of different microbial origin. Most of the available EPSs are non-toxic, biocompatible, biodegradable, and obtain from renewable resources. Microbial EPSs display unique functional properties due to their nature and structural composition. The demand for natural microbial EPSs utilization in the food industry due to their unique properties, including emulsifier, gelling agent, and stabilizers. Microbial EPSs and their derivatives have found a wide range of applications in food systems, including fermented dairy products, bakery products, cereal-based products, beverages, delivery of active agents, coatings, and films. This chapter will present a comprehensive overview of the recent developments of EPSs and their potential utilization in the food industry.

Keywords Exopolysaccharides · Hydrocolloids · Functional food · Food ingredients · Delivery systems

---

D. K. Arserim Ucar
Department of Nutrition and Dietetics, Faculty of Health Sciences, Bingöl University, 12000 Bingöl, Turkey

D. Konuk Takma
Department of Food Engineering, Faculty of Engineering, Aydın Adnan Menderes University, 09010 Aydın, Turkey

F. Korel (✉)
Department of Food Engineering, Faculty of Engineering, İzmir Institute of Technology, 35430 İzmir, Turkey
e-mail: figenkorel@iyte.edu.tr

A. K. Nadda et al. (eds.), *Microbial Exopolysaccharides as Novel and Significant Biomaterials*, Springer Series on Polymer and Composite Materials,
https://doi.org/10.1007/978-3-030-75289-7_8

1 Introduction

Microbial exopolysaccharides (EPSs) attract extensive attention in the food industry due to the growing interest in renewable sources. EPSs are considered potentially sustainable alternatives to chemical polymers because they are considered cost-effective, eco-friendly, non-toxic, high efficient, and biodegradable [119]. EPSs are naturally occurring extracellular polymers that are synthesized during the metabolic process of microorganisms, including bacteria, yeast, molds, and algae [12, 43, 45]. EPSs based on their compositions are divided into two groups; (1) homopolysaccharides and (2) heteropolysaccharides. The homopolysaccharides are created by a single type of monosaccharide including, D-glucose or L-fructose. The heteropolysaccharides are composed of several types of monosaccharides, including D-glucose, L-fructose, D-galactose, D-glucuronic acid, L- glucuronic acid, D-mannuronic acid [43, 119]. Cellulose, dextran, curdlan, and levan are examples of homopolysaccharides, while gellan, xanthan gum, and hyaluronic acid are examples of heteropolysaccharides [55]. The EPS production yield and structure depend on culture type, inoculum volume, including carbon and nitrogen source, substrate composition, airflow rate, temperature and pH of cultivating medium, agitation or mixing speed of incubation condition [41, 119]. Microbial EPSs are promising hydrocolloids, used as food ingredients such as stabilizing, gelling, thickening, and binding agents. For instance, pullulan enhanced the pasting and rheological properties of rice starch [23], xanthan gum improved the egg white foaming properties [34], xanthan gum addition into potato starch improved the physicochemical characteristics of gels [42], inulin and xanthan gum in the formulation of custard desserts increased the viscoelastic characteristic of product [131], xanthan gum in developing of low-fat food preparations [48], xanthan gum as gluten replacement [114], xanthan gum for development of low glycemic index food formulations [139], dextran and levan as bacterial EPSs in kefir beverage formulation [51], curdlan in set yogurt formulation [189], curdlan as fat mimetics ingredient for meat products [56], and alginate in low-fat mayonnaise fabrications [101], meat buffers were formulated by using gellan [166], hyaluronic acid and carrageenans for films or hydrogels preparations [30, 192], levan for nanoengineered structures [40] and welan in emulsion stabilization systems [111].

Prebiotics are the indigestible food ingredients that enhanced beneficial microorganisms activity and growth in the gastrointestinal tract [86]. Microbial EPSs are the most promising polysaccharides with prebiotic properties, and pullulan enhanced *Lactobacillus* and *Bifidobacterium* viability in low-fat yogurt [94] as well as; in another study, the prebiotic activity of pullulan were proved in wheat bread [128]. Antimicrobial agents such as metallic nanoparticles [2], bacteriocins [158], plant extracts [89], and essential oils [29] can be added or encapsulated into microbial EPSs for designing food-grade polymers to create an active food packaging system for maintaining food safety and quality during food storage. In contrast to synthetic polymers, biopolymers have the advantage of sustainability and environmentally friendly features. Recently, researchers have focused on biodegradable polymers [8]. Microbial exopolysaccharides are good candidates for the fabrication of high-performance non-toxic polymernanocomposites as well [5]. EPSs in polymer matrix resulted in significant improvement of mechanical, thermal, and barrier properties such as polyvinylalcohol (PVA)-bacterial cellulose nanocrystals (BCNC) [60], hydroxypropyl methylcellulose (HPMC)-BCNC [59], Konjac glucomannan-pullulan [181], pullulan-lysozyme nanofibers [153], packaging paper coated with curdlan and chitosan [17] and nanocomposite of cellulose nanocrystals and kefiran [149] are some examples of engineered nanocomposites with microbial EPSs.

Microbial EPSs also have the potential to produce biofilms and other bioengineered micro-nano structures for use in food applications [4] such as bacterial cellulose olive oil Pickering emulsions [180], pullulan based cinnamon essential oil nanoemulsions [29], astaxanthin encapsulated whey protein isolate-xanthan gum emulsions [15] and probiotics in dextran nanoparticles [86]. Microbial EPSs produced from bacteria as bacterial EPSs including bacterial cellulose [108], and xanthan gum [41], and from yeast-like fungus as fungal exopolysaccharide such as pullulan [155] by fermentation of a wide variety of substrates including organic and inorganic nutrients. The growth media substrates are expensive and increased the cost of microbial EPSs production on industrial scale. Alternative low-cost process substrates have been used for industrial-scale production. Agro-based wastes are composed of food constituents, which can be economically feasible raw materials in growth media with or without additional nutrients [72]. Agri-industrial residues including soya bean oil cake, mustard seed oil cake, rice bran oil cake, and corn steep liquor were used for pullulan production as growth media substrates [150], citrus peels including mandarin, orange, grapefruit, and lemon used for bacterial cellulose [72], waste bread hydrolysate consumed for xanthan gum production [41]. By the way, using industrial wastes utilization in this way could contribute to the waste management systems. This chapter provides an overview of the unique properties and the potential food applications of widely used microbial EPSs in the food industry. The food applications of microbial EPSs are also presented in Table 1.

2 Cellulose

Bacterial cellulose (BC) (β-1,4 linked, D-glucose) is a natural, edible, non-toxic, biodegradable, and biocompatible microbial exopolysaccharide [152]. BC can be produced from different kinds of microorganisms, including *Gluconacetobacter* [142], recently named as *Komagataeibacter, Aerobacter, Azotobacter, Achromobacter, Rhizobium, Alcaligenes, Escherichia, Salmonella, Pseudomonas* [123], *Sarcinia*, and *Agrobacterium* [32]. *Komagataeibacter xylinus* has been used as a model organism for industrial bacterial cellulose production [123]. The choice of bacterial strain is a significant parameter to produce a high yield of bacterial cellulose. The other factor was the growth media; Hestrin-Schramm (HS) medium [70] which was the most used medium for the BC production and alternative carbon (sugar, molasses) and nitrogen (peptone, yeast extract, corn steep liquor, peanut sprout extract) sources were used for the production [32, 162]. The high cost of growth media limited the BC production on an industrial scale. BC can be produced from industry by-products wastes for reducing the cost, such as whey protein from dairy industry [142], wastewater of candied jujube processing industry [105], and Colombian agro-industry waste pineapple peel and sugar cane juice [33]. BC has the same molecular formula $(C_6H_{10}O_5)_n$ of plant cellulose, while BC is pure material without lignin, hemicellulose, and pectin, making the purification easy for application [8]. BC can be used as a multifunctional food ingredient to improve the rheology of food, as a thickening, stabilizing, gelling, suspending agent, as well as in the formulation of low calorie and low cholesterol food product formulations, and immobilization of enzymes, as a nano carrier for encapsulation of food additives and food packaging material [152, 167]. BC is considered as "generally recognized as safe" (GRAS) by the Food and Drug Administration (FDA) in 1992 [167]. The nata de coco was accidentally discovered by Pagsanjan, Laguna in Luzon, Philippines. The first use of BC in food to manufacturing Philippine's traditional sweet candy dessert is called "*nata de coco*". It is prepared from coconut water through a fermentation process by *Acetobacter xylinum. Nata de coco is* also produced and consumed in Indonesia as a healthy diet and in East Asian countries [73]. The Philippines was the primary producer of nata de coco with an international market volume of about 6350 tons, with $6.63 billion worth in 2011. The nata de coco exported to the international markets, the major importing countries are Japan, the United States, Canada, Malaysia, and other 40 countries, including the European and Middle East countries [135]. Also, nata de coco industry had export value chains in Vietnam, Thailand, and Indonesia [134]. BC (nata) has been used as a fat replacement in different foods. Lin and Lin [109] were investigated the potential use of BC (nata) as a functional ingredient with Chinese-style meatball. Meatballs were produced with 10, 20, and 30% nata and 20% fat as control. Meatballs containing 10% nata had the same sensory and texture acceptability as control meatballs. A study was carried out by Halib et al. [66], evaluated nata de coco dessert as a possible source of pure bacterial cellulose for research study. In this study, the extracted BC characterized by Fourier transform infrared (FTIR) spectroscopy, thermogravimetric analysis (TGA) and scanning electron microscopy (SEM). The purified BC powder possessed similar FTIR spectra and degradation DTG peak that reported for BC. The food-grade material, nata de coco from local food industries, could be used as a source of BC.

Bacterial cellulose is a natural microbial extracellular polysaccharide. It has unique properties including low density, large surface area, high aspect ratio, high crystallinity, high purity, high water holding capacity, high tensile strength and nanoscale dimension, unique morphology, and 3D nanofibrillar cellulosic network [156, 184], these properties enable BC in a wide range of specific applications in food packaging and food-grade emulsion formulations specifically Pickering emulsions. Pickering emulsions are formed from solid colloidal particles, irreversibly attached at the between oil-water interface [80], depending on the solid particles wettability properties. Pickering emulsions can be either oil-in-water (o/w) or water-in-oil (w/o) emulsions [26]. Several investigations have reported the use of bacterial cellulose nanofiber (BCNF) and bacterial cellulose nanocrystals (BCNC) for stabilization of Pickering emulsions [79, 80, 180] and also Pickering emulsions of BC with other polymers such as BCNC-gelatin for cinnamon essential oil [140]. Bacterial cellulose nanocrystals can be produced from BC nanofibers under controlled acid hydrolysis conditions by chemical treatments [3] and by enzymatic treatments as green nanomaterials [60, 144]. Acid hydrolysis favors the removal of amorphous regions, provides crystal regions, and changes nanocrystals charge density [3]. Hydrochloric acid treatment results in weak surface charge, sulfuric acid-treated samples result in negatively charged sulfate esters on the crystal surface; this charge density affects formation of stable colloidal suspensions and the wettability of nanocrystals at oil-water interface [80].

The BCNCs amphiphilic character, hydrophilicity and hydrophobicity balance, nanocrystals wettability at the oil-water interface, emulsifying properties, crystals morphology, crystallinity, and size variation strongly affect the stability of Pickering emulsions [80, 133]. The emulsifying performance of hydroxypropyl methylcellulose (HPMC), carboxymethyl cellulose (CMC), and bacterial cellulose (BC) for extra virgin olive oil Pickering emulsions formulations were tested via using high shear mixer and ultrasound methods [133]. The BC and extra virgin olive oil Pickering emulsions showed better stability than the other commercial cellulose and were not affected by pH, temperature, and ionic strength changes. However, HPMC and CMC extra virgin olive oil Pickering emulsions were more sensitive to environmental stresses [133].

In another study, the possible use of BC and BCNCs in the stabilization of olive oil Pickering emulsion were evaluated [180]. Regarding this study results, the BCNCs with 259.6 nm particle size, −34.8 mV zeta potential, and 89.6% crystallinity index were hydrolyzed with sulfuric acid, acid treatment followed by the hydrogen peroxide oxidation. The olive oil Pickering emulsion formation with BC and BCNCs was proved by optical and fluorescence microscope. The BC oil Pickering emulsions showed better stability than the BCNCs towards the change of pH and ionic strength. Results proved that BC and BCNCs particles could be adsorbed at the oil and water interface to form the Pickering emulsions. The formed particle-stabilized emulsions showed better colloidal stability against the coalescence [180].

Apart from the BC fibers and crystals outstanding unique properties, BC was non-toxic, edible, available from renewable sources, biocompatible, and biodegradable polymer for encapsulation purposes. The BC engineered structures can be used as delivery systems for the encapsulation of bioactive substances and provide stability against environmental degradation, improving the stability of encapsulated food ingredients for food applications [8]. The BC and BCNCs as carriers for bioactive agents as Pickering emulsion could be used directly into the food matrix to produce functional foods and into the films and coatings solutions for food packaging applications. BC has great potential to use as the support material for films [49], foams, coatings, and rapid and simple sensing devices [92], due to its three-dimensional nanostructure, the high specific surface area, high water holding capacity, high tensile strength, besides bacterial cellulose films has the characteristic of transparency, flexibility, and hydrophilicity. In addition to BCs renewability and sustainability, its appealing characteristics such as high crystallinity and high mechanical strength enable to use BC as reinforcing agents in high-performance nanocomposite materials and eco-friendly materials for various applications, especially food packaging applications [9].

Most microbial contamination and deterioration reactions occur on the food surface [63]. Food packaging materials should exhibit good barrier and mechanical properties for preserving food quality and safety. Biobased polymers have not fulfilled these requirements, such as commercially available plastic packagings. BC and BCNCs are used as nanofiller, nano reinforcements with a wide range of polymers to develop nanocomposites to improve the polymers barrier and mechanical properties for packaging materials [8, 28]. George et al. [60] developed a reinforced Polyvinylalcohol (PVA) matrix with BCNC to generate polymer nanocomposites. BCNC was manufactured from enzyme hydrolysis with desirable properties, 100–300 nm length, and 10–15 nm diameter. BC-PVA nanocomposite display improved thermal stability and mechanical properties [60]. BCNC obtained by hydrochloric acid hydrolyzes is used as reinforcing hydroxypropyl methylcellulose (HPMC) polymer material. BCNC resulted in remarkable improvement in the tensile strength and modulus of HPMC, while the incorporation of 2–4% BCNC reduced the elongation properties [59]. The nanocomposite of bacterial cellulose nanofibrils (BCNs) supported zein nanoparticles (ZN) was developed by Li et al. [102]. The resulting BCN-ZN nanocomposites possessed improved mechanical and thermal properties. The incorporation/encapsulation/embedded antimicrobials into the food packaging polymers can reduce or retard the spoilage or pathogenic microorganism [2].

BC was used in a wide range of applications because of BCs outstanding properties, whereas BC has no antimicrobial property. Gao et al. [57] developed an antimicrobial BC film with nisin via the co-culturing method with nisin producing strain *Lactococcus lactis*. Jebel and Almasi [75] described a novel monolayer and multilayer of BC films with zinc oxide (ZnO) nanoparticles. Bacterial cellulose-containing antimicrobial composite films were also developed with a wide range of organic and inorganic antimicrobial agents such as PVA-BC with potassium sorbate [77] and silver nanoparticles incorporated chitosan-BC [146]. The bacterial cellulose films with antimicrobial properties are obtained with silver nitrate ($AgNO_3$). BC matrix was used for the stabilization of silver nanoparticles (AgNPs). An antimicrobial nanocomposite of bacterial cellulose silver extended tomatoes shelf-life up to 30 days [2]. The efficacy of nisin immobilized bacterial cellulose (BC) films was tested on processed meat [126]. Bacterial cellulose films were obtained by *Gluconacetobacter xylinus* K3 with Corn Steep Liquor-mannitol medium. BC-nisin composite films were prepared to immerse BC films into nisin solution for the absorption of nisin into the films. Nisin incorporated bacterial cellulose films that possessed antimicrobial activity on agar media and significantly reduced the artificially inoculated *L. monocytogenes* population on frankfurters as a food model. In addition, Zhu et al. [191] fabricated BC embedded ε-polylysine (ε-PL) casing for sausage packaging. The composite of BC/ε-PL was obtained soaking of BC into ε-PL solution to allow absorption of ε-PL through cellulose film. The composite film had remarkable antimicrobial activity against *E. coli* and *S. aureus* on agar media and sausages. The population of bacteria on sausage samples treated by BC/ε-PL was significantly lower than the BC films during 18 days storage at 4 °C. Bandyopadhyay et al. [11] fabricated films from BC and guar gum (GG) based polyvinyl pyrrolidone-carboxymethyl cellulose (PVP-CMC). The films were tested for shelf-life analysis of berries. Berries were packed with PVP-CMC, PVP-CMC-BC, PVP-CMC-GG, and PVP-CMC-BC-GG films. The weight loss of berries packed with PVP-CMC-BC-GG films was lower than the other films due to the least water vapor permeability (WVP) and oxygen permeability (OP) values. Yordshahi et al. [183] developed antimicrobial BC film as a carrier for postbiotics lactic acid bacteria. The antimicrobial effect of nanopaper with BC and postbiotics lactic acid bacteria tested on ground meat, lactic acid bacteria immobilized BC films decreased the *L. monocytogenes* growth on ground meat and total mesophilic and psychrophilic count during 9 days of storage at 4 °C. Because of BC outstanding properties with the high specific area and high porosity, BC would be used for intelligent food packaging applications as intelligent labels or sensors for freshness monitoring. In the study of Kuswandi et al. [92], edible pH sensor developed by BC membrane immobilized red cabbage anthocyanins for intelligent food packaging systems. The developed system can distinguish fresh milk from spoilage. Moreover, Mohammadalinejhad et al. [115] proved that *Echium amoenum* anthocyanins immobilized BC films may be used for development of novel non-destructive intelligent food packaging systems for monitoring the freshness or spoilage of shrimp as colorometric pH indicator. There are a wide range of promising applications for utilization of BC in food industry due to its purity, biocompatibility, high water holding capacity, mechanical strength, food-grade material, network structure, and better emulsifying capacity; however, production yield, quality, and the demand properties of end product, price of growth medium are the challenges of BC to be overcome for food applications.

3 Pullulan

Pullulan is a non-toxic, biodegradable, biocompatible water-soluble neutral edible, extracellular exopolysaccharide commercially produced by a yeast-like fungus *Aureobasidium pullulans* [21, 67]. Pullulan has a linear glucan structure consisting of maltotriose repeating units [155]. Pullulan can be produced from other microbial sources, including *Cytaria harioti, Cytaria darwinii, Cryphonectria parasitica, Teloschistes flavicans, Rhodosporidium paludigenum,* and *Rhodotorula bacarum* [154]. Pullulan has a potential application for food applications such as a thickener, stabilizer, binder, dietary fiber, texture improver, prebiotic, low-calorie food ingredient [94, 155]. Pullulan dietary fiber effects have been tested on fried potato starch [24] and rice starch digestibility [22]. Pullulan, a natural, odorless, tasteless polymer, is used for food packaging applications in developing edible film and coating formulations. Pullulan films and coatings with antimicrobial agents have great potential to reduce microbial spoilage and extend food products shelf-life. Chu et al. [29] prepared antimicrobial pullulan coatings with cinnamon essential oil nanoemulsion. Pullulan-cinnamon essential oil nanoemulsions coatings significantly decreased the total aerobic counts and yeast and mold counts of strawberries during room storage. Kraśniewska et al. [89] manufactured a pullulan coating enriched with leather bergenia leaves extract. Pullulan-leather bergenia leaves extract films antimicrobial efficacy was tested on artificially inoculated peppers with *Aspergillus niger* and *Staphylococcus aureus* and apples with *Aspergillus niger*. Pullulan-leather bergenia leaves extract coatings showed a high inhibitory effect on fungal contamination of apples, fungal, and peppers bacterial contamination. In another shelf-life study, carried out by Kumar et al. [91], chitosan-pullulan composite edible coatings supplemented with pomegranate peel was developed to enhance the quality and shelf-life of green bell paper at room (23 ± 3 °C, RH: 40–45%) and cold temperatures (4 ± 3 °C, RH: 90–95%) for 18 days of storage and no adverse effects were observed on sensory attributes during storage. Wu et al. [177] used *Laminaria japonica*-derived oligosaccharides in pullulan coating for cherry tomatoes preservation. The study performed by Yan et al. [181] improved strawberries' qualities during the storage period with Konjac glucomannan and pullulan composite films.

The efficacy of nisin embedded amaranth protein isolate-pullulan electrospun nanofibers was assessed on apple juice and fresh cheese as a real food model [158]. Fresh cheese and apple juice artificially inoculated with *Salmonella Typhimurium, L. monoctogenes*, and *L. mesenteroides*. Amaranth protein isolate-pullulan nanofibers showed satisfactory antibacterial effects against all bacteria. This study confirmed that nisin encapsulated amaranth protein isolate-pullulan nanofibers potential in controlling the post contamination.

Pullulan potential prebiotic properties were investigated in low-fat yogurt. The viability of *Streptococcus thermophilus*, *Lactobacillus*, and *Bifidobacterium* in the presence of pullulan at 0.5–2% in low-fat yogurt and pH changes during storage at 4 °C for 28 days were investigated [94]. Pullulan addition to reduced-fat yogurt enhanced the viability of *Lactobacillus* and *Bifidobacterium*, however, had no effect on the viability of *Streptococcus thermophilus*. Pullulan addition improved the texture properties but had an adverse impact on sensory attributes in low-fat yogurt. This study demonstrated the pullulan protective effect on the viability of probiotic bacteria for health-promoting properties.

4 Xanthan Gum

Xanthan gum is a charged heteropolysaccharide polymer, which is produced by the *Xanthomonas campestris*. The other Xanthomonas strain types such as *X. axonopodis pv. vesicatoria, X. Hortorum pv. pelargonii, X. Axonopodis pv. begoniae* and

X. campestris have been used for xanthan gum production [41]. Global production of xanthan gum exceeds 50000 tons/year with 600 and 800 million dollars/year market value [161].

Xanthan gum is an important microbial exopolysaccharide and widely used in the food industry due to its unique properties, including thickening properties [27], film-forming properties [90], and other applications, including a stabilizing agent for emulsions [170], nano-micro carriers for active agents [122] and as functionality enhancers [139]. Xanthan gum addition to the pasteurized egg white foam with Persian gum enhanced the foam texture and stability [34].

Edible coatings can be applied in different methods including, dipping, spraying, and coacervation. The study of Lara et al. [97] investigated the effect of xanthan gum-based edible coatings in the spraying method to improve the storage stability of fresh-cut lotus root. Xanthan gum, citric acid, and glycerol contained spraying solutions that prevent microbial growth, decreased the enzymatic browning, and enhanced the shelf-life of fresh-cut lotus root post-harvest storage. Cho and Yoo [27] determined the impact of commercially available food thickeners, including xanthan gum, guar gum, dextrin, and carboxymethyl cellulose, in cold beverage preparations. The thickened beverages prepared with xanthan gum possessed desired rheological properties and improved the swallowing ability. The study of Espert et al. [48] evaluated the palm oil in vitro digestion in the presence of xanthan gum. The impact of xanthan gum was significant in vitro digestion system, and the xanthan gum matrix remarkably reduced fat digestion.

The texture of the cake, pasta, and bread is considered a critical quality characteristic of these products; hydrocolloids such as xanthan gum are widely used to formulate bakery products. Milde et al. [114] developed a gluten-free pasta with cassava starch, cornflour, and xanthan gum. In the pasta formulation, the presence of 0.6% xanthan gum enhanced pasta dough handling, decreased cooking loss, and improved physical and textural properties. Mohammadi et al. [116] optimized gluten-free flatbread formulations, which included rice flour, corn starch, soy flour, xanthan gum, and also xanthan gum-carboxymethyl cellulose. The highest dough yield, bread yield, and lowest bread weight loss obtained with xanthan gum and xanthan gum-carboxymethyl cellulose combination bread formulations. In another study, the presence of 1% xanthan gum in sponge cake formulation with wheat flour and corn starch improved the hardness of the product but decreased the overall acceptance scores for sensory evaluation [130]. The other food applications of xanthan gum are; Zhao et al. [190] showed the potential of xanthan gum as sodium salt substitute in the low sodium meat products formulations, Santos et al. [147] developed food grade Pickering emulsions formulations with zein-xanthan gum, and sunflower oil, Sharma and Rao [151] developed the xanthan gum embedded cinnamic acid edible coatings for prevented browning and prolonged the shelf-life of fresh-cut pears.

5 Dextran

Dextran (α-1,6-linked glucan) is a homopolysaccharide composed of D-glucose units [176], produced by *Leuconostoc* spp, *Streptococcus* spp, *Weissella*, *Pediococcus* and *Lactobacillus* genera of lactic acid bacteria [13, 69, 88, 119]. Lactic acid bacteria produced a wide variety of different EPSs, including homopolysaccharides and heteropolysaccharides. Dextran, levan, kefiran, and hyaluronic acid are the LAB belonging EPSs [117]. Dextrans texturizing, emulsifying, prebiotics, cloud forming, edible coating properties are the potential applications of dextrans in the food industry [39, 46, 164]. In the baking industry, hydrocolloids have been used to enhance the dough and bread-making process textural and rheological properties [164] and bakery products shelf-life [176, 187]. Microbial EPSs from lactic acid bacteria during sourdough fermentation has the ability to act as hydrocolloid to improve the rheology, texture, and mouth feel properties of fermented dairy products and bakery products in the food industry [14, 44, 164].

Many studies have been conducted on the application of dextran in the bakery industry [13] to determine the prebiotic function of dextran from *Weissella cibaria* RBA12 isolated from pummelo, and its potential application in sourdough fermentation [145] tested the dextran as an alternative of commercially available hydrocolloids in gluten-free bread. Dextran produced by *Weissella confusa* (*W. confusa*) improved quality. It extended the shelf-life of Chinese steam bread with acceptable sensory attributes [164], and in the study of Wang et al. [174] it was reported that the improvement of wheat-fab a bean composite bread quality was due to dextran synthesized in situ *Weissella confusa*, and also wholegrain sorghum bread formulations improved with *W. confusa* [172]. In addition, dextran could be purified from a wide variety of sources with multifunctional properties in different kinds of food applications. Du et al. [44] obtained dextran from *Leuconostoc pseudomesenteroides* from the homemade vine with antioxidant property.

6 Kefiran

Kefiran is an extracellular polysaccharide from lactic acid bacteria during kefir production [149]. Kefiran is available in kefir grains and Lactobacillus species such as *L. kefir*, *L. parakefir*, *L. kefiranofaciens*, *L. brevis,* and *L. delbrueckii subsp. bulgaricus* [95, 118]. Kefiran has potential applications in producing food products, including edible films [61], a delivery system such as doxycycline antimicrobial agent embedded kefiran nanofibers [35], and green nanocomposites with cellulose nanocrystals and kefiran [149].

7 Curdlan

Curdlan is a water-soluble bacterial linear homopolysaccharide composed of repeating unit of (1 → 3)-β-glucan produced by *Agrobacterium* spp. (formerly *Alcaligenes faecalis var. Myxogenes*) [113, 119, 186], *Rhizobium* spp., *Cellulomonas* spp. are the other curdlan producing microorganisms [58].

Curdlan is widely used in the food industry as a stabilizer, thickener, and texturizer and approved by FDA (Food and Drug Administration) as a food additive [186]. Due to its unique properties, including high water holding capacity, rheological, gel-forming, textural improving properties, and freeze-thawed properties, curdlan has been used as a critical ingredient in the food industry to improve the quality of the various type of food products [25]. Textural and cooking qualities are the key quality for the noodles. Gao et al. [58] evaluated the noodle quality that was prepared with curdlan. Regarding the results, the use of curdlan in noodles formulations significantly improved the eating quality and textural properties. In another study, the addition of curdlan into potato starch noodles enhanced the textural properties and increased the syneresis [173]. Liang et al. [106] described how curdlan addition could minimize the effect of curdlan on the quality of frozen cooked noodles during frozen storage. The other important food applications are, curdlan and chitosan coatings improved the mechanical and barrier properties of packaging paper [17], curdlan has been used as a fortifier in a set of yogurt [189], the study of the Funami et al. [56] assessed the possibility of curdlan as fat mimetics ingredient for sausages.

8 Emulsan

Emulsan is an extracellular polysaccharide composed of sugar backbone with fatty acids produced by *Acinetobacter* spp., including *A. venetianus*, *A. calcoaceticus* [20, 78]. Emulsan has potential applications as biosurfactants [129] in the food industry.

9 Alginate

As a popular exopolysaccharide, alginate structure includes two main compounds: *β-D*-mannuronic acid units and α-*L*-guluronic acid units. These acid units are linked by α-1,4 glycosidic bonds. Alginate is collected from various kinds of brown seaweeds and species of *Pseudomonas* and *Azetobacter* also synthesize alginate as an exopolysaccharide [68]. *Pseudomonas* species that are able to produce alginate are known as *P. aeruginosa*, *P. fluorescens* and *P. syringae* [16].

Bacteria cells can adhere to solid surfaces and adhesion process occurs easily in the presence of some EPSs such as alginate. For instance, the attachment of *P. aeruginosa* to solid surfaces was improved by alginate synthesized in the medium. Biofilms created by exopolysaccharide can contribute to overcoming electrostatic repulsion between the bacteria cells and the surface. Hence, alginate has a significant function in the development of biofilms by species of *P. aeruginosa* [16]. Alginates are purified from these medium created by various bacteria species and the production of alginate is commercially performed by using different species of brown algae such as *Laminaria digitata*, *Laminaria japonica*, *Ascophyllum nodosum*, and *Macrocystis pyrifera*. Extracts obtained from these species were treated with alkali solutions such as NaOH. At the end of purification steps, sodium alginate is obtained in powder form [99].

Alginates are able to produce gels when the medium including divalent cations such as calcium ions. Interaction between alginate and divalent cations is called crosslinking and that is prefered between G blocks and divalent cations. Therefore, guluronate content of alginate is proportional to the gel strength [52]. Alginate has considerable properties as gelling agent, thickening, stabilizing, emulsifying agents, and encapsulation. Besides that, it is a biocompatible and biodegradable material used as a biopolymer, especially for food coating exopolysaccharide [141].

Alginates are commonly used in many industrial processes. In food industry, alginates have been utilized as thickener, emulsifier, and stabilizing agents. The applications of alginate in food products are related to its physical properties. These physical properties come from its unique chemical structure. The α-*L*-guluronic acid units known as G blocks and β-*D*-mannuronic acid units known as M blocks defined the functional properties of alginate. In terms of gelling property, G blocks play an important role by binding Ca^{2+} or H^+ binding that results in gel formation. In the food industry, alginate is well known as thickening agent because of its ability to form gels, which are heat stable, at low temperature in the absence of heating or at low pH medium or in the presence of calcium [141]. Ca-dependent gelation and gel properties of alginate are significantly influenced by intrinsic factors including molecular weight, guluronic acid percentage, the length and distribution of Ca-binding blocks, and extrinsic factors such as concentration of alginate, concentration of calcium ions, ion strength, pH, and temperature. Alginate having high molecular weight offers more calcium-binding sites that develop rheological properties of the gels and improve viscosities. Increase in guluronic acid percentage increased the stability and mechanical strength in terms of Ca-dependent gels [19].

Considering the role of alginate in the viscosity properties of foods, diverse food formulations like ice cream, jam, jellies, mayonnaise, salad dressing, desserts, cakes, and candies have been produced. In low-fat mayonnaise formulations, fat reduction affected the textural properties of product and the application of alginate-based gel systems were investigated as alternatives to produce low-fat mayonnaise. Li et al. [101] performed the production of low-fat mayonnaise by emulsifying the oil molecules in the gel medium created by using alginate and alginate stabilized gel-based emulsion produced by using alginate (2%) and Tween

80 (0.5%). Fat droplets in these kinds of products have an important structural function and thus development of low-fat formulation is a challenging issue. Alginates take part in formulations as an alternative ingredient to reduce the amount of fat in the product while maintaining the textural properties of the product. Main advantage of alginate-based gels over other polysaccharide gels is thixotropy, shear-thinning property, but one disadvantage is the formation of weak hydrogels [182]. In the study, it was mentioned that mechanical strength of alginate-based gels was enhanced by using with different polysaccharides such as cellulose, chitosan, pectin, and others. Thus, alginate was combined with glucomannan in different ratios in order to produce low-fat mayonnaise. It was investigated glucomannan addition to alginate at high concentration as 4% considerably enhanced the strength of alginate-glucomannan matrix that forms a complex gel structure. Moreover, the prepared low-fat emulsion gels including alginate and glucomannan indicated good thermal stability after heating at 100 °C for 30 min and freeze-thaw stability after freezing the gels at −18 °C for 24 h [182]. Alginate is also used to replace milk fat droplets to obtain low-fat dairy products due to the higher water-binding capacity of alginate. Sodium alginate was used in the production of Cheddar cheese having low-fat content and its usage improved textural properties. Four levels of sodium alginate 0.12, 0.17, 0.18, and 0.23% were investigated as an ingredient to replace fat content of Cheddar cheese until 91% fat reduction. During ripening of 180 days, fat reduction by addition of alginate resulted in improved textural properties by increasing hardness and microstructural properties, but poor color development [84]. In recent years, low-fat meat products have become more popular and are preferred by consumers. Alginate is one of the hydrocolloids used in low-fat emulsion meat products in order to replace fat by maintaining the quality [82]. The effect of substitution of pork back-fat with alginate solution at 0, 25, 50, 75, and 100% ratios on the quality characteristics, protein conformation, and sensory attributes of frankfurters were evaluated [82]. When the use of sodium alginate solution was 25 and 50% ratios, the cooking yield, emulsion stability, and color values of frankfurters were not significantly different. Texture properties including hardness, springiness, cohesiveness, and chewiness of frankfurters produced by 50% pork back-fat replacement with sodium alginate solution had significantly higher values than others [82].

As a gelling agent, alginate forms stable gels at a wide range of temperatures and low pH conditions. Alginate is widely used in the production of ice cream products for functions such as thickening, stabilizing, controlling viscosity, shrinkage, and ice crystal formation [137]. Moreover, alginate has also been used to obtain hydrogels maintaining viability of probiotic bacteria in low acid medium that simulates gastric juice. Encapsulation of probiotic bacteria, *Bifidobacterium breve*, was carried out by alginate-based capsules and viability of probiotics was enhanced in various pH environments simulating gastric and intestinal medium [104]. Addition of *Lactobacillus rhamnosus* and *Lactobacillus casei* into ice cream in free and encapsulated form by using alginate and chitosan were investigated. Encapsulated *L. rhamnosus* was preserved at low temperatures, but the *L. casei* indicated greater viability during the encapsulation as well as in the gastrointestinal environment [50].

Alginate is being an attractive polysaccharide used for edible films and coatings of foods because of its film-forming ability with non-toxicity and low price. Among the different kinds of alginate salts, sodium alginate is commonly used for producing water-soluble, tasteless, odorless, glossy, and flexible edible films with low permeability to oxygen [165]. Alginate-based biodegradable coatings present an alternative to substitute synthetic coatings for maintaining the quality of fresh fruit and vegetables after harvesting. One disadvantage is poor moisture barriers of edible films produced from alginate due to its hydrophilic structure. However, addition of calcium ions improves the water-resistance of alginate films [143]. A number of studies in the literature have presented the potential of alginate-based edible coatings for maintaining quality attributes of fruit and vegetables such as grapes [87], peach [103], pineapple [7], plum [168], mushroom [76], and arbutus berry [62]. Alginate based coatings have been commonly used as carriers of active ingredients to increase shelf-life and improve quality properties of foods [165]. Recent investigations in literature focus on nanoemulsion alginate coatings in which active ingredients exhibit better properties. Effectiveness of alginate-based nanoemulsion coatings incorporated with sweet orange essential oil were evaluated in terms of antibacterial and antibiofilm activity against *Salmonella typhi* and *Listeria monocytogenes* as well as coating effect on quality attributes of tomatoes during storage at 22 °C. Alginate nanoemulsion indicated antibacterial property against *S. typhi* and *L. monocytogenes* and increased the shelf-life of tomatoes by delaying spoilage caused by the bacteria. In addition, the coating significantly enhanced firmness and reduced weight loss of tomatoes by delaying ripening [38]. Application of alginate-based nanoemulsion coating incorporating lemongrass essential oil at the concentrations of 0.1 and 1% were investigated on fresh-cut apples stored at 4 °C for 14 days. In terms of the quality parameters including weight loss, pH, and acidity, formulation incorporating 0.1% essential oil content was found to have positive effects compared to high essential oil content (1%) [31]. Cinnamon essential oil nanoemulsions were incorporated into alginate-based biocomposite films [54]. In this study, developed biocomposite films including 20% cinnamon essential oil nanoemulsion indicated antibacterial activities against *Salmonella typhimurium, Bacillus cereus, Escherichia coli, and Staphylococcus aureus.*

10 Gellan

Gellan is a kind of gum formed by bacterium *Sphingomonas elodea* (ATCC31461), also called as *Pseudomonas elodea*, through aerobic fermentation. Gellan is one of the EPSs which has ability to form gel in a wide range of conditions such as low concentrations and acidic medium [110]. Structurally, gellan could be in two different forms which are high acyl (HA) and low acyl (LA) gellan. The structure of HA and LA gellan is mainly a linear tetrasaccharide repeat unit of two glucose, one glucuronate, and one rhamnose units [185]. In the tetrasaccharide repeating sequence of gellan, three of the four glycosidic linkages have equatorial bonds at first and fourth carbon atoms of the participating residues. The rest of linkages in the gellan exhibits a systematic "twist" in direction of the chain which ensures helix structure of gellan. The water-based solution of anionic polysaccharides such as gellan present cations which are counter ions to the charged groups of the polymer chains. The balance of positively-charged ions in the solution is provided by interaction with negatively-charged polymer chains. Linear polyanion charge intensity and the charge of individual cations state the strength of this interaction [120].

Gel formation ability of polysaccharides differs from each other. Some of that, such as carrageenan, gellan, and curdlan, are able to form a gel by means of heating and then cooling. Others, such as alginate, LA gellan, and high methoxyl pectin, form a gel at specific conditions including temperature, types of cations, and pH [185]. Different factors should be considered in terms of gelation mechanisms. Gelation process takes place by aggregation of double helices of gellan. It is explained that pH reduction assists aggregation and gelation by means of decreasing the negative charge on the polymer which results in the decrease in electrostatic repulsion between the helices. Some of the cations can bind the helices in defined sites related to carboxylate groups of polymer, and thus decreasing electrostatic repulsion. Increase in ionic size of cations increases the strength of binding ($Li^+ < Na^+ < K^+ < Rb^+ < Cs^+$) [120]. Due to its rheological properties, gellan has been commonly used in food products. Generally, food gels are classified as fluid gels, soft gels, and hard gels. Even though xanthan gum is commonly used in weakly gelled foods, such as salad dressings, gellan is used as a gelling agent to form soft and firm hydrogels in foods such as beverages, desserts, jams, and jellies. Low concentration of gellan is applied to form fluid gels in diverse types of fruit juices and beverages. In order to form soft and firm gels in foods such as desserts and jellies, higher concentration of gellan are required [185].

Regarding the food applications of gellan, studies in the literature have investigated the effect of gellan addition in food formulations from different perspectives. Schelegueda et al. [148] evaluated low-sugar content food model systems developed by using natural additives having different roles. One of the additives was gellan, used as gelling agent, xylitol, used as water activity depressor, and natamycin, used as an antimicrobial agent. Addition of gellan at two different concentrations(0.90 and 1.80 g/100 g) was investigated with the combination of other additives. In the model food system, growth of *Zygosaccharomyces bailii* was inhibited by highest gellan concentration at the beginning of storage. However, consumption of nutrients in the medium leads to the utilization of gellan as an energy source by yeasts. It was also explained that higher level of gellan application with natamycin resulted in weak structure in the gel. Effects of low acyl (LA) and high acyl (HA) gellan on the thermal stability of anthocyanins in model beverage systems was investigated [179]. It was indicated that gellan addition maintained thermal stability of anthocyanins during heat treatment and HA gellan provided significantly higher stability improvement than LA gellan. It is probably due to higher degree of acylation in the HA gellan. Acyl

Table 1 Food applications of microbial EPSs

EPS	Sources	Applications in food industry	References
Cellulose	*Komagataeibacter, Aerobacter, Azotobacter, Achromobacter, Rhizobium, Alcaligenes, Escherichia, Salmonella, Pseudomonas, Sarcinia, Agrobacterium, Dickeya, Rhodobacter*	Gelling agent, emulsifying agents, delivery of bioactive agents, coatings, and films	[32, 92, 107, 123]
Pullulan	*Aureobasidium pullulans, Cytaria harioti, Cytaria darwinii, Cryphonectria parasitica, Teloschistes flavicans, Rhodosporidium paludigenum, and Rhodotorula bacarum*	Delivery of bioactive agents, coatings films, and prebiotic properties	[21, 67, 154]
Xanthan gum	*Xanthomonas axonopodis pv. vesicatoria, Xanthomonashortorum pv. pelargonii, Xanthomonas axonopodis pv. begoniae and Xanthomonas campestris*	Gelling agent, thickening, delivery of bioactive agents, coatings, and films, sodium salt substitute, gluten-free food formulations	[41, 47, 98, 151, 190]
Dextran	*Leuconostoc spp, Streptococcus, Weissella, Pediococcus, and Lactobacillus genera of lactic acid bacteria (LAB)*	Functional food applications, texturizing, emulsifying, prebiotics, cloud forming, edible coating	[13, 39, 46, 88, 164]
Kefiran	*Lactobacillus species such as Lactobacillus kefir, Lactobacillus parakefir, Lactobacillus kefiranofaciens, Lactobacillus brevis, and Lactobacillus delbrueckii subsp. bulgaricus*	Delivery of bioactive agents, coatings, and films	[35, 61, 95, 118]
Curdlan	*Agrobacterium sp., Alcaligenes faecalis var. Myxogenes, Rhizobium spp, Cellulomonas spp*	Texturizing, emulsifying, delivery of bioactive agents, coatings, films, and fat mimetics	[56, 58, 113, 173, 186, 188]
Emulsan	*Acinetobacter sp, including A. venetianus, A. calcoaceticus*	Adsorption, biosurfactants	[20, 78, 129]
Alginate	Species of *Pseudomonas* and *Azetobacter*	Gelling agent, thickening, stabilizing, emulsifying agents, encapsulation, biodegradable coating	[68, 141]
Gellan	*Sphingomonas elodea, Pseudomonas elodea*	Gelling agent to form soft and firm hydrogels in foods, edible films for food packaging applications	[110, 185]

(continued)

Table 1 (continued)

EPS	Sources	Applications in food industry	References
Carragenan	Seaweeds of the class Rhodophyceae	Gelling, thickening, emulsifying agent, texture enhancers, and stabilizers in food products	[18]
Hyaluronic acid	*Streptococcus* species such as *Streptococcus equisimilis, Streptococcus pyogenes, and Streptococcus equi*	Hydrating agent in cosmetics and pharmaceuticals, biocompatible films, or hydrogels	[74, 163]
Levan	Lactobacillus johnsonii and Lactobacillus gasserii, Bacillus subtilis, Aerobacter levanicum, S. salivarius	Emulsifier, stabilizer, thickener, encapsulating agent, and carrier for flavors, inhibition hyperglycemia induced by diabetes	[36, 159]
Colanic acid	*E. coli* and also other species of Enterobacteriaceae family	Gelling agent in cosmetics and personal care products	[132, 160]
Welan	*Sphingomonas* sp., *Alcaligenes* sp.	Thickening, binding, and emulsifying agent in food products	[83]

groups in gellan molecules tend to regulation of structure into more extended frame. Thus, LA gellan molecules showed more strict arrangement. In another study, meat buffers were formulated by using gellan in order to reduce fat and sodium content [166]. Texture properties of reduced fat and sodium meat batters were evaluated by usage of gellan in the study. Hardness of samples was highest in the control samples while rigid structure decrease with the reduction in fat content and increased gellan content. Moreover, gellan is newly used gum in edible films developed for the applications of food packaging. For instance, gellan is used as an edible film matrix for carrying of ascorbic acid for improving food quality by antioxidant activity. Ascorbic acid degradation in the developed gellan based edible film indicated a pseudo-first order kinetics and it was indicated that ascorbic acid 100%-retained in film matrix after film casting [100]. Production and storage of gellan based edible films with the gellan concentrations of 0,0.02, 0.04, 0.06, 0.08, and 0.10% were investigated by Xiao et al. [178] and it was found that concentration of 0.08% gellan in edible films presented a desirable tensile strength and perfect film barrier properties. Storage of gellan based films was also investigated under four different conditions including refrigerated conditions at 0 °C, supermarket storage environment at 6 °C, room temperature at 25 °C, and high temperature at 35 °C [178]. Under storage conditions of room temperature and high temperature, higher water vapor permeability was observed due to the increase in velocity of gellan and other molecules. In another study, LA and HA gellan based edible coatings were optimized for application on ready-to-eat mango bars. As an independent variable, gellan concentration considerably affected the coating thickness. Gellan coating of mango bars improved sensory characteristics, color, and volatiles of fruits during storage [37].

Oil-in-water emulsions including 30% sunflower oil were stabilized by using HA gellan at different concentrations changing from 0.01 to 0.2%. HA gellan was effective to stabilize emulsions at concentration above 0.05% and it was compared with LA gellan at the same conditions. LA gellan was not found to be effective in stabilization of emulsions. It was suggested that HA gellan can be used as a potential ingredient to stabilize food emulsions [171]. Gellan was used with microcrystalline cellulose in order to develop physical and thermal stability of ginkgo beverages. The stability of ginkgo beverage was evaluated by characteristics of particle size, size distribution, zeta potential, and rheological properties. The study indicated that gellan with a small amount of 0.05 or 0.08% contibuted effectively to beverage stability compared to the larger amount of gellan by decreasing particle size and increasing zeta potential [127]. In yogurt and yogurt-based beverages, serum separation is one of the quality defects delayed by using hydrocolloid stabilizers. In this respect, EPSs can be able to reduce sedimentation rate by increasing the product viscosity. In a yogurt-based Iranian drink, effects of addition of gellan at the percentages of 0.01, 0.03, and 0.05, alone as well as its combination with high-methoxy pectin at 0.25% on the quality characteristics of viscosity, particle-size, and serum separation were investigated. Gellan was effective against syneresis by forming strong networks, thus serum volume decreased. Moreover, sensory analysis performed after 1 day of storage at 5 °C indicated consumer acceptance of yogurt-based drink was not affected by addition of 0.05% gellan in the presence of high-methoxy pectin at 0.25% while stability of product was maintained [85].

11 Carrageenan

Carrageenan is one of the gel-forming and viscosifying EPSs, extracted from a number of seaweeds of the class Rhodophyceae. Carrageenan has a molecular mass above 100 kDa and is structurally sulfated polygalactan with 15–40% of ester-sulfate. Its structure contains D-galactose and 3,6-anhydro-galactose units and the units are linked by the glycosidic linkage of α-1,3 and β-1,4. Based on the sulfate content, there are different kind of carrageenans such as lambda (λ), kappa (κ), iota (ι), epsilon (ε), and mu (μ), changing 22–35% sulphate. Increase in ester sulfate content results in the decreasing solubility temperature and gel strength [124]. From a commercial perspective, iota (ι), kappa (κ), and lambda (λ) are important three main varieties of carrageenans. The κ-carrageenan is comprised of 3-linked β-D-galactose 4-sulfate and 4-linked 6-anhydro-α-galactopyranose which has one negative charge per disaccharide repeating unit while ι-carrageenan includes two sulfate groups per disaccharide repeating unit. In the medium containing cations such as K^+, Ca^{2+}, ι- and κ-carrageenans form highly viscous aqueous solutions, thus indicate gelling properties. However, λ-carrageenan is incapable to produce gel at all temperatures, and gelation of λ-carrageenan is feasible with trivalent ions [192].

EPSs have different functional properties and each has various application areas. Carrageenan is useful to produce gels in the presence of potassium and calcium ions in certain foods, i.e., meat products, dairy products, and desserts. Gelation of carrageenan has resulted from the helix formation that develops when a 3, 6 anhydro bridge is present on the B unit of the carrageenan molecule [71]. Three main forms of carrageenan including kappa, iota and lambda differentiate in the number of sulphate groups as one, two, and three per disaccharide in the structure, respectively. In an aqueous solution, ι- and κ-carrageenan change form from a temperature-dependent and disordered structure to ordered helix transition. The helix structure is related with the gelation property of the carrageenans [96]. Thus, the carrageenans are used mainly as gelling agents, but also as fat substitutes, stabilizing, and thickening agents in the food industry. Among three most important types of carrageenan, except trisulfated λ-carrageenan, only monosulfated κ-carrageenan and bisulfated ι-carrageenan are able to form gels. Hydrogels obtained by using κ-carrageenan have thermo-sensitive structural and thus, κ-carrageenan is preferred for developing delivery agents in which targeted compounds' release depends on the temperature. In encapsulation of active compounds such as antioxidants, enzymes, and probiotics, κ-carrageenan hydrogels have been used to protect compounds during production and storage of the food products [93].

Carrageenans are widely used in food industry for properties of gelling, thickening, and emulsifying. Carrageenans are added as texture enhancers or stabilizers in many types of food products like frozen desserts, sauces, ice cream, yogurt, chocolate milk, cheese, and meat products [18]. Synergistic effect of carrageenan with locust bean gum is applied in order to increase the gel strength and combination with other hydrocolloids have an effect on gel strength and cohesiveness. In dairy products, λ- and κ-carrageenan are commonly used to improve solubility and texture due to easy combination with milk proteins [124]. In the study performed by Günter et al. [64], gel microparticles were produced by using pectin and κ-carrageenan. Higher pectin concentration in the microparticle gels resulted in lower swelling degree in simulated gastric, intestinal, and colonic fluids. Addition of carrageenan into the gel formulation provided an increase in the swelling degree. Influence of the carrageenan types including iota, kappa, and lambda on the casein micelle and carrageenan interactions in milk were studied and all three types of carrageenan showed adsorption onto casein micelles. The most highly charged form of carrageenan was found as λ-carrageenan which indicates attractive interactions with casein micelles at the temperature of 60 °C [96]. Hydrocolloids have an important role in controlling ice recrystallization. The usage of κ-carrageenan as a secondary stabilizer was investigated in terms of stabilization of ice cream during storage [10]. It was indicated that κ-carrageenan significantly reduce the hardness and iciness and supported the functionality of primary stabilizer. κ-carrageenan was added to lactose-free frozen yogurts as a stabilizer with three different concentrations of 0.05, 0.1, and 0.15%. Quality characteristics of lactose-free frozen yogurts including acidity, texture, viscosity, overrun, melting properties, and color attributes were improved with 0.15% κ-carrageenan addition. Sensory properties were also better in the yogurt samples incorporated with κ-carrageenan [157]. Foerster et al. [53] studied on a milk emulsion system in order to determine the optimum amount of λ-carrageenan based on the size of fat globules and stability of emulsion. Emulsions prepared with a carrageenan concentration range of 0.3 and 0.4% indicated the highest stability and minimal fat globule size. Moreover, the addition of carrageenan into milk emulsions also provided stabilization and reduction of surface fat in powders obtained by spray drying. Thus, the λ-carrageenan is to be useful in powders of dairy-based emulsion by stabilizing emulsions in the presence of milk protein. In a viscous food model systems containing 10% of sucrose, the addition of λ-carrageenan was investigated in terms of the release of aroma compounds including aldehydes, esters, ketones, and alcohols. The difference between release of the aroma compounds in water and in viscous λ-carrageenan solution was compared by using dynamic headspace gas chromatography. Mass transfer of aroma compounds demonstrated a decrease in release rates was changed depending on the physicochemical characteristics of the aroma compounds, but the effect of λ-carrageenan on decreasing the release rates was observed in the most volatile compounds [18].

12 Hyaluronic Acid

Hyaluronan is glycosaminoglycan with a high molecular weight and it is formed from repetition of disaccharide units which comprise of D-glucuronic acid and D-N-acetylglucosamine linked by β-1,4 and β-1,3 glycosidic bonds. Dissachharide units found in the hyaluronan molecule changes in the number of 10000 or more. Synthesis of hyaluronic acid is provided by internal membrane proteins called as hyaluronan synthases [125]. Hyaluronic acid was discovered in 1934 and its extraction was obtained from animal tissue, especially from rooster coombs. However, now production of hyaluronic acid is performed by recombinant bacteria [121].

Hyaluronic acid has a role as key molecule in the regulation of various cellular and biological processes. Thus, hyaluronic acid has been used for many biomedical applications not only due to the biocompatible, biodegradable, and nonimmunogenic properties but also because of its biological functions [136]. Synthesis of hyaluronic acid is occurred by many *Streptococcus* species such as *Streptococcus equi, Streptococcus equisimilis*, *Streptococcus pyogenes.* Among the species, *S. equi* subsp. *zooepidemicus* achieved industrial production of hyaluronic acid, but these strains are known as pathogenic. Therefore, a recombinant strain of *Bacillus subtilis* was found to be able to produce hyaluronic acid on a laboratory scale [74]. Recently, *S. thermophilus* isolated from traditional dairy food products was investigated for production of hyaluronic acid. It was found to be effective to produce EPSs including hyaluronic acid with a wide range of molecular masses [74]. Hyaluronic acid is found to be useful natural material in surgery of human eye due to the compatibility with human immune system. Another property of the hyaluronic acid which makes it useful is high water-binding capacity. The water-binding capacity of hyaluronic acid was remarked as up to 6 litres of water per gram of the polysaccharide, that associated with molecular mass of the polysaccharide [163].

Hyaluronic acid is mainly used as a hydrating agent in cosmetics and pharmaceuticals, and also used in eye surgery due to its biological properties [163]. The usage of hyaluronic acid in food industry is limited. Hyaluronic acid is also known with the ability to form biocompatible films or hydrogels in defined conditions. Galactomannans are a kind of polysaccharides used to increase the viscosity of food products and in films/coatings for foods. Combination of galactomannan and hyaluronic acid was investigated in order to observe physical and chemical characteristics of hyaluronic acid in solution. Two types of galactomannans including guar gum and locust bean gum improved the viscoelastic behavior in hyaluronic acid mixture solution compared to pure solution. Synergism between hyaluronic acid and locust bean gum was obtained successfully at 50% in which hydrogel with the best viscoelastic behavior and desired properties was obtained [112]. Nano-delivery system based on oligo-hyaluronic acid and loaded with curcumin and resveratrol was produced successfully and the hyalurosomes containing both curcumin and resveratrol were suggested to improve stability, bioavailability, and antioxidant activity of functional compounds used into juice, yogurt, and nutritional supplements. Coencapsulation of curcumin and resveratrol in hyalurosomes formulation was obtained with the average particle nano-size of 134.5 ± 5.1 nm and stability observed in vitro gastrointestinal release test [65]. In another study, caffeic acid as an antioxidant agent was incorporated into biopolymer hydrogels composed of hyaluronic acid, hydrolyzed collagen, and chitosan. Increase in the amount of hyaluronic acid resulted in excellent swelling behavior. Release of caffeic acid in the produced hydrogels was initially 70% within 60 min and followed by a release of 80% in 480 min [30].

13 Levan

Levan is a homopolysaccharide comprised of fructose, which makes it a unique carbohydrate. A wide range of microorganisms are capable to produce levan as an exopolysaccharide while limited plant-based sources are found to be the storage of levan. Bacterial strains producing levansucrase include *Lactobacillus johnsonii* and *Lactobacillus gasserii, Bacillus subtilis, Aerobacter levanicum, S. salivarius*, etc. Levansucrase plays a part in catalysation of the transferring D-fructosyl residues from fructose to yield levan. Thus, levansucrase is a catalyzer of microbial levan biosynthesis [159]. Among the bacteria, *Bacillus subtilis* Takahashi is known as the most efficient strain in terms of levan production. Structurally, levan is a fructose polymer including β-(2, 6)-linkages and also about 12% branched with β-(1, 2)-linkages. Its molecular weight is around 2×10^6 Da [169]. Levan can differ from the molecular weight and the fraction of residues incorporated in side chains according to the source obtained and conditions produced. It is an exopolysaccharide with a variety of usage in cosmetics, foods, and pharmaceuticals [1]. In terms of solubility, levan is a water-soluble carbohydrate polymer, but insoluble in all organic solvents like methanol, ethanol, isopropanol, n-propanol, acetone, toluene, etc. [159]. Intrinsic viscosity of levan is lower than other polysaccharides having similar molecular weight. Its low intrinsic viscosity is affiliated with the compact and spherical molecular conformation of the polysaccharide. Intrinsic viscosity values for levan in water was found in a range of 0.07 and 0.18 dL/g [6].

Levan is an exopolysaccharide used with a variety of usage in cosmetics, foods, and pharmaceuticals. Potential applications of levan have been mentioned as an emulsifier, stabilizer, thickener, encapsulating agent, and carrier for flavors. In addition, it is found efficient in inhibiting hyperglycemia and oxidative stress induced by diabetes [36]. Rheological properties of the solutions of levan obtained from *Bacillus* spp. were investigated at 20 °C and different concentrations. The intrinsic viscosity of obtained levan was compared with the levans produced by other bacteria. Viscosity values demonstrated shear-thinning behavior appears at a higher concentration than for other levans [6]. Levan has been used in the production of nanostructured system. Encapsulation of nanoparticles incorporated with vitamin E was performed by using levan and effect of parameters such as type of homogenization, speed of homogenization, and concentration of vitamin E on encapsulation efficiency was investigated. In the study, nanoparticules having spherical particles between 50 and 100 nm were successfully produced by homogenizer at the conditions of 16000 rpm speed and vitamin E concentrations ranging from 2 to 10% [40].

14 Colanic Acid

Colanic acid is an important exopolysachharide in terms of the survival of *E. coli* for outside the host. Fundamentally, lipopolysaccharides produced by Gram-negative bacteria includes O antigen which is a polysaccharide component. Antigens including total of 173 O, 80 capsular, and K antigens are determined in *Escherichia coli*. Among these antigens, colanic acid, in other words M antigen, is widely found exopolysaccharide produced by *E. coli* and also other species of Enterobacteriaceae family [160]. Colanic acid is found as a major component in sugar composition of EPSs obtained from isolated bacteria including *Enterobacter* sp., *Klebsiella* sp., *Enterobacter amnigenus, Citrobacter* sp., *Enterobacter cloacae* [138]. Different factors based on genetic and environmental conditions such as membrane integrity, cell envelope stress, osmotic shock, metabolic stress, growth pH, or temperature are determined to activate the synthesis of colanic acid in cells [175].

Colanic acid is important forms of EPSs located on the cell surface of *Escherichia coli*, *Shigella* spp., *Salmonella* spp. and *Enterobacter* spp. In the literature, any information has not been found in the applications in food products. Applications of colanic acid are mentioned in cosmetics and personal care products with its gelling property [132].

15 Welan

Welan is an exopolysaccharide produced by *Sphingomonas* sp. which is also known as *Alcaligenes* sp. and gram-negative bacteria. The production of welan is carried out by submerged fermentation from *Alcaligenes* species and using a medium including glucose, ammonium nitrate, di-potassium hydrogen phosphate, magnesium sulphate, and ferrous ions. The yield of welan depends on the factors such as source of carbon, source of nitrogen, temperature, agitation speed, pH, and others [83]. Welan and gelan have similar structures with same repeating unit. However, welan has a single glycosyl side-chain substituent that can be an a-L-rhamnopyranosyl or an a-L-mannopyranosyl unit linked (1 → 3) to the 4-O-substituted β-D-glucopyranosyl unit in the main structure [81].

Welan is well known with its high viscosity even at very low concentrations and its high viscosity retained stable at high temperatures for long time. For instance, compared to xanthan which is another thermostable polysaccharide, while viscosity of xanthan solutions disappear at 135 °C, viscosity of welan solutions maintained up to 163 °C [81]. Rheological properties of welan solutions were explained in fresh water, sea water, and 3% KC1 solutions. The rheological properties obtained by welan ensure higher penetration rates, lower formation damage, and good suspensions as an advantage. It has also an important role to prevent phase separation in cementitious products. Welan gum has been used as thickening, binding, or emulsifying agent in food products such as jellies, beverages like citric acid based drinks, dairy products like ice creams, yogurt, and salad dressings [83]. However, in the literature, studies focused on the rheological properties of welan gum solutions and applications in food products were limited. Effect of welan gum addition on viscoelastic properties, flow behavior, droplet size distribution, and physical stability of thyme oil in water emulsions was investigated by Martin-Piñero et al. [111]. While emulsion without welan gum indicated Newtonian behavior, emulsions including welan gum showed a weak gel-like behavior with greater viscosity. It was pointed out that welan gum played an important role as a rheological modifier for thyme oil in water emulsions. Thus, it is a natural polysaccharide which can control the rheological properties of these kinds of emulsions by Martin-Piñero et al. [111] (Table 1).

16 Conclusion

The microbial polysaccharides can open up new opportunities in the food industry that are environmentally friendly, sustainable, and multifunctional. Microbial EPS are used in the food industry due to their unique characteristics such as emulsifier, stabilizer, thickener, texturizer, encapsulating and carrying of bioactive agents, and coating and film-forming ability. Microbial EPSs have the potential to be utilized as food additives and functional food ingredients. Microbial EPSs structural complexity, low production yield, and production cost are the most critical challenges for microbial polysaccharides utilization in the near future; however, agro-industrial wastes could be considered as an alternative for reducing the microbial polysaccharides production cost. The new microbial EPSs engineered structures with multifunctional properties have the potential to broaden the food applications. Considering the utilization of microbial EPSs as the introduction of green technology in the food industry offers great potential for clean label products and promotes health benefits.

References

1. Abdel-Fattah AM, Gamal-Eldeen AM, Helmy WA, Esawy MA (2012) Antitumor and antioxidant activities of levan and its derivative from the isolate Bacillus subtilis NRC1aza. Carbohydr Polym 89:314–322. https://doi.org/10.1016/j.carbpol.2012.02.041
2. Adepu S, Khandelwal M (2018) Broad-spectrum antimicrobial activity of bacterial cellulose silver nanocomposites with sustained release. J Mater Sci 53:1596–1609. https://doi.org/10.1007/s10853-017-1638-9
3. Arserim-Uçar DK, Korel F, Liu LS, Yam KL (2021) Characterization of bacterial cellulose nanocrystals: effect of acid treatments and neutralization. Food Chem 336:127597. https://doi.org/10.1016/j.foodchem.2020.127597
4. Arserim-Uçar DK (2020) Nanocontainers for food safety. In: Smart Nanocontainers, pp 105–117. Elsevier. https://doi.org/10.1016/B978-0-12-816770-0.00007-1
5. Arserim-Uçar DK, Çabuk B (2020) Emerging antibacterial and antifungal applications of nanomaterials on food products. In: Nanotoxicity, pp 415–453. Elsevier. https://doi.org/10.1016/B978-0-12-819943-5.00027-0
6. Arvidson SA, Rinehart BT, Gadala-Maria F (2006) Concentration regimes of solutions of levan polysaccharide from Bacillus sp. Carbohydr Polym 65:144–149. https://doi.org/10.1016/j.carbpol.2005.12.039
7. Azarakhsh N, Osman A, Ghazali HM et al (2014) Lemongrass essential oil incorporated into alginate-based edible coating for shelf-life extension and quality retention of fresh-cut pineapple. Postharvest Biol Technol 88:1–7. https://doi.org/10.1016/j.postharvbio.2013.09.004
8. Azeredo HMC, Barud H, Farinas CS et al (2019) Bacterial cellulose as a raw material for food and food packaging applications. Front Sustain Food Syst 3:7. https://doi.org/10.3389/fsufs.2019.00007
9. Azeredo HMC, Rosa MF, Mattoso LHC (2017) Nanocellulose in bio-based food packaging applications. Ind Crops Prod 97:664–671. https://doi.org/10.1016/j.indcrop.2016.03.013

10. BahramParvar M, Tehrani MM, Razavi SMA (2013) Effects of a novel stabilizer blend and presence of κ-carrageenan on some properties of vanilla ice cream during storage. Food Biosci 3:10–18. https://doi.org/10.1016/j.fbio.2013.05.001
11. Bandyopadhyay S, Saha N, Brodnjak UV, Sáha P (2019) Bacterial cellulose and guar gum based modified PVP-CMC hydrogel films: characterized for packaging fresh berries. Food Packag Shelf Life 22:100402. https://doi.org/10.1016/j.fpsl.2019
12. Barcelos MCS, Vespermann KAC, Pelissari FM, Molina G (2020) Current status of biotechnological production and applications of microbial exopolysaccharides. Crit Rev Food Sci Nutr 60:1475–1495. https://doi.org/10.1080/10408398.2019.1575791
13. Baruah R, Maina NH, Katina K et al (2017) Functional food applications of dextran from Weissella cibaria RBA12 from pummelo (Citrus maxima). Int J Food Microbiol 242:124–131. https://doi.org/10.1016/j.ijfoodmicro.2016.11.012
14. Bejar W, Gabriel V, Amari M et al (2013) Characterization of glucansucrase and dextran from Weissella sp. TN610 with potential as safe food additives. Int J Biol Macromol 52:125–132. https://doi.org/10.1016/j.ijbiomac.2012.09.014
15. Boonlao N, Shrestha S, Sadiq MB, Anal AK (2020) Influence of whey protein-xanthan gum stabilized emulsion on stability and in vitro digestibility of encapsulated astaxanthin. J Food Eng 272:109859. https://doi.org/10.1016/j.jfoodeng.2019.109859
16. Boyd A, Chakrabarty AM (1995) Pseudomonas aeruginosa biofilms: role of the alginate exopolysaccharide. J Ind Microbiol 15:162–168. https://doi.org/10.1007/BF01569821
17. Brodnjak UV (2017) Experimental investigation of novel curdlan/chitosan coatings on packaging paper. Prog Org Coatings 112:86–92. https://doi.org/10.1016/j.porgcoat.2017.06.030
18. Bylaite E, Ilgunaité Ž, Meyer AS, Adler-Nissen J (2004) Influence of λ-carrageenan on the release of systematic series of volatile flavor compounds, from viscous food model systems. J Agric Food Chem 52:3542–3549. https://doi.org/10.1021/jf0354996
19. Cao L, Lu W, Mata A et al (2020) Egg-box model-based gelation of alginate and pectin: a review. Carbohydr Polym 242:116389. https://doi.org/10.1016/j.carbpol.2020.116389
20. Castro GR, Chen J, Panilaitis B, Kaplan DL (2009) Emulsan-alginate beads for protein adsorption. J Biomater Sci Polym Ed 20:411–426. https://doi.org/10.1163/156856209X416449
21. Catley BJ (1970) Pullulan, a relationship between molecular weight and fine structure. FEBS Lett 10:190–193. https://doi.org/10.1016/0014-5793(58)80450-1
22. Chen L, Tian Y, Zhang Z et al (2017) Effect of pullulan on the digestible, crystalline and morphological characteristics of rice starch. Food Hydrocoll 63:383–390. https://doi.org/10.1016/j.foodhyd.2016.09.021
23. Chen L, Tong Q, Ren F, Zhu G (2014) Pasting and rheological properties of rice starch as affected by pullulan. Int J Biol Macromol 66:325–331. https://doi.org/10.1016/j.ijbiomac.2014.02.052
24. Chen L, Zhang H, McClements DJ et al (2019) Effect of dietary fibers on the structure and digestibility of fried potato starch: a comparison of pullulan and pectin. Carbohydr Polym 215:47–57. https://doi.org/10.1016/j.carbpol.2019.03.046
25. Chen Y, Wang F (2020) Review on the preparation, biological activities and applications of curdlan and its derivatives. Eur Polym J 110096. https://doi.org/10.1016/j.eurpolymj.2020.110096
26. Chevalier Y, Bolzinger MA (2013) Emulsions stabilized with solid nanoparticles: pickering emulsions. Colloids Surfaces A Physicochem Eng Asp 439:23–34. https://doi.org/10.1016/j.colsurfa.2013.02.054
27. Cho HM, Yoo B (2015) Rheological characteristics of cold thickened beverages containing xanthan gum e based food thickeners used for dysphagia diets. J Acad Nutr Diet 115:106–111. https://doi.org/10.1016/j.jand.2014.08.028
28. Choi SM, Shin EJ (2020) The nanofication and functionalization of bacterial cellulose and its applications. Nanomaterials 10:406. https://doi.org/10.3390/nano10030406

29. Chu Y, Gao CC, Liu X et al (2020) Improvement of storage quality of strawberries by pullulan coatings incorporated with cinnamon essential oil nanoemulsion. LWT Food Sci Technol 122:109054. https://doi.org/10.1016/j.lwt.2020.109054
30. Chuysinuan P, Thanyacharoen T, Thongchai K et al (2020) Preparation of chitosan/hydrolyzed collagen/hyaluronic acid based hydrogel composite with caffeic acid addition. Int J Biol Macromol 162:1937–1943. https://doi.org/10.1016/j.ijbiomac.2020.08.139
31. Cofelice M, Lopez F, Cuomo F (2019) Quality control of fresh-cut apples after coating application. Foods 8:189. https://doi.org/10.3390/foods8060189
32. Costa AFS, Almeida FCG, Vinhas GM, Sarubbo LA (2017) Production of bacterial cellulose by Gluconacetobacter hansenii using corn steep liquor as nutrient sources. Front Microbiol 8:2027. https://doi.org/10.3389/fmicb.2017.02027
33. Castro C, Zuluaga R, Putaux JL et al (2011). Structural characterization of bacterial cellulose produced by Gluconacetobacter swingsii sp. from Colombian agroindustrial wastes. Carbohydr Polyme 84:96–102. https://doi.org/10.1016/j.carbpol.2010.10.072
34. Dabestani M, Yeganehzad S (2019) Effect of Persian gum and Xanthan gum on foaming properties and stability of pasteurized fresh egg white foam. Food Hydrocoll 87:550–560. https://doi.org/10.1016/j.foodhyd.2018.08.030
35. Dadashi S, Boddohi S, Soleimani N (2019) Preparation, characterization, and antibacterial effect of doxycycline loaded kefiran nanofibers. J Drug Deliv Sci Technol 52:979–985. https://doi.org/10.1016/j.jddst.2019.06.012
36. Dahech I, Belghith KS, Hamden K et al (2011) Antidiabetic activity of levan polysaccharide in alloxan-induced diabetic rats. Int J Biol Macromol 49:742–746. https://doi.org/10.1016/j.ijbiomac.2011.07.007
37. Danalache F, Beirão-da-Costa S, Mata P et al (2015) Texture, microstructure and consumer preference of mango bars jellified with gellan gum. LWT Food Sci Technol 62:584–591. https://doi.org/10.1016/j.lwt.2014.12.040
38. Das S, Vishakha K, Banerjee S et al (2020) Sodium alginate-based edible coating containing nanoemulsion of Citrus sinensis essential oil eradicates planktonic and sessile cells of food-borne pathogens and increased quality attributes of tomatoes. Int J Biol Macromol 162:1770–1779. https://doi.org/10.1016/j.ijbiomac.2020.08.086
39. Davidović S, Miljković M, Tomić M et al (2018) Response surface methodology for optimisation of edible coatings based on dextran from Leuconostoc mesenteroides T3. Carbohydr Polym 184:207–213. https://doi.org/10.1016/j.carbpol.2017.12.061
40. de Siqueira EC, de Rebouças JS, Pinheiro IO, Formiga FR (2020) Levan-based nanostructured systems: an overview. Int J Pharm 580:119242. https://doi.org/10.1016/j.ijpharm.2020
41. Demirci AS, Palabiyik I, Apaydın D et al (2019) Xanthan gum biosynthesis using Xanthomonas isolates from waste bread: process optimization and fermentation kinetics. LWT Food Sci Technol 101:40–47. https://doi.org/10.1016/j.lwt.2018.11.018
42. Dobosz A, Sikora M, Krystyjan M et al (2020) Influence of xanthan gum on the short- and long-term retrogradation of potato starches of various amylose content. Food Hydrocoll 102:105618. https://doi.org/10.1016/j.foodhyd.2019
43. Donot F, Fontana A, Baccou JC, Schorr-Galindo S (2012) Microbial exopolysaccharides: main examples of synthesis, excretion, genetics and extraction. Carbohydr Polym 87:951–962. https://doi.org/10.1016/j.carbpol.2011.08.083
44. Du R, Qiao X, Zhao F et al (2018) Purification, characterization and antioxidant activity of dextran produced by Leuconostoc pseudomesenteroides from homemade wine. Carbohydr Polym 198:529–536. https://doi.org/10.1016/j.carbpol.2018.06.116
45. Du R, Xing H, Yang Y et al (2017) Optimization, purification and structural characterization of a dextran produced by *L. mesenteroides* isolated from Chinese sauerkraut. Carbohydr Polym 174:409–416. https://doi.org/10.1016/j.carbpol.2017.06.084
46. Eckel VPL, Vogel RF, Jakob F (2019) In situ production and characterization of cloud forming dextrans in fruit-juices. Int J Food Microbiol 306:108261. https://doi.org/10.1016/j.ijfoodmicro.2019.108261

47. Encina-Zelada CR, Cadavez V, Monteiro F et al (2018) Combined effect of xanthan gum and water content on physicochemical and textural properties of gluten-free batter and bread. Food Res Int 111:544–555. https://doi.org/10.1016/j.foodres.2018.05.070
48. Espert M, Constantinescu L, Sanz T, Salvador A (2019) Food Hydrocolloids Effect of xanthan gum on palm oil in vitro digestion. Application in starch-based fi lling creams. Food Hydrocoll 86:87–94. https://doi.org/10.1016/j.foodhyd.2018.02.017
49. Fabra MJ, López-Rubio A, Ambrosio-Martín J, Lagaron JM (2016) Improving the barrier properties of thermoplastic corn starch-based films containing bacterial cellulose nanowhiskers by means of PHA electrospun coatings of interest in food packaging. Food Hydrocoll 61:261–268. https://doi.org/10.1016/j.foodhyd.2016.05.025
50. Farias TGS de, Ladislau HFL, Stamford TCM et al (2019) Viabilities of Lactobacillus rhamnosus ASCC 290 and Lactobacillus casei ATCC 334 (in free form or encapsulated with calcium alginate-chitosan) in yellow mombin ice cream. LWT Food Sci Technol 100:391–396. https://doi.org/10.1016/j.lwt.2018.10.084
51. Fels L, Jakob F, Vogel RF, Wefers D (2018) Structural characterization of the exopolysaccharides from water kefir. Carbohydr Polym 189:296–303. https://doi.org/10.1016/j.carbpol.2018.02.037
52. Fernando IPS, Lee WW, Han EJ, Ahn G (2020) Alginate-based nanomaterials: fabrication techniques, properties, and applications. Chem Eng J 391:123823. https://doi.org/10.1016/j.cej.2019.123823
53. Foerster M, Liu C, Gengenbach T et al (2017) Reduction of surface fat formation on spray-dried milk powders through emulsion stabilization with λ-carrageenan. Food Hydrocoll 70:163–180. https://doi.org/10.1016/j.foodhyd.2017.04.005
54. Frank K, Garcia CV, Shin GH, Kim JT (2018) Alginate biocomposite films incorporated with cinnamon essential oil nanoemulsions: physical, mechanical, and antibacterial properties. Int J Polym Sci 2018:1519407. https://doi.org/10.1155/2018/1519407
55. Freitas F, Alves VD, Reis MAM (2011) Advances in bacterial exopolysaccharides: from production to biotechnological applications. Trends Biotechnol 29:388–398. https://doi.org/10.1016/j.tibtech.2011.03.008
56. Funami T, Yada H, Nakao Y (1998) Curdlan properties for application in fat mimetics for meat products. J Food Sci 63:283–287. https://doi.org/10.1111/j.1365-2621.1998.tb15727.x
57. Gao G, Fan H, Zhang Y et al (2021) Production of nisin-containing bacterial cellulose nanomaterials with antimicrobial properties through co-culturing Enterobacter sp. FY-07 and Lactococcus lactis N8. Carbohydr Polym 251:117131. https://doi.org/10.1016/j.carbpol.2020
58. Gao H, Xie F, Zhang W et al (2020) Characterization and improvement of curdlan produced by a high-yield mutant of Agrobacterium sp. ATCC 31749 based on whole-genome analysis. Carbohydr Polym 245:116486. https://doi.org/10.1016/j.carbpol.2020
59. George J, Kumar R, Sajeevkumar VA et al (2014) Hybrid HPMC nanocomposites containing bacterial cellulose nanocrystals and silver nanoparticles. Carbohydr Polym 105:285–292. https://doi.org/10.1016/j.carbpol.2014.01.057
60. George J, Ramana KV, Bawa AS, Siddaramaiah (2011) Bacterial cellulose nanocrystals exhibiting high thermal stability and their polymer nanocomposites. Int J Biol Macromol 48:50–57. https://doi.org/10.1016/j.ijbiomac.2010.09.013
61. Ghasemlou M, Khodaiyan F, Oromiehie A (2011) Physical, mechanical, barrier, and thermal properties of polyol-plasticized biodegradable edible film made from kefiran. Carbohydr Polym 84:477–483. https://doi.org/10.1016/j.carbpol.2010.12.010
62. Guerreiro AC, Gago CML, Faleiro ML et al (2015) The effect of alginate-based edible coatings enriched with essential oils constituents on Arbutus unedo L. fresh fruit storage. Postharvest Biol Technol 100:226–233. https://doi.org/10.1016/j.postharvbio.2014.09.002
63. Guillard V, Issoupov V, Redl A, Gontard N (2009) Food preservative content reduction by controlling sorbic acid release from a superficial coating. Innov Food Sci Emerg Technol 10:108–115. https://doi.org/10.1016/j.ifset.2008.07.001

64. Günter EA, Martynov VV, Belozerov VS et al (2020) Characterization and swelling properties of composite gel microparticles based on the pectin and κ-carrageenan. Int J Biol Macromol 164:2232–2239. https://doi.org/10.1016/j.ijbiomac.2020.08.024
65. Guo C, Yin J, Chen D (2018) Co-encapsulation of curcumin and resveratrol into novel nutraceutical hyalurosomes nano-food delivery system based on oligo-hyaluronic acid-curcumin polymer. Carbohydr Polym 181:1033–1037. https://doi.org/10.1016/j.carbpol.2017.11.046
66. Halib N, Amin MCIM, Ahmad I (2012) Physicochemical properties and characterization of nata de coco from local food industries as a source of cellulose. Sains Malaysiana 41:205–211
67. Hassan AHA, Cutter CN (2020) Development and evaluation of pullulan-based composite antimicrobial films (CAF) incorporated with nisin, thymol and lauric arginate to reduce foodborne pathogens associated with muscle foods. Int J Food Microbiol 320:108519. https://doi.org/10.1016/j.ijfoodmicro.2020.108519
68. Hay ID, Rehman ZU, Moradali MF et al (2013) Microbial alginate production, modification and its applications. Microb Biotechnol 6:637–650. https://doi.org/10.1111/1751-7915.12076
69. Heperkan ZD, Bolluk M, Bülbül S (2020) Structural analysis and properties of dextran produced by Weissella confusa and the effect of different cereals on its rheological characteristics. Int J Biol Macromol 143:305–313. https://doi.org/10.1016/j.ijbiomac.2019.12.036
70. Hestrin S, Schramm M (1954) Synthesis of cellulose by Acetobacter xylinum. II. Preparation of freeze-dried cells capable of polymerizing glucose to cellulose. Biochem J 58:345–352. https://doi.org/10.1042/bj0580345
71. Hotchkiss S, Brooks M, Campbell R et al (2016) The use of carrageenan in food. In: Leonel P (eds) Carrageenans, pp 229–243. Nova Science Publishers
72. Hussain Z, Sajjad W, Khan T, Wahid F (2019) Production of bacterial cellulose from industrial wastes: a review. Cellulose 26:2895–2911. https://doi.org/10.1007/s10570-019-02307-1
73. Iguchi, Yamanaka S, Budhiono A (2000) Bacterial cellulose—a masterpiece of nature's arts. J Mater Sci 35:261–270. https://doi.org/10.1023/A
74. Izawa N, Hanamizu T, Iizuka R et al (2009) Streptococcus thermophilus produces exopolysaccharides including hyaluronic acid. J Biosci Bioeng 107:119–123. https://doi.org/10.1016/j.jbiosc.2008.11.007
75. Jebel FS, Almasi H (2016) Morphological, physical, antimicrobial and release properties of ZnO nanoparticles-loaded bacterial cellulose films. Carbohydr Polym 149:8–19. https://doi.org/10.1016/j.carbpol.2016.04.089
76. Jiang T, Feng L, Wang Y (2013) Effect of alginate/nano-Ag coating on microbial and physicochemical characteristics of shiitake mushroom (Lentinus edodes) during cold storage. Food Chem 141:954–960. https://doi.org/10.1016/j.foodchem.2013.03.093
77. Jipa IM, Dobre L, Stroescu M et al (2012) Preparation and characterization of bacterial cellulose-poly(vinyl alcohol) films with antimicrobial properties. Mater Lett 66:125–127. https://doi.org/10.1016/j.matlet.2011.08.047
78. Johri AK, Yalpani M, Kaplan DL (2003) Incorporation of fluorinated fatty acids into emulsan by Acinetobacter calcoaceticus RAG-1. Biochem Eng J 16:175–181. https://doi.org/10.1016/S1369-703X(03)00034-2
79. Kalashnikova I, Bizot H, Bertoncini P et al (2013) Cellulosic nanorods of various aspect ratios for oil in water Pickering emulsions. Soft Matter 9:952–959. https://doi.org/10.1039/c2sm26472b
80. Kalashnikova I, Bizot H, Cathala B, Capron I (2012) Modulation of cellulose nanocrystals amphiphilic properties to stabilize oil/water interface. Biomacromol 13:267–275. https://doi.org/10.1021/bm201599j
81. Kang KS, Pettitt DJ (1993) Xanthan, Gellan, Welan, and Rhamsan. Industrial gums, pp 341–397, 3rd edn. Academic Press

82. Kang ZL, Wang TT, Li YP et al (2020) Effect of sodium alginate on physical-chemical, protein conformation and sensory of low-fat frankfurters. Meat Sci 162:108043. https://doi.org/10.1016/j.meatsci.2019
83. Kaur V, Bera MB, Panesar PS et al (2014) Welan gum: microbial production, characterization, and applications. Int J Biol Macromol 65:454–461. https://doi.org/10.1016/j.ijbiomac.2014.01.061
84. Khanal BKS, Bhandari B, Prakash S et al (2018) Modifying textural and microstructural properties of low fat Cheddar cheese using sodium alginate. Food Hydrocoll 83:97–108. https://doi.org/10.1016/j.foodhyd.2018.03.015
85. Kiani H, Mousavi ME, Razavi H, Morris ER (2010) Effect of gellan, alone and in combination with high-methoxy pectin, on the structure and stability of doogh, a yogurt-based Iranian drink. Food Hydrocoll 24:744–754. https://doi.org/10.1016/j.foodhyd.2010.03.016
86. Kim WS, Han GG, Hong L et al (2019) Novel production of natural bacteriocin via internalization of dextran nanoparticles into probiotics. Biomaterials 218:119360. https://doi.org/10.1016/j.biomaterials.2019.119360
87. Konuk Takma D, Korel F (2017) Impact of preharvest and postharvest alginate treatments enriched with vanillin on postharvest decay, biochemical properties, quality and sensory attributes of table grapes. Food Chem 221:187–195. https://doi.org/10.1016/j.foodchem.2016.09.195
88. Kothari D, Tingirikari JMR, Goyal A (2015) In vitro analysis of dextran from Leuconostoc mesenteroides NRRL B-1426 for functional food application. Bioact Carbohydrates Diet Fibre 6:55–61. https://doi.org/10.1016/j.bcdf.2015.08.001
89. Krasńiewska K, Gniewosz M, Synowiec A et al (2015) The application of pullulan coating enriched with extracts from Bergenia crassifolia to control the growth of food microorganisms and improve the quality of peppers and apples. Food Bioprod Process 94:422–433. https://doi.org/10.1016/j.fbp.2014.06.001
90. Kurt A, Toker OS, Tornuk F (2017) Effect of xanthan and locust bean gum synergistic interaction on characteristics of biodegradable edible film. Int J Biol Macromol 102:1035–1044. https://doi.org/10.1016/j.ijbiomac.2017.04.081
91. Kumar N, Neeraj P, Ojha A et al (2020) Effect of active chitosan-pullulan composite edible coating enrich with pomegranate peel extract on the storage quality of green Bell pepper. LWT Food Sci Technol 138:110435. https://doi.org/10.1016/j.lwt.2020.110435
92. Kuswandi B, Asih NPN, Pratoko DK et al (2020) Edible pH sensor based on immobilized red cabbage anthocyanins into bacterial cellulose membrane for intelligent food packaging. Packag Technol Sci 33:321–332. https://doi.org/10.1002/pts.2507
93. Kwiecień I, Kwiecień M (2018) Application of polysaccharide-based hydrogels as probiotic delivery systems. Gels 4:47. https://doi.org/10.3390/gels4020047
94. Kycia K, Chlebowska-Śmigiel A, Szydłowska A et al (2020) Pullulan as a potential enhancer of Lactobacillus and Bifidobacterium viability in synbiotic low fat yoghurt and its sensory quality. LWT Food Sci Technol 128:109414. https://doi.org/10.1016/j.lwt.2020
95. la Riviére JWM, Kooiman P, Schmidt K (1967) Kefiran, a novel polysaccharide produced in the kefir grain by Lactobacillus brevis. Arch Mikrobiol 59:269–278. https://doi.org/10.1007/BF00406340
96. Langendorff V, Cuvelier G, Michon C et al (2000) Effects of carrageenan type on the behaviour of carrageenan/milk mixtures. Food Hydrocoll 14:273–280. https://doi.org/10.1016/S0268-005X(99)00064-8
97. Lara G, Yakoubi S, Mae C et al (2020) Spray technology applications of xanthan gum-based edible coatings for fresh-cut lotus root (Nelumbo nucifera). Food Res Int 137:109723. https://doi.org/10.1016/j.foodres.2020.109723
98. Lee H, Yoo B (2020) Agglomerated xanthan gum powder used as a food thickener: effect of sugar binders on physical, microstructural, and rheological properties. Powder Technol 362:301–306. https://doi.org/10.1016/j.powtec.2019.11.124

99. Lee KY, Mooney DJ (2012) Alginate: properties and biomedical applications. Prog Polym Sci 37:106–126. https://doi.org/10.1016/j.progpolymsci.2011.06.003
100. León PG, Rojas AM (2007) Gellan gum films as carriers of l-(+)-ascorbic acid. Food Res Int 40:565–575. https://doi.org/10.1016/j.foodres.2006.10.021
101. Li A, Gong T, Hou Y et al (2020) Alginate-stabilized thixotropic emulsion gels and their applications in fabrication of low-fat mayonnaise alternatives. Int J Biol Macromol 146:821–831. https://doi.org/10.1016/j.ijbiomac.2019.10.050
102. Li Q, Gao R, Wang L et al (2020) Nanocomposites of bacterial cellulose nanofibrils and zein nanoparticles for food packaging. ACS Appl Nano Mater 3:2899–2910. https://doi.org/10.1021/acsanm.0c00159
103. Li XY, Du XL, Liu Y et al (2019) Rhubarb extract incorporated into an alginate-based edible coating for peach preservation. Sci Hortic (Amsterdam) 257:108685. https://doi.org/10.1016/j.scienta.2019.108685
104. Li Y, Feng C, Li J et al (2017) Construction of multilayer alginate hydrogel beads for oral delivery of probiotics cells. Int J Biol Macromol 105:924–930. https://doi.org/10.1016/j.ijbiomac.2017.07.124
105. Li Z, Wang L, Hua J et al (2015) Production of nano bacterial cellulose from waste water of candied jujube-processing industry using Acetobacter xylinum. Carbohydr Polym 120:115–119. https://doi.org/10.1016/j.carbpol.2014.11.061
106. Liang Y, Qu Z, Liu M et al (2020) Effect of curdlan on the quality of frozen-cooked noodles during frozen storage. J Cereal Sci 95:103019. https://doi.org/10.1016/j.jcs.2020.103019
107. Lin D, Liu Z, Shen R et al (2020) Bacterial cellulose in food industry: current research and future prospects. Int J Biol Macromol 158:1007–1019. https://doi.org/10.1016/j.ijbiomac.2020.04.230
108. Lin D, Lopez-Sanchez P, Li R, Li Z (2014) Production of bacterial cellulose by Gluconacetobacter hansenii CGMCC 3917 using only waste beer yeast as nutrient source. Bioresour Technol 151:113–119. https://doi.org/10.1016/j.biortech.2013.10.052
109. Lin KW, Lin HY (2004) Quality characteristics of chinese-style meatball containing bacterial cellulose (nata). J Food Sci 69:SNQ107–SNQ111. https://doi.org/10.1111/j.1365-2621.2004.tb13378.x
110. Lorenzo G, Zaritzky N, Califano A (2013) Rheological analysis of emulsion-filled gels based on high acyl gellan gum. Food Hydrocoll 30:672–680. https://doi.org/10.1016/j.foodhyd.2012.08.014
111. Martin-Piñero MJ, García MC, Muñoz J, Alfaro-Rodriguez MC (2019) Influence of the welan gum biopolymer concentration on the rheological properties, droplet size distribution and physical stability of thyme oil/W emulsions. Int J Biol Macromol 133:270–277. https://doi.org/10.1016/j.ijbiomac.2019.04.137
112. Martin AA, Sassaki GL, Sierakowski MR (2020) Effect of adding galactomannans on some physical and chemical properties of hyaluronic acid. Int J Biol Macromol 144:527–535. https://doi.org/10.1016/j.ijbiomac.2019.12.114
113. McIntosh M, Stone BA, Stanisich VA (2005) Curdlan and other bacterial $(1 \rightarrow 3)$-β-D-glucans. Appl Microbiol Biotechnol 68:163–173. https://doi.org/10.1007/s00253-005-1959-5
114. Milde LB, Chigal PS, Olivera JE, González KG (2020) Incorporation of xanthan gum to gluten-free pasta with cassava starch. Physical, textural and sensory attributes. LWT Food Sci Technol 131:109674. https://doi.org/10.1016/j.lwt.2020.109674
115. Mohammadalinejhad S, Almasi H, Moradi M (2020) Immobilization of Echium amoenum anthocyanins into bacterial cellulose film: a novel colorimetric pH indicator for freshness/spoilage monitoring of shrimp. Food Control 113:107169. https://doi.org/10.1016/j.foodcont.2020.107169
116. Mohammadi M, Sadeghnia N, Azizi MH et al (2014) Development of gluten-free flat bread using hydrocolloids: Xanthan and CMC. J Ind Eng Chem 20:1812–1818. https://doi.org/10.1016/j.jiec.2013.08.035

117. Moradi M, Guimarães JT, Sahin S (2021) Current applications of exopolysaccharides from lactic acid bacteria in the development of food active edible packaging. Curr Opin Food Sci 40:33–39. https://doi.org/10.1016/j.cofs.2020.06.001
118. Moradi Z, Kalanpour N (2019) Kefiran, a branched polysaccharide: preparation, properties and applications: a review. Carbohydr Polym 223:115100. https://doi.org/10.1016/j.carbpol.2019.115100
119. More TT, Yadav JSS, Yan S et al (2014) Extracellular polymeric substances of bacteria and their potential environmental applications. J Environ Manage 144:1–25. https://doi.org/10.1016/j.jenvman.2014.05.010
120. Morris ER, Nishinari K, Rinaudo M (2012) Gelation of gellan—a review. Food Hydrocoll 28:373–411. https://doi.org/10.1016/j.foodhyd.2012.01.004
121. Moscovici M (2015) Present and future medical applications of microbial exopolysaccharides. Front Microbiol 6:1012. https://doi.org/10.3389/fmicb.2015.01012
122. Muhammad DRA, Sedaghat Doost A, Gupta V et al (2020) Stability and functionality of xanthan gum–shellac nanoparticles for the encapsulation of cinnamon bark extract. Food Hydrocoll 100:105377. https://doi.org/10.1016/j.foodhyd.2019.105377
123. Naloka K, Matsushita K, Theeragool G (2020) Enhanced ultrafine nanofibril biosynthesis of bacterial nanocellulose using a low-cost material by the adapted strain of Komagataeibacter xylinus MSKU 12. Int J Biol Macromol 150:1113–1120. https://doi.org/10.1016/j.ijbiomac.2019.10.117
124. Necas J, Bartosikova L (2013) Carrageenan: a review. Vet Med (Praha) 58:187–205. https://doi.org/10.17221/6758-VETMED
125. Necas J, Bartosikova L, Brauner P, Kolar J (2008) Hyaluronic acid (hyaluronan): a review. Vet Med (Praha) 53:397–411. https://doi.org/10.17221/1930
126. Nguyen VT, Gidley MJ, Dykes GA (2008) Potential of a nisin-containing bacterial cellulose film to inhibit Listeria monocytogenes on processed meats. Food Microbiol 25:471–478. https://doi.org/10.1016/j.fm.2008.01.004
127. Ni Y, Tang X, Fan L (2021) Improvement in physical and thermal stability of cloudy ginkgo beverage during autoclave sterilization: Effects of microcrystalline cellulose and gellan gum. LWT Food Sci Technol 135:110062. https://doi.org/10.1016/j.lwt.2020.110062
128. NithyaBalaSundari S, Nivedita V, Chakravarthy M et al (2020) Characterization of microbial polysaccharides and prebiotic enrichment of wheat bread with pullulan. LWT Food Sci Technol 122:109002. https://doi.org/10.1016/j.lwt.2019.109002
129. Nitschke M, Costa SGVAO (2007) Biosurfactants in food industry. Trends Food Sci Technol 18:252–259. https://doi.org/10.1016/j.tifs.2007.01.002
130. Noorlaila A, Hasanah HN, Asmeda R, Yusoff A (2020) The effects of xanthan gum and hydroxypropylmethylcellulose on physical properties of sponge cakes. J Saudi Soc Agric Sci 19:128–135. https://doi.org/10.1016/j.jssas.2018.08.001
131. Noreña CPZ, Bayarri S, Costell E (2015) Effects of Xanthan gum additions on the viscoelasticity, structure and storage stability characteristics of prebiotic custard desserts. Food Biophys 10:116–128. https://doi.org/10.1007/s11483-014-9371-2
132. Nwodo UU, Green E, Okoh AI (2012) Bacterial exopolysaccharides: functionality and prospects. Int J Mol Sci 13:14002–14015. https://doi.org/10.3390/ijms131114002
133. Paximada P, Tsouko E, Kopsahelis N et al (2016) Bacterial cellulose as stabilizer of o/w emulsions. Food Hydrocoll 53:225–232. https://doi.org/10.1016/j.foodhyd.2014.12.003
134. Phisalaphong M, Tran TK, Taokaew S et al (2016) Nata de coco Industry in Vietnam, Thailand, and Indonesia. In: Bacterial nanocellulose, pp 231–236. Elsevier. https://doi.org/10.1016/B978-0-444-63458-0.00014-7
135. Piadozo MES (2016) Nata de Coco Industry in the Philippines. In: Bacterial nanocellulose, pp 215–229. Elsevier. https://doi.org/10.1016/B978-0-444-63458-0.00013-5
136. Pitarresi G, Palumbo FS, Calabrese R et al (2007) Crosslinked hyaluronan with a protein-like polymer: novel bioresorbable films for biomedical applications. J Biomed Mater Res, Part A 84:413–424. https://doi.org/10.1002/jbm.a

137. Qin Y, Jiang J, Zhao L et al (2018) Applications of alginate as a functional food ingredient. Biopolymers for food design, pp 409–429. Academic Press. Elsevier
138. Rättö M, Verhoef R, Suihko ML et al (2006) Colanic acid is an exopolysaccharide common to many enterobacteria isolated from paper-machine slimes. J Ind Microbiol Biotechnol 33:359–367. https://doi.org/10.1007/s10295-005-0064-1
139. Raungrusmee S, Shrestha S, Sadiq MB, Anal AK (2020) Influence of resistant starch, xanthan gum, inulin and defatted rice bran on the physicochemical, functional and sensory properties of low glycemic gluten-free noodles. LWT Food Sci Technol 126:109279. https://doi.org/10.1016/j.lwt.2020.109279
140. Razavi MS, Golmohammadi A, Nematollahzadeh A et al (2020) Preparation of cinnamon essential oil emulsion by bacterial cellulose nanocrystals and fish gelatin. Food Hydrocoll 109:106111. https://doi.org/10.1016/j.foodhyd.2020.106111
141. Rehm BHA (2015) Alginates: biology and applications: biology and applications, vol 13. Springer Science & Business Media
142. Revin V, Liyaskina E, Nazarkina M et al (2018) Cost-effective production of bacterial cellulose using acidic food industry by-products. Brazilian J Microbiol 49:151–159. https://doi.org/10.1016/j.bjm.2017.12.012
143. Rhim JW (2004) Physical and mechanical properties of water resistant sodium alginate films. LWT Food Sci Technol 37:323–330. https://doi.org/10.1016/j.lwt.2003.09.008
144. Rovera C, Ghaani M, Santo N et al (2018) Enzymatic hydrolysis in the green production of bacterial cellulose nanocrystals. ACS Sustain Chem Eng 6:7725–7734. https://doi.org/10.1021/acssuschemeng.8b00600
145. Rühmkorf C, Rübsam H, Becker T et al (2012) Effect of structurally different microbial homoexopolysaccharides on the quality of gluten-free bread. Eur Food Res Technol 235:139–146. https://doi.org/10.1007/s00217-012-1746-3
146. Salari M, Sowti Khiabani M, Rezaei Mokarram R et al (2018) Development and evaluation of chitosan based active nanocomposite films containing bacterial cellulose nanocrystals and silver nanoparticles. Food Hydrocoll 84:414–423. https://doi.org/10.1016/j.foodhyd.2018.05.037
147. Santos J, Alcaide-González MA, Trujillo-Cayado LA et al (2020) Development of food-grade Pickering emulsions stabilized by a biological macromolecule (xanthan gum) and zein. Int J Biol Macromol 153:747–754. https://doi.org/10.1016/j.ijbiomac.2020.03.078
148. Schelegueda LI, Zalazar AL, Herbas ET et al (2020) Effect of gellan gum, xylitol and natamycin on Zygosaccharomyces bailii growth and rheological characteristics in low sugar content model systems. Int J Biol Macromol 164:1657–1664. https://doi.org/10.1016/j.ijbiomac.2020.07.277
149. Shahabi-Ghahfarrokhi I, Khodaiyan F, Mousavi M, Yousefi H (2015) Green bionanocomposite based on kefiran and cellulose nanocrystals produced from beer industrial residues. Int J Biol Macromol 77:85–91. https://doi.org/10.1016/j.ijbiomac.2015.02.055
150. Sharma N, Prasad GS, Choudhury AR (2013) Utilization of corn steep liquor for biosynthesis of pullulan, an important exopolysaccharide. Carbohydr Polym 93:95–101. https://doi.org/10.1016/j.carbpol.2012.06.059
151. Sharma S, Rao TVR (2015) Xanthan gum based edible coating enriched with cinnamic acid prevents browning and extends the shelf-life of fresh-cut pears. LWT Food Sci Technol 62:791–800. https://doi.org/10.1016/j.lwt.2014.11.050
152. Shi Z, Zhang Y, Phillips GO, Yang G (2014) Utilization of bacterial cellulose in food. Food Hydrocoll 35:539–545. https://doi.org/10.1016/j.foodhyd.2013.07.012
153. Silva NHCS, Vilela C, Almeida A et al (2018) Pullulan-based nanocomposite films for functional food packaging: exploiting lysozyme nanofibers as antibacterial and antioxidant reinforcing additives. Food Hydrocoll 77:921–930. https://doi.org/10.1016/j.foodhyd.2017.11.039
154. Singh RS, Kaur N, Kennedy JF (2019) Pullulan production from agro-industrial waste and its applications in food industry: a review. Carbohydr Polym 217:46–57. https://doi.org/10.1016/j.carbpol.2019.04.050

155. Singh RS, Saini GK, Kennedy JF (2008) Pullulan: microbial sources, production and applications. Carbohydr Polym 73:515–531. https://doi.org/10.1016/j.carbpol.2008.01.003
156. Singhsa P, Narain R, Manuspiya H (2018) Bacterial cellulose nanocrystals (BCNC) preparation and characterization from three bacterial cellulose sources and development of functionalized BCNCs as nucleic acid delivery systems. ACS Appl Nano Mater 1:209–221. https://doi.org/10.1021/acsanm.7b00105
157. Skryplonek K, Henriques M, Gomes D et al (2019) Characteristics of lactose-free frozen yogurt with κ-carrageenan and corn starch as stabilizers. J Dairy Sci 102:7838–7848. https://doi.org/10.3168/jds.2019-16556
158. Soto KM, Hernández-Iturriaga M, Loarca-Piña G et al (2019) Antimicrobial effect of nisin electrospun amaranth: pullulan nanofibers in apple juice and fresh cheese. Int J Food Microbiol 295:25–32. https://doi.org/10.1016/j.ijfoodmicro.2019.02.001
159. Srikanth R, Reddy CHSSS, Siddartha G et al (2015) Review on production, characterization and applications of microbial levan. Carbohydr Polym 120:102–114. https://doi.org/10.1016/j.carbpol.2014.12.003
160. Stevenson G, Andrianopoulos K, Hobbs M, Reeves PR (1996) Organization of the Escherichia coli K-12 gene cluster responsible for production of the extracellular polysaccharide colanic acid. J Bacteriol 178:4885–4893. https://doi.org/10.1128/jb.178.16.4885-4893.1996
161. Subhash M, Jadhav A, Jana S (2015) Sustainable production of microbial polysaccharide xanthan gum from supplemental substrate. Magar Subhash B Abhijit S Jadhav Sumitkumar Jana. Int J Sci Res 4:9–11
162. Surya E, Fitriani Ridhwan M et al (2020) The utilization of peanut sprout extract as a green nitrogen source for the physicochemical and organoleptic properties of Nata de coco. Biocatal Agric Biotechnol 29:101781. https://doi.org/10.1016/j.bcab.2020.101781
163. Sutherland IW (1998) Novel and established applications of microbial polysaccharides. Trends Biotechnol 16:41–46. https://doi.org/10.1016/S0167-7799(80)01139-6
164. Tang X, Liu R, Huang W et al (2018) Impact of in situ formed exopolysaccharides on dough performance and quality of Chinese steamed bread. LWT Food Sci Technol 96:519–525. https://doi.org/10.1016/j.lwt.2018.04.039
165. Tavassoli-Kafrani E, Shekarchizadeh H, Masoudpour-Behabadi M (2016) Development of edible films and coatings from alginates and carrageenans. Carbohydr Polym 137:360–374. https://doi.org/10.1016/j.carbpol.2015.10.074
166. Totosaus A, Pérez-Chabela ML (2009) Textural properties and microstructure of low-fat and sodium-reduced meat batters formulated with gellan gum and dicationic salts. LWT Food Sci Technol 42:563–569. https://doi.org/10.1016/j.lwt.2008.07.016
167. Ullah H, Santos HA, Khan T (2016) Applications of bacterial cellulose in food, cosmetics and drug delivery. Cellulose 23:2291–2314. https://doi.org/10.1007/s10570-016-0986-y
168. Valero D, Díaz-Mula HM, Zapata PJ et al (2013) Effects of alginate edible coating on preserving fruit quality in four plum cultivars during postharvest storage. Postharvest Biol Technol 77:1–6. https://doi.org/10.1016/j.postharvbio.2012.10.011
169. Venugopal V (2016) Marine polysaccharides: food applications. CRC Press
170. Viana VR, Silva MBF, Azero EG et al (2018) Assessing the stabilizing effect of xanthan gum on vitamin D-enriched pecan oil in oil-in-water emulsions. Colloids Surfaces A Physicochem Eng Asp 555:646–652. https://doi.org/10.1016/j.colsurfa.2018.07.052
171. Vilela JAP, Da Cunha RL (2016) High acyl gellan as an emulsion stabilizer. Carbohydr Polym 139:115–124. https://doi.org/10.1016/j.carbpol.2015.12.045
172. Wang C, Zhang H, Wang J et al (2020) Colanic acid biosynthesis in Escherichia coli is dependent on lipopolysaccharide structure and glucose availability. Microbiol Res 239:126527. https://doi.org/10.1016/j.micres.2020.126527
173. Wang M, Chen C, Sun G et al (2010) Effects of curdlan on the color, syneresis, cooking qualities, and textural properties of potato starch noodles. Starch/Staerke 62:429–434. https://doi.org/10.1002/star.201000007

174. Wang Y, Sorvali P, Laitila A et al (2018) Dextran produced in situ as a tool to improve the quality of wheat-faba bean composite bread. Food Hydrocoll 84:396–405. https://doi.org/10.1016/j.foodhyd.2018.05.042
175. Wang Y, Trani A, Knaapila A et al (2020) The effect of in situ produced dextran on flavour and texture perception of wholegrain sorghum bread. Food Hydrocoll 106:105913. https://doi.org/10.1016/j.foodhyd.2020.105913
176. Wolter A, Hager AS, Zannini E et al (2014) Influence of dextran-producing Weissella cibaria on baking properties and sensory profile of gluten-free and wheat breads. Int J Food Microbiol 172:83–91. https://doi.org/10.1016/j.ijfoodmicro.2013.11.015
177. Wu S, Lu M, Wang S (2016) Effect of oligosaccharides derived from Laminaria japonica-incorporated pullulan coatings on preservation of cherry tomatoes. Food Chem 199:296–300. https://doi.org/10.1016/j.foodchem.2015.12.029
178. Xiao G, Zhu Y, Wang L et al (2011) Production and storage of edible film using gellan gum. Procedia Environ Sci 8:756–763. https://doi.org/10.1016/j.proenv.2011.10.115
179. Xu XJ, Fang S, Li YH et al (2019) Effects of low acyl and high acyl gellan gum on the thermal stability of purple sweet potato anthocyanins in the presence of ascorbic acid. Food Hydrocoll 86:116–123. https://doi.org/10.1016/j.foodhyd.2018.03.007
180. Yan H, Chen X, Song H et al (2017) Synthesis of bacterial cellulose and bacterial cellulose nanocrystals for their applications in the stabilization of olive oil pickering emulsion. Food Hydrocoll 72:127–135. https://doi.org/10.1016/j.foodhyd.2017.05.044
181. Yan Y, Duan S, Zhang H et al (2020) Preparation and characterization of Konjac glucomannan and pullulan composite films for strawberry preservation. Carbohydr Polym 243:116446. https://doi.org/10.1016/j.carbpol.2020.116446
182. Yang X, Gong T, Lu YH et al (2020) Compatibility of sodium alginate and konjac glucomannan and their applications in fabricating low-fat mayonnaise-like emulsion gels. Carbohydr Polym 229:115468. https://doi.org/10.1016/j.carbpol.2019.115468
183. Yordshahi AS, Moradi M, Tajik H, Molaei R (2020) Design and preparation of antimicrobial meat wrapping nanopaper with bacterial cellulose and postbiotics of lactic acid bacteria. Int J Food Microbiol 321:108561. https://doi.org/10.1016/j.ijfoodmicro.2020.108561
184. Zhai X, Lin D, Liu D, Yang X (2018) Emulsions stabilized by nanofibers from bacterial cellulose: new potential food-grade Pickering emulsions. Food Res Int 103:12–20. https://doi.org/10.1016/j.foodres.2017.10.030
185. Zhang H, Zhang F, Yuan R (2019) Applications of natural polymer-based hydrogels in the food industry. Hydrogels based on natural polymers, pp 357–410, Elsevier
186. Zhang R, Edgar KJ (2014) Properties, chemistry, and applications of the bioactive polysaccharide curdlan. Biomacromol 15:1079–1096. https://doi.org/10.1021/bm500038g
187. Zhang Y, Li D, Yang N et al (2018) Comparison of dextran molecular weight on wheat bread quality and their performance in dough rheology and starch retrogradation. LWT Food Sci Technol 98:39–45. https://doi.org/10.1016/j.lwt.2018.08.021
188. Zhang Y, Zhou L, Zhang C et al (2020) Preparation and characterization of curdlan/polyvinyl alcohol/ thyme essential oil blending film and its application to chilled meat preservation. Carbohydr Polym 247:116670. https://doi.org/10.1016/j.carbpol.2020.116670
189. Zhao Y, Fu R, Li J (2020) Effects of the β-glucan, curdlan, on the fermentation performance, microstructure, rheological and textural properties of set yogurt. LWT Food Sci Technol 128:109449. https://doi.org/10.1016/j.lwt.2020.109449
190. Zhao Z, Wang S, Li D, Zhou Y (2020) Effect of xanthan gum on the quality of low sodium salted beef and property of myofibril proteins. Food Sci Hum Wellness 10:112–118. https://doi.org/10.1016/j.fshw.2020.09.003
191. Zhu H, Jia S, Yang H et al (2010) Characterization of bacteriostatic sausage casing: a composite of bacterial cellulose embedded with ε-polylysine. Food Sci Biotechnol 19:1479–1484. https://doi.org/10.1007/s10068-010-0211-y
192. Zia KM, Tabasum S, Nasif M et al (2017) A review on synthesis, properties and applications of natural polymer based carrageenan blends and composites. Int J Biol Macromol 96:282–301. https://doi.org/10.1016/j.ijbiomac.2016.11.095

Microbial EPS as Immunomodulatory Agents

K. V. Jaseera and Thasneem Abdulla

Abstract The rich diversity of microorganisms represents a potential reservoir of bioactive compounds with valuable pharmaceutical and nutraceutical applications. Microbial EPS signifies one of the major biomolecules which are much explored in pharmaceutical and food industries. They have added advantage of production as they can be cultivated in a controlled condition and process can be optimised for large-scale production. Microbial EPS as immunomodulatory agent is of great interest due to its effectiveness in treating infectious diseases in the absence of suitable antimicrobial therapy and the international market for immunomodulators is believed to increase exponentially. Immunomodulatory activity is the pharmacological effect to influence the cellular and/or humoral immune system, either through stimulation or suppression, to maintain immune homeostasis. Such agents, which possess activity to modulate pathophysiological processes, are addressed as immunomodulatory agents. Immunostimulation or immunopotentiation augments immunological reactions to prevent or to control diseases either through humoral immunity or cell-mediated immunity, whereas immune suppression impedes immune responses. Numerous investigations have been conducted to evaluate the immunomodulatory potencies of these microbial EPS. This chapter gives an insight on the available literature on microbial EPS-based immune modulation, how immunomodulators work, factors influencing immunomodulation and evaluation of immunomodulatory activity of EPS.

Keywords Extracellular polysaccharides · Immunomodulatory agents · Immune cells · Immune effector mechanisms · Immunomodulatory assays

---

K. V. Jaseera (✉)
ICAR-Central Marine Fisheries Research Institute, Ernakulam North (PO), Kochi, Kerala 682018, India

T. Abdulla
Department of Biotechnology, Sir Syed Institute for Technical Studies, Kannur, Taliparamba, Kerala 670142, India

A. K. Nadda et al. (eds.), *Microbial Exopolysaccharides as Novel and Significant Biomaterials*, Springer Series on Polymer and Composite Materials,
https://doi.org/10.1007/978-3-030-75289-7_9

1 Introduction

Polysaccharides are biological polymers having vital physiological functions. They are an integral part of the cell wall material and serve as energy storage material in all living organisms [3]. Besides these cell wall components many eukaryotes and prokaryotes, namely, bacteria, algae, archaea and fungi excrete a mixture of high molecular weight biomolecules termed as extracellular polymers into their surroundings [96, 104, 116, 151, 159, 169] . Even though three-dimensional polymeric network of extracellular polymeric substance consist of diverse biopolymers mainly polysaccharides, proteins, nucleic acids and lipids, exopolysaccharides (EPS) constitute major per cent of extracellular polymeric substances of almost all organisms [31, 169]. EPS may be either capsular polysaccharides (CPS), found tightly anchored to the microbial membrane or exopolysaccharides (EPS) that are loosely bound to the microbes or completely secreted out [8]. They may be either homopolymers or heteropolymers [8, 71, 104].

Functional role of EPS includes, protecting cells from unfavourable, limiting or toxic conditions such as desiccation, antibiotics, phagocytosis and phage attacks [160], thereby improving microbial competition in different environments [74] and also aid in adherence of the biofilm to substratum and helps to hold cells in close vicinity resulting in a synergistic community [31, 34, 114]. Its unique physio-chemical properties are effectively utilised in the food industry as stabilising, viscosifzying, and emulsifying agents [64]. Bacterial alginates, yeast glucans, dextrans, gellan, xanthan and pullulan are few commercially produced microbial EPS applied to improve the physicochemical properties of food formulations in the food industry [137]. Being the renewable hydrocolloids, EPS are also explored in industrial, pharmaceutical, nutraceutical, biomedical and agricultural sectors, as a replacement for petroleum-based polymer [1, 86, 89, 129].

2 Immunomodulatory Activity

Current reports on the ability of EPS to stimulate and modulate the immune system and their role as antitumour, antiviral, anti-inflammatory and antioxidant agents have promoted curiosity among researchers [82, 107]. Immunomodulatory activity is the pharmacological effects of some bioactive compounds to influence the cellular and/or humoral immune responses [161] either through stimulation or suppression to maintain immune homeostasis [36, 55] and such agents, which modulate pathophysiological processes are addressed as immunomodulators [5]. The immunomodulatory activity may be either specific (single antigen) or non-specific (multiple antigens) [27]. Immunostimulation or immunopotentiation augments immunological reactions to prevent or to control diseases either through humoral immunity or cell-mediated immunity, whereas immune suppression impede immune responses and is adopted particularly in case of autoimmune disorders, allergy, organ transplantation, etc. Immunomodulators act against wide-range of bacteria, virus and fungi and hence can be offered against new pathogen outbreaks [33]. Besides the vaccines or therapeutic antibodies for the effective clearance of SARS-CoV-2, immunomodulatory agents have also been a therapeutic arsenal to treat the life-threatening cytokine storm syndrome [131]. Certain conventional Chinese medicines, especially the lung cleansing and detoxifying decoction named 'Qing Fei Pai Du Tang', composed of mainly different polysaccharide combination have shown therapeutic effects on mild and ordinary COVID-19 patients [11].

3 Microbial Polysaccharides and Their Immunomodulatory Activities

Though EPS from higher species are abundantly studied as immunomodulatory agents, microbial EPS offer added advantage as their cultivation and thus EPS production can be achieved under controlled, optimised and contamination-free conditions. Moreover, the demand for natural renewable sources has increased which mooted an increasing interest in EPSs synthesised by microorganisms and it can be suggested as a substitute to polysaccharide extraction from other sources like plant, animal and seaweed [64, 65, 98, 105, 112, 146, 155]. Another benefit of microbial EPS is that optimisation of cultural conditions, media engineering, and scaling up enables the industrial level production of microbial EPS [12]. Smirnou et al. [141] could be achieved in producing microbial EPS from *Cryptococcus laurentii* using bioreactor, with EPS production of 4.3 g/L in 144 h of incubation under optimised condition of pH 3 at 25 °C with low aeration, $1\% < PO_2 < 10\%$. He also concluded that the medium composition and culturing condition determine the composition of EPS produced. A temperature of 25 °C resulted in EPS with glucuronic acid whereas low temperature causes production of glucose containing EPS. Media optimisation for immunomodulatory Sphingobactan, α-mannan EPS from *Sphingobacterium* sp. IITKGP-BTPF3 resulted in production of 1.42 g/L EPS in a medium containing 4% sucrose, 1.5% tryptone at 25 °C and pH 8.0 on sixth day of incubation [13].

3.1 Bacterial EPS as Immunomodulators

Bacteria comprise a major class of microbial EPS producers. Bacterial EPS have been a main focus of study due to their industrially promising physiochemical and biological properties [3]. Lactic acid bacteria, key members of gastrointestinal tract (GIT), are important EPS secreting microbes. LAB-derived EPS remains the most investigated bacterial EPS so far in regard to medical application, owing to its widespread use as probiotics and moreover it is Generally Regarded as Safe (GRAS). LAB synthesises both homopolysaccharides such as dextran and levan and heteropolymeric EPS composed primarily of sugar monomers glucose, galactose and rhamnose [26]. Halaas et al. [46] revealed the immunomodulatory potential of an EPS (mannuronan) from *Pseudomonas aeruginosa* to enhance natural cytotoxicity by increasing Fas ligand expression on NK cells. The immunomodulatory action of V2-7, a sulfated EPS produced by *Halomonas eurihalina*, exerted via induction of lymphocyte proliferation was demonstrated by Pérez-Fernández et al. [119]. An EPS from *Paenibacillus jamilae* CP-7 was revealed to enhance the resistance of mice to intracellular pathogen *Listeria monocytogenes* via macrophage stimulation [128]. TA-1-an EPS from *Thermus aquaticus* YT1-was reported to stimulate macrophages via TLR2 and enhance production of cytokines TNF-α and IL-6 [80]. 101 EP and 102 EP, EPS from *Lactobacillus paracasei subsp. paracasei* NTU 101and *Lactobacillus plantarum* NTU 102 respectively was shown elicit secretion of proinflammatory cytokines TNF-α, IL-6 and IL-1β in murine macrophage cell line, Raw 264.7 [83]. Evidences supporting the elevated production of pro- and anti-inflammatory cytokines by murine macrophages challenged by different EPS from *Lactobacillus* and *Bifidobacterial* strains indicate the immunomodulatory potential of enteric bacterial EPS [109]. EPS from *Bacillus licheniformis* strain T14 was shown to prevent HSV-2 replication in human peripheral blood mononuclear cells by stimulating Th1 cell-mediated immunity [43]. Furthermore, Jones et al. [60] confirmed the ability of EPS from *Bacillus subtilis* to prevent intestinal inflammation via a TLR-4 dependant mechanism [60]. Porcine intestinal epithelial (PIE) cells when stimulated with LPS produces inflammatory cytokines via Toll-like receptor (TLR4) mediated signalling and EPS from LAB was shown to lower this effect and generate an anti-inflammatory response [75]. Experiments by Pérez Ramos et al. [113], on 2-substituted (1, 3)-β-D-glucan-an EPS produced by *Pediococcus parvulus* 2.6-revealed its ability to downregulate expression of genes involved in inflammatory response, in zebrafish models. EPS from *Lactobacillus fermentum*, in its purified form, was revealed to boost production of proinflammatory cytokine TNF-α in THP-1 cell line. Both purified and crude version of the EPS was indicated to increase synthesis of regulatory cytokine IL-10 [2]. In a recent study, Mizuno et al. [95] established the ability of *Streptococcus thermophilus* ST538 EPS to modulate innate antiviral immune response in PIE cells.

Xanthan gum (XG), EPS from *Xanthomonas campestris*, was shown to enhance the secretion of IL-1β, IL-6, and TNFα in mouse RAW264.7 macrophage cells. The same study also demonstrated the ability of XG to lower production of these proinflammatory molecules in LPS-stimulated RAW264.7 cells [85, 135] demonstrated the ability of xanthan gum, when used with ovalbumin, to stimulate both cellular and humoral immune response suggesting the possibility of using it as a vaccine adjuvant.

3.2 Microalgal EPS as Immunomodulators

EPS from microalgae are promised to be a potential immunomodulatory agent, especially sulphated polysaccharides [121, 169]. EPS from *Porphyridium* sp., *Chlorella* sp. and *Ellipsoidon* sp. showed inhibitory action against viral haemorrhagic septicemia virus (VHSV) of salmonid fish and the African swine fever virus (ASFV) [29]. EPS from diatoms, *Phaeodactylum tricornutum*, is recognised to inhibit carrageenan-induced paw edema and thus can be utilised as an anti-inflammatory agent against it [44]. Sulfated polysaccharide obtained from red microalgae *Porphyridium* exhibited anti-inflammatory activity in polymorphonuclear leukocytes in vitro and inhibited the development of erythema in vivo by inhibiting the migration/adhesion of polymorphonuclear leukocytes (PMNs) [90]. The sulphated exopolysaccharides (p-KG03) from the marine dinoflagellate *Gyrodinium impudicum* (strain KG03) showed immunomodulatory activity by stimulating the release of nitric oxide and cytokines in macrophages via the JNK dependent pathway through the NF-κB transcription factor activation [4]. Similarly, Yim et al. [175] explored immunostimulatory effects of sulphated exopolysaccharide p-KG03 (*G. impudicum* KG03) and demonstrated that it can induce tumouricidal activities of macrophages and NK cells in vivo and also enhance the release of cytokines, such as interleukins-1b and -6, and TNF–α by macrophages. The in vitro (macrophages) and in vivo (S180-tumour-bearing mouse) studies of the extracellular polysaccharides (EPSs) derived product from *Porphyridium cruentum* showed strong immunostimulating potency by activating macrophages and EPS substantially enhanced the proliferation lymphocyte, representing the antitumour effect of EPS [144]. He also proved that molecular weight of EPS significantly influences immune stimulatory effect and showed that low molecular weight product showed stronger activity when compared to the high molecular weight product. Raposo et al. [122] showed that proteoglycans of *Porphyridium* (sulphated glycosaminoglycans) elicit anti-inflammatory properties. EPS from mutant *Thraustochytriidae* sp. GA strain stimulated B cell growth, contributing to support humoral immune system. It decreased the secretion of interferon-γ (IFN-γ) and interleukin-6 (IL-6) by T cells, but the production of tumour necrosis factor-α (TNF-α) was not affected [108]. Sulfated polysaccharides from *Tribonema* sp. named TSP a heteropolysaccharide comprising galactose as main constituent exhibited potential immune stimulatory activity in macrophage cells and enhanced the release of interleukin 6 (IL-6), interleukin 10 (IL-10), and tumour necrosis factor α (TNF-α) [17, 14]. Ethyl acetate fraction of EPS from *Dunaliella salina* enhanced PBMC proliferative index and cytokines (IFN-γ, TNF-α, TGF-β) secretion in a dose-dependant manner and showed dose-dependant inhibitory effect on macrophages and thus NO production [41]. Spirulina is a potent response modifier, elevating activation of macrophages, NK cells, by accelerating secretion of antibodies and cytokines, and by activating and mobilising T and B cells [32, 66]. Shokri et al. [139] reported that treatment with *Spirulina platensis* gave better result in mice suffering from systemic candidiasis and breast cancer, where the IL-4 and IL-10 expression was considerably decreased and the expression of IL-17, TNF-α and IFN-γ were elevated. Chen et al. [15] reported that phycocyanin was the active biomolecule present in *Spirulina platensis*, to induce immunomodulation by promoting the secretion of TNF-α, IL-1β, and IL-6 in macrophage cell line J774A cells. Supplementation of Spirulina for 12 weeks in senior volunteers (>50) resulted in increased indoleamine 2,3-dioxygenase enzyme activity, a sign of immune function, along with increased white blood cell count [136].

3.3 Fungal EPS as Immunomodulators

Glucan is the most abundant polysaccharides from fungus and yeast. Extracellular heteropolysaccharide fraction, PS-F2, from *Ganoderma formosanum* exhibits immunostimulatory activities by triggering MAPK and NF-κB pathway which in turn activate TNF-α expression [162]. β-Glucan is a pathogen-associated molecular pattern (PAMP) that is detected upon fungal infection to trigger the host's immune responses in both vertebrates and invertebrates and pre-administration causes innate immune memory [47]. EPS having immune-enhancing and anticancer activities was reported from *Rhizopus nigricans* [179, 181]. Immunomodulatory EPS from *Aspergillus terreus* activated macrophages causing production of superoxide anion, nitric oxide (NO), interleukin (IL)-6 and tumour necrosis factor (TNF)-α [20]. The in vivo studies of microbial EPS from *Cryptococcus laurentii* showed that it significantly improved wound healing in healthy rats [141]. Low molecular weight (1→3)-β-D-glucans (90–100 kDa) from baker's yeast *Saccharomyces cerevisiae* showed immunomodulatory role [132]. Hromádková et al. [51] reported the immunomodulatory role of particulate (1→3)-β-D-glucan isolated from *S. cerevisiae.* Pelizon et al. [111] investigated role of β-glucan, extracted from *S. cerevisiae,* on cytokine production, natural killer activity and fungicidal activity. In vivo β-glucan administration resulted in higher production of IL-12 and TNF-α by spleen cells, when *S. aureus* was used as a stimulus. In addition, β-glucan increased NK spleen cells activity against YAC target cells. Intravenous administration of particulate β-glucan was effective in treatment of Paracoccidioidomycosis (PCM), an endemic disease [93]. The immunomodulatory effects of *Aureohasidium pullulans* SM-2001 exopolymers containing β-1,3/1,6-glucan were evaluated in cyclophosphamide (CPA)-treated immunosuppressed mice and established that it can successfully avert immunosuppression through the recruitment of T cells and TNF-α+cells or by enhancing their activity [177]. In vitro assessment of Botryosphaeran (BR), β-glucan comprising β-(1→3) backbone and β-(1→6) branched glucose residues, revealed its immunomodulatory role in macrophages by improving phagocytosis and production of nitricoxide and TNF-α [166]. Tamegai et al. [152] reported, co-treatment of R-848, an analog of imidazoquinoline derivative and an immune response modifier, with the culture supernatant of *Aureobasidium pullulans* significantly augmented TNF-α and IL-12p40 cytokine expression and phagocytosis of apoptotic Jurkat cells.

4 Factors Influencing the Immunomodulatory Activity of EPS

Owing to its diverse structural and functional properties, EPS offers potential bioactivities which are significantly influenced by source, molecular weight (MW), composition, functional group, type of linkage and chain length of EPS [30, 38, 149]. Characteristics of EPS vary greatly depending on the source from where it is produced [104]. EPS from *Tetraselmis suecica* helped attachment of *Helicobacter pylori* to HeLa S3 cells, contradicting the report of suppressive role of EPS from *Tetraselmis* spp. owing to dissimilarity in EPS structure and/or chemical composition indicating even EPS from the same genus act differently [45].

Molecular weight and size of EPS greatly influence immunomodulatory activity. Lake et al. [76], reported that the 3-D conformations of low MW polysaccharides facilitate the ligand–receptor interaction and thus improve binding capacity. Suryot et al. [150] revealed that partially hydrolysed EPS having a low molecular weight of values $\leq 70 \times 10^3$ g/mol comprising homopolymer of D-glucose linked by α-(1→6) glycosidic linkages from *Lactobacillus confusus* TISTR 1498 showed immunomodulatory activity in macrophage cell lines (RAW264.7) and induced the expression responses (NF-α, IL-1β, IL-6 and IL-10) whereas intact EPS with high MW couldn't initiate immunological responses. Similarly product produced as a result of oxidative degradation of EPS from *P. cruentum* with H_2O_2 under ultrasonic wave generated 6 fragments with different molecular weights. The in vitro studies on macrophages and in vivo on S180-tumour-bearing mouse of these products showed that, even though all the fragments showed immune stimulating potency by activating macrophages, low molecular weight products showed stronger activity indicating molecular weight has strong influence on immunomodulation [144].

Surayot and You [147] revealed that EPS dose, position of functional group, sulphates and its composition significantly alter its potency and can be attributed to alteration in binding capacity with cell surface receptor. As in the case of sulphated *polysaccharides (SP-F2) from Codium fragile, an adequate dose or concentration is required* to elicit immune responses, and lower and higher level diminishes its effect. He also reported that both the content (protein) and group (sulphate) of SP-F2 have profound effect on its NK cell activation as deproteination (DP-F2) and desulfation (DS-F2) significantly reduced expression of IFN, granzyme-B, NKp30 and FasL.

Monosaccharide composition of EPS affects its potency on immune cells. Three unpurified EPSs from *Lactobacillus reuteri* Mh-001 exhibited anti-inflammatory affect differently, EPS with galactose showed highest activity followed by rhamnose containing EPS and then glucose containing [17, 14]. The ability to evoke immune responses by EPS is positively influenced by sulphate content [144].

The immunomodulatory activity by microbial EPS depends on its structure and charge it carries. The bacterial polysaccharides with a zwitterionic charge evoke in vitro and in vivo CD4 T cell responses and its action is dependent on specific structure and free amino and carboxyl groups on the repeating units of these polysaccharides. Degradation of structure and charge neutralisation resulted in loss of its activity [159]. Drying methods used in EPS preparation such as solvent exchange (GE), lyophilisation (GL), and spray drying (GS) influence its microstructure leading to changes in physiochemical properties and hence its immunostimulatory role. Spray dried (1→3)-β-D-glucan from *S. cerevisiae* was suggested as potent immunomodulator [51].

The bioactivities can be effectively altered by suitable modifications at structural level. Alteration may be performed by biological, physical or chemical ways. These modifications may alter the molecular weight, structure, composition, etc. of EPS [178]. In physical modifications, polysaccharides are degraded with radiation-induced reactions, ultrasonic disruption, and microwave exposure and generally bring about changes in molecular weight [146]. Chemical modification may add functional groups to enhance the bioactivities. Several methods are employed in which selenisation, sulfation and sulfonation are most common. These modifications may alter the ionic charge of the molecule and thus its solubility. Modification of EPS purified from *Lactococcus lactis* subsp. *Lactis* culture to selenium-exopolysaccharide (Se-EPS) by adding selenium chloride oxide ($SeCl_2O$) improved the immunomodulatory activity of EPS [106]. In biological modification, specific enzymes are used to precisely and efficiently digest the EPS to alter its molecular weight. Xu et al. [171] reported that activation of macrophages by enzymatically digested alginate oligosaccharides (AOS) was influenced by its structure, molecular size, and M/G ratio. In addition to these, mutation at molecular level is also employed, which significantly alter its structure and thus the function [108].

Mannosylated dextran particles improve antigen presentation via MHC class I molecules by dentritic cells in vitro [22]. An investigation on immunomodulatory potential of pullulan, and its acidic and alkaline derivatives, revealed the ability of alkaline pullulan to more strongly induce secretion of proinflammatory cytokines by plasmacytoid dendritic cell line, CAL-1 via TLR-mediated MyD88-dependent pathway and NF-κB pathway [163]. Comparison of immunomodulatory potential of particulate and soluble β-glucans from *S. cerevisiae* revealed that particulate β-glucans modulates immune response via dectin-1 dependent pathway, but not soluble β-glucans [120]. Low molecular weight sulphated β-glucan from *S. cerevisiae* exhibited better immunological potential than high molecular weight glucans [77]. Phosphorylated derivates of β-D-Glucan from *Poria cocos* were demonstrated to exhibit stronger anti-tumour activities both in vivo and in vitro [16]. Mannuronate-rich alginate induces production of proinflammatory cytokines via TLR4/2 receptor signalling along with CD14 [28]. Iwamoto et al. [57] demonstrated that enzymatically depolymerised guluronate and mannuronate oligomers could induce cytotoxic cytokine production in human mononuclear cells, whereas depolymerised oligomers could not bring about this response. AOS prepared enzymatically could enhance NO synthase expression leading to nitric oxide production in RAW 264.7 cells [84].

5 Mechanism of Action of EPS Immunomodulators

In order to understand the underlying mechanism of EPS elicited immunomodulation, one must understand the immune system and cells of the immune system, as well as and how our immune system operates. Immune system is indispensable in protecting the host against foreign evading agents through numerous mechanisms. Innate and adaptive immunity forms the two pillars of immune system. Innate immune response is non-specific, rapid and the first line of defence against foreign particles, while, adaptive immune response is more specific, slow but long-lasting and generates immunologic memory. The two systems work in concert and exert their effects through a large number of immune cells mainly blood monocytes, tissue macrophages, neutrophils, T and B lymphocytes, antigen presenting cells (dendritic cells), natural killer cells and others and these cells express diverse pattern recognition receptors (PRR) that initiate an array of responses via different intracellular signalling cascades [100].

EPS exerts immunomodulatory effect by activating and regulating multiple signal transduction pathways. Even though very little is explored about the microbial EPS, advanced research on EPS as immunomodulatory agents have disclosed the underlying mechanism of action of some of these molecules. Microbial EPS elicit humoral and cell-mediated immunity via activation of B cells, T cells, macrophages and natural killer cells and thus improve responses to certain infections in host. Further, this approach can be efficiently exploited to augment existing antibiotic treatments that are becoming less efficient with the increase in antibiotic resistance. Immunomodulatory therapy is found effective in treating autoimmune encephalitis [49]. Nowak et al. [101] demonstrated that oral administration of probiotic bacteria *Lactobacillus rhamnosus* KL37 reduces several types of experimental arthritis, including collagen-induced arthritis (CIA) and inhibits arthritogenic autoantibodies. It was found that EPS from *Bacillus subtilis* encourages development of anti-inflammatory M2 macrophages in a TLR4-dependent manner, and these cells hinder T cell stimulation in vitro and in *Citrobacter* rodentium-infected mice and hence it was suggested that administration of these EPS can be used to regulate T cell-mediated immune responses in many inflammatory diseases [110]. Kocuran, EPS from *Kocuria rosea* strain BS-1 have powerful immunosuppressive properties and demonstrated in vitro suppression of macrophage stimulation and macrophage-derived inflammatory cytokines and complement mediated haemolysis indicating its in vitro immunosuppressive activity [73].

5.1 *Activation of Macrophage and Immune Modulatory Cascade*

Macrophage stimulation is a crucial part of immune responses and includes several receptors and activators and involves interaction of EPS with multiple pattern recognition receptors on the surface of macrophage followed by an array of signalling events to control expression of immune response genes [78, 146]. Polysaccharides activate macrophages on interaction with pattern recognition molecules, such as Toll-like receptor-2/4 (TLR-2/4), scavenger receptor, dectin-1, complement receptor 3 (CR3), CD14, and mannose receptor. Most receptors interact synergistically, e.g., TLR4 and CD14, Dectin-1 and TLR2, and CD14 and CR3 form complexes which triggers intracellular signalling [162, 187]. Macrophages exert its effects generally through reactive oxygen species (ROS) generation, release of cytokines, lymphocyte proliferation, and phagocytosis [133]. Investigation by Surayot et al. [146] on the immunomodulatory activity of partially hydrolysed EPS from the culture broth of *L. confusus* TISTR 1498 confirmed NF-κB (nuclear factor kappa-light-chain-enhancer of activated B cells) and JNK (c-Jun N-terminal kinase) pathways as the underlying mechanism of macrophage stimulation.

5.1.1 Toll-Like Receptors 2/4-Mediated Signalling Cascade

Toll-like receptors (TLRs) are type I transmembrane glycoproteins, a group of pattern recognition receptors (PRRs), which are extensively found on the macrophage, lymphocyte and neutrophil surfaces [125]. TLRs are vital in pathogen recognition and initiation of the innate immune response. Activation of TLRs induces expression of various genes for multiple effector molecules like chemokines, cytokines, MHC, co-stimulatory molecules, antimicrobial peptides and nitric oxide synthase (iNOS) which effectively abolishes infectious agents [156]. Two types of TLRs, TLR2 and TLR4 recognise and interact with carbohydrate moieties. Polysaccharides indirectly interact with TLR4 in association with CD14, RP105, MD1 or MD2 [59, 138]. TLRs interact with EPS and stimulates TNF receptor-associated factor 6 (TRAF6) via myeloid differentiation factor 88 (MyD88)-mediated signalling events or MyD88 independent pathway (TLR-related interferon factor (TRIF)-mediated signalling pathway) [133, 173]. MyD88, have a C-terminal TIR domain and a cytoplasmic *N*-terminal death domain, which bind with the TIR domain of TLRs by MyD88 adaptor like (MAL) protein. Upon activation, MyD88 recruits a kinase called IRAK-4 (IL-1 receptor-associated kinase-4) to TLRs which phosphorylates IRAK-1. Phosphorylated IRAK-1 interacts with TRAF6, [94] stimulating two distinct signalling cascades, leading to activation of MAP kinase and TAK1/TAB complex formation. MAP kinase pathway activates AP-1 transcription factors and TAK1/TAB complex activates Inhibitor kappa B kinase (IκK) complex. On activation, the IκK complex causes phosphorylation and successive destruction of IκBα. In resting cells, NF-κB (P50, p65/c-Rel) noncovalently interacts with IκBα and remains inactive in the cytosol. Degradation of IκBα causes nuclear transportation of transcription factor NF-κB (Takeda 2004; Klinman 2004, [58, 72, 94]) and activation of expression of immune response genes [182]. NF-κB, an important family of eukaryotic transcription factors mediates various normal cell and tissue responses, including immune and inflammatory responses [182].

In MyD88-independent pathways, TRAM, which is functionally similar to MAL, combines with TRIF and TLR [67, 173]. TRAM/TRIF complex recruit TRAF-6, which regulate the synthesis of correlated cytokines and interferon types I/ II [127] and hence TRAF-6 is the junction of two types of TLR4 signalling pathway, which activate NF-κB and MAPK reaction cascade [87]. MAPKs are serine/ threonine protein kinases that phosphorylates ERK, JNK and p38 which activate AP1 (activator protein-1) transcription factors [134]. AP-1, a c-Fos and c-Jun heterodimer, [52] whose binding regulate expression of immune response genes [130]. The mechanism of action of microbial EPS-induced macrophage activation is illustrated in Fig. 1.

TLR2/4 receptors are also expressed on T and B lymphocytes [174]. Ligand binding to TLR4 mediates activation and proliferation of B cells and their differentiation into plasma B cells via activation of MAPKs which subsequently stimulates transcription factors, including NF-τB, and AP-1 [92, 103, 167, 185]. TLR4

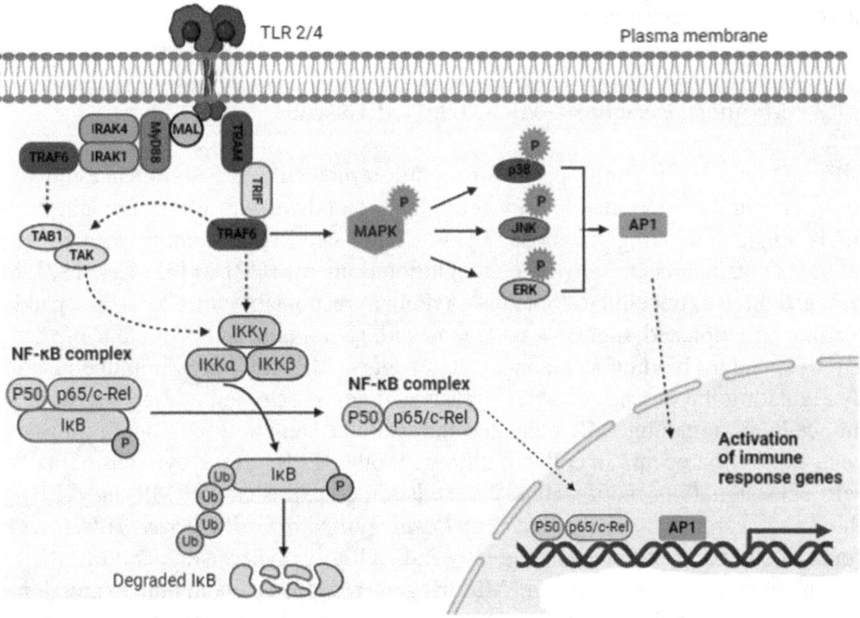

Fig. 1 Mechanism of action of macrophage activation by microbial EPS. Created with BioRender.com

is responsible for sulphated polysaccharide-induced B cell activation via calcium signalling cascade [81].

5.1.2 Complement Receptor 3 (CR3) Mediated Cascade

CR3 is a heterodimer of CD11b and CD18 and belongs to integrin family. CD11b is exclusive to CR3 and it binds ligands via lectin site [39, 157]. CR3 is expressed particularly on almost all T cell surfaces, polymorphonuclear leukocytes, macrophages, and monocytes and natural killer cells [126] and act as a receptor for complement proteins and β-glucan [180]. Concurrent ligation of iC3b binding and glucan binding domains are responsible for CR3-mediated phagocytosis and degranulation [158]. CR3 signal transduction regulates the synthesis of interleukin-12 (IL-12), a prime mediator of cell-mediated immunity (CMI) [88]. CD14 and CR3, arranged as a transmembrane complex, activate phospholipase (PLC), which in turn triggers both phosphatidylinositol 3-kinase (PI3–K) and protein kinase (PKC) via NF-κB or MAPK cascade and controls the activation of related genes [97].

Jung et al. [61] reported that *Bacillus* EPS exerts its immunomodulatory activity via CR3 dependant pathway. Extracellular heteropolysaccharide fraction, PS-F2, from the submerged mycelia culture of *G. formosanum* elicited macrophage activation mediated by Dectin-1 and CR3 receptors [162]. Tzianabos [158] reported that β-glucan induces macrophage, neutrophils and eosinophils proliferation via increased phosphatase activity.

5.1.3 Mannose Receptors (MR)-Mediated Cascade

MR, a C-type lectin family pattern recognition molecule, is a significant endocytic receptor found predominantly on macrophages and dendritic cells. It recognises and binds sugars including mannose, fucose, and acetyl glucosamine with different affinities through its carbohydrate recognition domains (CRDs) [35, 154, 183]. MR contains three extracellular domains—a domain responsible for Ca^{2+}—independent binding to sulphated sugars, a collagen binding domain and a domain mediating Ca^{2+} dependent binding to mannosylated sugars. MR modulates immune response by regulating the process of phagocytosis, antigen processing and presentation and intracellular signalling. MR-mediated intracellular signalling culminating in target gene expression occurs in collaboration with other PPRs as the cytoplasmic tails of MR lacks signalling motifs. HEK 293 cells, on co expression of MR and TLR was shown to secrete IL-8 in response to *Pnemocystis carinnii* (*jirovecci*) but neither could elicit the response independently indicating its synergistic action [35].

On recognition of pathogen, MR triggers responses, including expression of inflammatory mediators, majorly via NF-κB and MAPK activation, lysozymal enzyme secretion and modulates receptors of other cells [168], which assists in elimination of pathogenic organisms [79]. Since MR is responsible for early sensing of infection and provides a link between innate and adaptive immunity, it is hailed as a critical regulator of immune response [35]. Wang et al. [164] established MR-mediated p38 MAPK/NF-kB pathway to be responsible for *Aspergillus fumigatus*-induced inflammatory response in human corneal epithelial cells. Mycobacterial LPS, lipoarabinomannan (LAM) on binding MR induces T cell proliferation indicating its role in antigen presentation [153].

5.1.4 Scavenger Receptors (SR)-Mediated Cascade

SR is a membrane bound glycoprotein which is expressed by diverse immune cells including macrophages and can bind a range of endogenous and exogenous ligands. Eight classes of SR (A-H) have been discovered so far [118]. On binding their ligands, SR initiates a cascade of reactions resulting in stimulation of MAPK and NF-κB, in a manner similar to CR3 receptor, eliciting transcription of numerous proinflammatory genes, including iNOS, leading to production of NO in macrophages and thus induces inflammatory response [40, 56, 176].

SR acts in co-operation with TLR2/4 receptors in mediating signal transduction, particularly through initiation of NF-κB or IRF3 pathway, which ultimately results in upregulation of genes coding for bioactive molecules like chemokines, cytokines, ROS/RNS and others [24].

Binding of fucoidan—a sulphated polysaccharide—to Macrophage scaveneger receptor A (MSR-A), in human THP-1 macrophages, regulates synthesis of inflammatory cytokines TNFα and IL-1β via differential modulation of specific protein kinase signal transduction pathways [53].

5.1.5 Dectin-1-Mediated Cascade

Dectin is a C-type lectin (CLR) family receptor, predominantly seen on monocyte/macrophage, and neutrophil surfaces [99] and possess carbohydrate recognition domain (CRD), comprising 18 conserved amino acid residues including two disulphide bonds, formed by four cysteine residues. It acts synergistically with TLR2/4. Ligand binds Dectin-1 in a Ca^{2+} dependent manner and triggers series of signalling events which trigger MAPK/NF-κB signalling pathway [50]. β-(1, 3)-glucan component of the EPS delivers immunomodulatory activity by binding to Dectin-1 via macrophages and dendritic cell activation which causes exclusion of the phosphatases CD45 and CD148 from the phagocytic cup, resulting in proinflammatory cytokine production and cell maturation through its canonical ITAM/Syk/CARD9 dependent pathway signalling. Downstream to Syk, CARD9 associate with Bcl10 and MALT1, triggering the activation of the NF-kB transcription factor [23]. Additionally, there are reports supporting the non-canonical Raf-1-dependent mechanism (Syk-independent pathway) of NF-kB activation. Furthermore, it induces re-polarisation of M2-like macrophages to M1-like macrophages leading to enhanced synthesis of NO culminating in efficient killing of pathogens [6, 23]. EPS from *Trichoderma pseudokoningii* stimulate the activation of macrophages and induces the production of nitric oxide (NO), interleukin (IL)-1β and tumour necrosis factor (TNF)-α and enhance phagocytic activity through NF-κB and MAPKs signalling pathways via TLR4 and Dectin-1 [165].

5.2 Activation of T/B Lymphocytes and Immune Modulatory Cascade

EPS triggers lymphocyte immune responses via series of signalling events [102]. The immunomodulatory role of EPS from *Thraustochytriidae* sp. GA is accomplished through B cell multiplication and cytokine release by T cells [108]. Acidic fraction of EPS from *Lactobacillus delbrueckii* ssp. *bulgaricus* 1073R-1 stimulated the multiplication of murine splenocytes and Peyer's patches but not thymocytes [68]. Kalka-moll et al. [63] suggested CD4 T cell response as a signalling pathway for the immunomodulatory activity of bacterial polysaccharides.

Very little literature is available regarding the mechanism of induction of T cell and B cell proliferation in response to microbial EPS.

5.2.1 Membrane Immunoglobulin (mIg) Mediated Cascade

B cells harbour mIg receptors on their surface that complexes with CD19 and CD79b to regulate B cell proliferation [184]. Here, mIgM and mIgD serve as signal transducers. Ligand binding to its receptor, causes the activation of a G-protein (*N* protein) leading to GTP binding that activates phospholipase-C (PLC) which in turn cleave, phosphatidylinositol bisphosphate (PIP2) into inositol trisphosphate (IP3) and diacylglycerol (DAG). IP3 causes the efflux of Ca^{2+} from intracellular stores while diacylglycerol causes translocation and activation of protein kinase C (PKC) [10, 42] (Fig. 2). Binding of polysaccharides to IgM/CD79 trigger protein tyrosine kinase (PTK) also [184]. PTK and PKC further activate MAPK and stimulate activator protein-1 (AP-1) that controls genes in B lymphocyte immune responses [48]. Polysaccharides with sulphate group can initiate spleen lymphocytes activation and differentiation into IgM secretory plasma cells, and enhance the expression of mIg receptor complex [186].

5.2.2 T Cell Receptor-Mediated Cascade

TCR interacts with major histocompatibility complex (MHC) molecules on T cells. TCR exists as a complex comprising variable αβ chains linked with the nonpolymorphic CD3 proteins noncovalently. Both CD3 and TCR act synergistically for the cascading event to occur. TCR docking leads to activation of Src (Lck and Fyn)

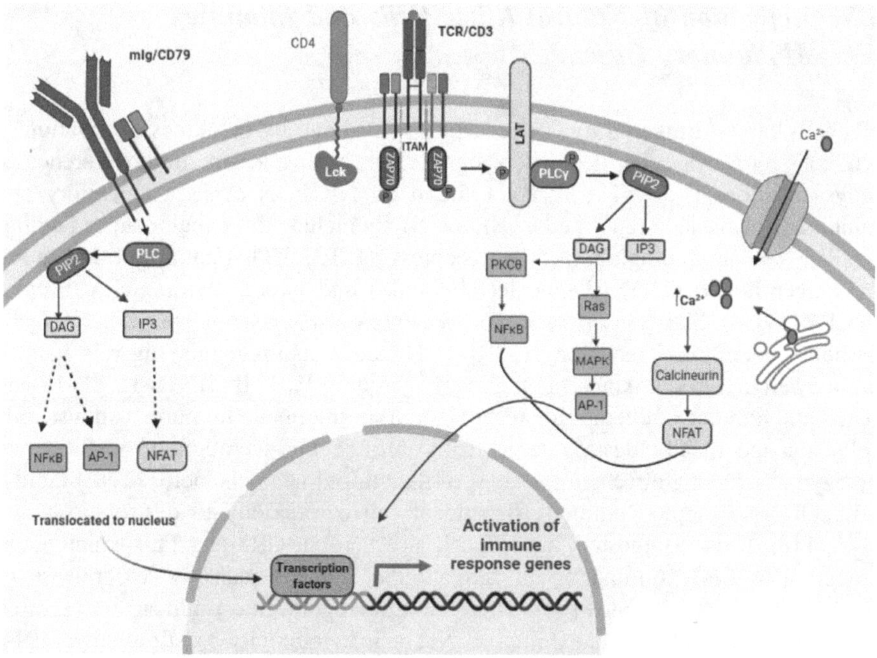

Fig. 2 Mechanism of activation of lymphocytes by microbial EPS. Created with BioRender.com

protein tyrosine kinases, inducing phosphorylation of the tyrosine residue within CD3 ITAMs which then serve as docking site and is followed by the recruitment of ZAP-70, which phosphorylate LAT (linker for the activation of T cells) and SH2 (Src homology 2) domain of leukocyte phosphoprotein. Once phosphorylated, LAT activates phospholipase-C isoform (PLCϒ), PI3K, GRB2 and Gads which induces the hydrolysis of membrane lipid PI(4, 5)P2, releasing IP3 and DAG, second messengers required for T cell functioning. DAG via Ras and PKCθ pathway trigger MAPK and NF-κB respectively and ultimately leads to AP-1/STAT3 activation. The IP3 initiate efflux and influx of Ca^{2+} ion from endoplasmic reticulum (ER) and cell exterior, respectively. Increased intracellular Ca^{2+} levels activate several downstream regulatory proteins and transcription factor NFAT that in turn activate several gene expression. Dephosphorylation of the NFAT (nuclear factor of activated T cells) family bycalcineurin members trigger its translocation to the nucleus where it complexes with a variety of other transcription factors and initiate signalling cascade in responses to TCR signal [142]. Figure 2 illustrates the mechanism of activation lymphocytes by microbial EPS.

5.3 Activation of Natural Killer Cells and Immune Modulatory Cascade

NK cells can be triggered by CR3 receptor, TLR4 and by cytokines. Activation of NK cells by polysaccharides differs significantly owing to structural difference of polysaccharides [147]. The receptor found on NK cells exist in activatory and inhibitory forms and are termed as NKRs. NKRs include three major super families —killer cell immunoglobulin-like receptors (KIR), CD94/natural killer group 2-member D (NKG2D) (C-type lectin family) and natural cytotoxicity receptor (NCR) [19]. Ig-like transcripts (ILTs) receptors also exist on NK. All NK cells harbour interleukin-2 receptor (IL-2Rβγ). NK cells also recognise ligands such as monocyte-derived cytokines (monokines), including IL-1, IL-10, IL-12, IL-15 and IL-18 and number of chemokine receptors during the innate immune response. NK cells non-specifically destroy target cells through the secretion of perforin and granzymes or using effector molecules of the tumour necrosis factor (TNF) family. NK cells also employ antibody dependant cell cytotoxicity to destroy its target [145, 148]. It was suggested that NK cell also harbour CR3 and TL4 which act as receptor for EPS culminating in activation of NF-κB pathway and release of immune effector molecules [12]. There are also reports that β-glucan polysaccharide from yeast and fungus exhilarated NK cells cytotoxicity by stimulating IFN-γ and perforin secretion [25]. Sulphated fucan (SF) from *Stichopusjaponicus* may enhance NK cells proliferation by secreting IFN-γ. SF activates signalling cascade by binding CR3 receptor culminating in the expression of NKp30 and release of FasL, IFN- γ, and granzyme-B [145]. Brusilovsky et al. [9] reported that KIR2DL4 stimulates cytokine production and cytolytic activity in resting NK cells, which were normally thought to activate NK cells by soluble HLA-G ligand, which can also be induced by heparin sulphate (HS). There are also report of recognition of sulphur containing polysaccharides, such as dextran sulfate, fucoidan, and κ-carrageenan, by NK cell receptors, Ly-49C containing C-type lectin domain [7]. Simplified depiction of mechanism of activation of NK cell by microbial EPS is presented in Fig. 3.

5.4 Activation of the Complement System and Immune Modulatory Cascade

Polysaccharides can activate complement system, and play role in primary host defence mechanism [18]. Complement system is composed of over 20 serum proteins, which are activated by classical, alternative and lectin pathway triggering a series of reaction culminating in the formation of membrane attack complex [37]. Sulfated polysaccharides can also exhibit their immunomodulatory effect through complement pathway [54]. The peptides generated by complement activation can facilitate several immune reactions, such as opsonization, activation of lymphocytes

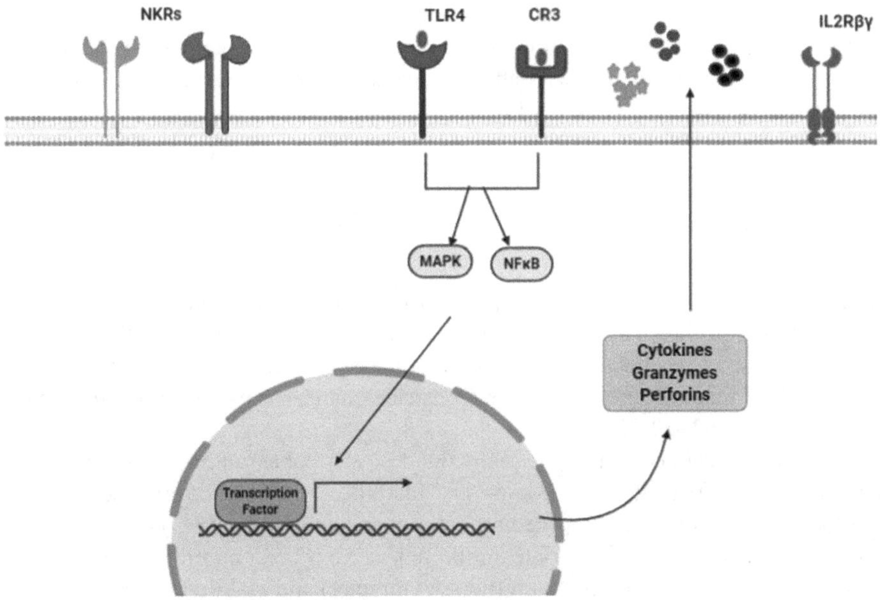

Fig. 3 Mechanism of activation of NK cells by microbial EPS. Created with BioRender.com

and mast cell degranulation [91]. Classical pathway is activated by antigen-antibody (IgM and IgG antibodies) complex, leading to activation of complement system by a series of mechanisms. In alternative pathway, microorganisms and some molecules such as lipopolysaccharide trigger C3 in an antibody independent mechanism. A third pathway of complement activation is a carbohydrate binding pathway-mannan binding lectin (MBL) pathway. The MBL pathway is stimulated by carbohydrate structures of microorganisms leading to elimination of microbes via lytic complement components or by enhancing phagocytosis. There are reports of polysaccharide-induced modulation of complement system [172], but less is known about microbial EPS-induced complement modulation. GY785 and HE800, EPS from *Alteromonas infernus* and *Vibrio diabolicus* respectively activated the classical complement system, while its over-sulphated LMW EPS (GY785 DROS and HE800 DROS) inhibited complement activation as indicated by a decrease in sheep erythrocyte lysis [21].

6 Evaluation of Immunomodulatory Activity

Biological evaluation methodologies are very crucial and indispensable in the assessment of the beneficial and detrimental potencies of candidate drugs. Bioactive compounds must be investigated by means of in vitro and in vivo pharmacological screening. Macrophages and monocytes are the most common and appropriate model for the in vitro evaluation of immunomodulatory potential of such compounds. For in vivo assays, Swiss albino mice, Wistar albino rats and BALB/c mice are commonly used [124, 140, 143].

6.1 *In Vitro Assays*

Macrophages isolated from peritoneal cavity of mice and commercially available macrophage cell lines (RAW 264.7, murine macrophage J774.A1, THP-1, U-937) are widely used to evaluate the immunomodulatory potential of biomolecules [69]. The yield of macrophages from mouse can be increased by thio-glycollate elicitation method [62, 123].

The macrophage cell lines are maintained in a suitable culture medium such as RPMI 1640 (Roswell Park Memorial Institute 1640 medium) and DMEM (Dulbecco's Modified Eagle Medium) supplemented with 10% heat-inactivated fetal bovine serum and appropriate antibiotics at 37 °C in a humidified atmosphere containing 5% CO_2. The cell count (hemocytometer) and viability (trypan-blue dye exclusion assay) are determined and the cells are further used to perform various in vitro assays [62].

One of the common assays is **analysis of nitric oxide (NO) production** in which the macrophage cells are incubated in different concentrations of test polysaccharide fractions and after 24 h, colorimetric determination at 540 nm of nitrite is performed using the Greiss reagent (Kauakou et al. 2013). Reactive oxygen species (ROS) production by macrophages is assessed by **NBT (Nitroblue tetrazolium) dye reduction assay**. Here, the macrophage cells treated with NBT are incubated with test polysaccharide fractions followed by spectrophotometric measurement at 630 nm and Stimulation Index (SI) is represented as the ratio of OD of the treated to the control macrophages. [62]. ELISA kits are used to quantitatively measure the **production of cytokines** including TNF-α, interleukins and GM-CSF [124]. Activation of NF-κB is measured by **alkaline phosphatase reporter gene assay** (Kauakou et al. 2013).

6.2 *In Vivo Assays*

The animals used for the assay should be housed under standard husbandry conditions (room temperature of 25 ± 1 °C; relative humidity 45–55% and a 12:12 h light/dark cycle) in a well-ventilated animal house approved by Committee for Control, and Supervision on Experiments on Animals (CPCSEA) and fed at ad libitum level. Prior to the experimental procedures, the mice are acclimatised to the laboratory conditions. All the experimental protocols performed must be in accordance with rules laid down by Institutional Animal Ethics Committee [140].

The test extract is administered (oral/injection) to one group of animals and the control group is administered with normal saline and different assays are performed.

Carbon clearance assay is a method to measure the effect of extract on phagocytosis [117].

After 48 h of last dose, colloidal carbon solution diluted with normal salinities injected via tail vein of each mouse and at 0 and 15 min after injection, blood samples are drawn from retro-orbital plexus and mixed with 0.1% sodium carbonate for RBC lysis. Absorbance is measured calorimetrically at 660 nm and Phagocytic Index (K) is expressed using the equation:

$$K = (\ln OD_1 - \ln OD_2)/(T_2 - T_1)$$

OD_1 and OD_2 are optical densities at times T_1 and T_2, respectively.

The phagocytic index indicates phagocytic function of mononuclear macrophage and thus non-specific immunity [62].

To detect **thymus and spleen indices**, mice are grouped into three-control, immunosuppressed and immunosuppressed mice administered with test extract. The bodyweight of mice in each group is noted and then on specific days, after the last dose, the animals are killed by cervical dislocation [170] Thymus and spleen are excised and immediately weighed. The thymus and spleen indices are calculated as follows:

$$\text{Thymus or spleen index} = [\text{weight of thymus or spleen(mg)}/\text{body weight (g)}] \times 100$$

The relative spleen and thymus weight is an indicator of non-specific immunity. If the extract has immune stimulatory effect, the immunosuppressant-induced immune organ weight reduction will be restored.

In **lymphocyte proliferation assay**, [124] the spleenocytes are isolated aseptically and cultured (5×10^5 cells/well) in the presence of concanavalin A in a 96-well plate. After 72 h of incubation, cells are treated with tetrazolium salt solution and incubated for 3 h. Then, dimethyl sulfoxide is used to dissolve the crystals formed and absorbance is measured at 570 nm (OD_{570}). Spleenocyte proliferation is measured as stimulation index (SI):

$$SI = OD_{570}\text{of stimulated cells}/OD_{570}\text{ of negative control.}$$

An increase in SI indicates enhanced lymphocyte proliferation [124].

Haemagglutinating titre (HA) [140] is a measure of activation of humoral immune response. Sheep RBCs (SRBC) are used as antigenic material for immunisation. Oral administration of the test extract is done 2 days prior to, on the day and 2 days after immunisation. On 7th day blood samples are collected and serum separated. Serially diluted serum is mixed with SRBC and cells incubated at 37 °C for an hour, after which agglutination is observed. Elevated HA titre value is an indication of induced antibody production (Shrestha et al. 2009).

Delayed-type hypersensitivity (DTH) response [62] is a measure of cell-mediated immune response and is assessed by footpad reaction test. The extract is orally administered before and after immunisation by subcutaneous injection of SRBCs (1×10^8 cells) in the right footpad. After 7 days, the left footpad is immunised with the same dose of SRBC which serve as control. At 4 and 48 h after injection, footpad thickness is measured. The difference in thickness before and after antigenic challenge is a measure of delayed-type hypersensitivity.

7 Conclusion

Microbial EPS are polysaccharides released into the close vicinity of the cells. Microbial EPS have important biotechnological, pharmaceutical and industrial applications. Microorganisms are easy to grow and are remarkable biological resources. Microbial EPS plays great role as immunomodulators by stimulating or suppressing host immune responses via interaction with specific receptors on immune cells. Various in vitro and in vivo studies have established its immunomodulatory potencies. Several factors influence the immunomodulatory effects of microbial EPS and its influence on immunomodulation, and its interaction with immune effector cells. The extensive research on immunomodulatory activities revealed the underlying mechanism of immunomodulation and stimulation of host immune responses. Immunomodulators can be utilised in complementary and alternative medicines. Immunomodulators might prove to be a great boon during disease outbreaks, where no vaccines and medicines are available.

References

1. Ahmad NH, Mustafa S, Che Man YB (2015) Microbial polysaccharides and their modification approaches: a review. Int J Food Prop 18:332–347
2. Ale EC, Bourin MJ, Peralta GH, Burns PG, Ávila OB, Contini L, Reinheimer J, Binetti AG (2019) Functional properties of exopolysaccharide (EPS) extract from *Lactobacillus fermentum* Lf2 and its impact when combined with *Bifidobacterium animalis* INL1 in yoghurt. Int Dairy J 96:114–125
3. Badel S, Bernardi T, Michaud P (2011) New perspectives for Lactobacilli exopolysaccharides. Biotechnol Adv 29(1):54–66
4. Bae SY, Yim JH, Lee HK, Pyo S (2006) Activation of murine peritoneal macrophages by sulfated exopolysaccharide from marine microalga *Gyrodinium impudicum* (strain KG03): involvement of the NF-κB and JNK pathway. Int Immunopharmacol 6(3):473–484
5. Bafna AR, Mishra SH (2005) Immunomodulatory activity of methanol extract of roots of *Cissampelospareira* Linn. Ars Pharm 46:253–262
6. Basso, AMM, De Castro RJA, de Castro TB, Guimarães HI, Polez VLP, Carbonero ER et al (2019). *I*mmunomodulatory activity of β-glucan-containing exopolysaccharides

from *Auricularia auricular* in phagocytes and mice infected with *Cryptococcus neoformans*. Med Mycol 58(2):227–239. https://doi.org/10.1093/mmy/myz042

7. Brennan J, Takei F, Wong S, Mager DL (1995) Carbohydrate recognition by a natural killer cell receptor, Ly-49C. J Biol Chem 28, 270(17):9691–9694
8. Broadbent JR, McMahon DJ, Welker DL, Oberg CJ, Moineau S (2003) Biochemistry, genetics, and applications of exopolysaccharide production in *Streptococcus thermophilus*: a review. J Dairy Sci 86(2):407–423
9. Brusilovsky M, Cordoba M, Rosental B, Hershkovitz O, Andrake MD, Pecherskaya A, Einarson MB, Zhou Y, Braiman A, Campbell KS, Porgador A (2013) Genome-wide siRNA screen reveals a new cellular partner of NK cell receptor KIR2DL4: heparan sulfate directly modulates KIR2DL4-mediated responses. J Immunol 191(10):5256–5267
10. Cambier J, Justement L, KarenNewell M, Chen Z, Harris L, Sandoval V, Ransom J (1987) Transmembrane signals and intracellular "second messengers" in the regulation of quiescent B-lymphocyte activation. Immunol Rev 95(1):37–57
11. Cao P, Wu S, Wu T, Deng Y, Zhang Q, Wang K, Zhang Y (2020) The important role of polysaccharides from a traditional Chinese medicine-Lung cleansing and detoxifying decoction against the COVID-19 pandemic. Carbohydr Polym 22:
12. Chaisuwan W, Jantanasakulwong K, Wangtueai S, Phimolsiripol Y, Chaiyaso T, Techapun C, Phongthai S, You S, Regenstein JM, Seesuriyachan P (2020) Microbial exopolysaccharides for immune enhancement: fermentation, modifications and bioactivities. Food Biosci 35:100564. https://doi.org/10.1016/j.fbio.2020.100564
13. Chatterjee S, Mukhopadhyay SK, Gauri SS, Dey S (2018) Sphingobactan, a new α-mannan exopolysaccharide from Arctic *Sphingobacterium* sp. IITKGP-BTPF3 capable of biological response modification. Int Immunopharmacol 60:84–95. https://doi.org/10.1016/j.intimp.2018.04.039
14. Chen YC, Wu YJ, Hu CY (2019) Monosaccharide composition influence and immunomodulatory effects of probiotic exopolysaccharides. Int J Biol Macromol 15(133):575–582
15. Chen HW, Yang TS, Chen MJ et al (2014) Purification and immunomodulating activity of C-phycocyanin from *Spirulina platensis* cultured using power plant flue gas. Process Biochem 49:1337–1344
16. Chen X, Xu X, Zhang L, Zeng F (2009) Chain conformation and anti-tumor activities of phosphorylated (1 → 3)-β-d-glucan from *Poria cocos*. Carbohydr Polym 15: 78(3):581–587
17. Chen X, Song L, Wang H, Liu S, Yu H, Wang X, Li R, Liu T, Li P (2019) Partial characterization, the immune modulation and anticancer activities of sulfated polysaccharides from filamentous microalgae *Tribonema* sp. Molecules 24(2):322
18. Chun H, Shin DH, Hong BS, Cho HY, Yang HC (2001) Purification and biological activity of acidic polysaccharide from leaves of *Thymus vulgaris* L. Biol Pharma Bull 24(8):941–946. https://doi.org/10.1248/bpb.24.941
19. Cooper MA, Fehniger TA, Caligiuri MA (2001) The biology of human natural killer-cell subsets. Trends Immunol 22(11):633–640
20. Costa CRLM, Menolli RA, Osaku EF, Tramontina R, de Melo RH, do Amaral AE, Duarte PAD, de Carvalho MM, Smiderle FR, Silva JLDC, Mello RG (2019) Exopolysaccharides from *Aspergillus terreus*: Production, chemical elucidation and immunoactivity. Int J Biol Macromol 139:654–664
21. Courtois A, Berthou C, Guézennec J, Boisset C, Bordron A (2014) Exopolysaccharides isolated from hydrothermal vent bacteria can modulate the complement system. PLoS ONE 9(4):
22. Cui L, Cohen JA, Broaders KE, Beaudette TT, Fréchet JM (2011) Mannosylated dextran nanoparticles: a pH-sensitive system engineered for immunomodulation through mannose targeting. Bioconjug Chem 18:22(5):949–957
23. Dambuza IM, Brown GD (2015) C-type lectins in immunity: recent developments. Curr Opin Immunol 32:21–27

24. De Stefano D, Maiuri MC, Carnuccio R (2010). Effects of hydroxytyrosol on macrophage activation. In: Olives and olive oil in health and disease prevention. Academic Press, pp 1275–1282. https://doi.org/10.1016/B978-0-12-374420-3.00141-8
25. Del Cornò M, Gessani S, Conti L (2020) Shaping the innate immune response by dietary glucans: any role in the control of cancer? Cancers 12(1):155
26. Dertli E (2013) Biochemistry and functional analysis of exopolysaccharide production in *Lactobacillus johnsonii*. Doctoral dissertation, School of Biological Sciences
27. Dhama K, Saminathan M, Jacob SS, Singh M, Karthik K, Amarpal et al (2015) Effect of immunomodulation and immunomodulatory agents on health with some bioactive principles, modes of action and potent biomedical applications. Int J Pharmacol 11(4):253–290
28. Draget KI, Taylor C (2011) Chemical, physical and biological properties of alginates and their biomedical implications. Food Hydrocolloids 1;25(2):251–256
29. Fabregas J, García D, Fernandez-Alonso M, Rocha AI, Gómez-Puertas P, Escribano JM, Coll JM (1999) In vitro inhibition of the replication of haemorrhagic septicaemia virus (VHSV) and African swine fever virus (ASFV) by extracts from marine microalgae. Antiviral Res 44(1):67–73. https://doi.org/10.1016/s0166-3542(99)00049-2
30. Ferreira SS, Passos CP, Madureira P, Vilanova M, Coimbra MA (2015) Structure-function relationships of immunostimulatory polysaccharides: a review. Carbohydr Polym 132:378–396
31. Flemming HC, Wingender J (2010) The biofilm matrix. Nat Rev Microbiol 8(9):623–633
32. Gad AS, Khadrawy YA, El-Nekeety AA et al (2011) Antioxidant activity and hepatoprotective effects of whey protein and Spirulina in rats. Nutrition 27(5):582–589
33. Gallois M, Oswald IP (2008) Immunomodulators as efficient alternatives to in-feed antimicrobials in pig production. Arch Zootech 11(3):15–32
34. Gauri SS, Mandal SM, Mondal KC, Dey S, Pati BR (2009) Enhanced production and partial characterization of an extracellular polysaccharide from newly isolated *Azotobacter* sp. SSB81. Bioresour Technol 100(18):4240–4243
35. Gazi U, Martinez-Pomares L (2009) Influence of the mannose receptor in host immune responses. Immunobiology 214:554–561
36. Ghule BV, Murugananthan G, Nakhat PD, Yeole PG (2006) Immunostimulant effects of *Capparis zeylanica* Linn. leaves. J Ethnopharmacol 108(2):311–315
37. Goldsby RA, Kindt TJ, Osborne BA, Kuby JI (2000) In: Kuby immunology. W H Freeman and Company, New York
38. Gong G, Dang T, Deng Y, Han J, Zou Z, Jing S, Zhang Y, Liu Q, Huang L, Wang Z (2018) Physicochemical properties and biological activities of polysaccharides from *Lycium barbarum* prepared by fractional precipitation. Int J Biol Macromol 109:611–618
39. Gordon S (2007) The macrophage: past, present and future. Eur J Immunol 37(S1):S9–S17
40. Gough PJ, Gordon S (2000) The role of scavenger receptors in the innate immune system. Microbes Infect 2:305–311
41. Goyal M, Baranwal M, Pandey SK et al (2019) Hetero-Polysaccharides Secreted from *Dunaliella salina* exhibit immunomodulatory activity against peripheral blood mononuclear cells and RAW 264.7 macrophages. Indian J Microbiol 59:428–435
42. Griffioen AW, Sanders L, Rijkers GT, Zegers BJ (1992) Cell biology of B lymphocyte activation by polysaccharides. J Infect Dis S71–S73
43. Gugliandolo C, Spanò A, Lentini V, Arena A, Maugeri TL (2014) Antiviral and immunomodulatory effects of a novel bacterial exopolysaccharide of shallow marine vent origin. J Appli Microbiol 116(4):1028–1034
44. Guzman S, Gato A, Lamela M, Freire-Garabal M, Calleja J (2003) Anti-inflammatory and immunomodulatory activities of polysaccharide from Chlorella stigmatophora and Phaeodactylum tricornutum. Phytother Res 17(6):665–670
45. Guzman-Murillo M, Ascencio F (2000) Anti-adhesive activity of sulphated exopolysaccharides of microalgae on attachment of red sore disease-associated bacteria and Helicobacter pylori to tissue culture cells. Lett Appl Microbiol 30(6):473–478

46. Halaas Ø, Vik R, Espevik T (1998) Induction of Fas ligand in murine bone marrow NK cells by bacterial polysaccharides. J Immunol 160(9):4330–4336
47. Han B, Baruah K, Cox E, Vanrompay D, Bossier P (2020) Structure-functional activity relationship of β-glucans from the perspective of immunomodulation: a mini-review. Front Immunol 11:658. https://doi.org/10.3389/fimmu.2020.00658
48. Han SB, Yoon YD, Ahn HJ, Lee HS, Lee CW, Yoon WK, … & Kim H M (2003). Toll-like receptor-mediated activation of B cells and macrophages by polysaccharide isolated from cell culture of Acanthopanaxsenticosus. Int immunopharmaco, 3(9):1301–1312
49. Hermetter C, Fazekas F, Hochmeister S (2018) Systematic review: syndromes, early diagnosis, and treatment in autoimmune encephalitis. Front Neurol 9. https://doi.org/10.3389/fneur.2018.00706
50. Herre J, Marshall AS, Caron E, Edwards AD, Williams DL, Schweighoffer E et al (2004) Dectin-1 uses novel mechanisms for yeast phagocytosis in macrophages. Blood 104:4038–4045
51. Hromádková Z, Ebringerová A, Sasinková V, Šandula J, Hříbalová V, Omelková J (2013) Influence of the drying method on the physical properties and immunomodulatory activity of the particulate (1 → 3)-β-D-glucan from *Saccharomyces cerevisiae*. Carbohydr Polym 1;51 (1):9–15
52. Hsieh HL, Lin CC, Shih RH, Hsiao LD, Yang CM (2012) NADPH oxidase-mediated redox signal contributes to lipoteichoic acid-induced MMP-9 upregulation in brain astrocytes. J Neuroinflamm 9(1):1–6
53. Hsu HY, Chiu SL, Wen MH, Chen KY, Hua KF (2001) Ligands of macrophage scavenger receptor induce cytokine expression via differential modulation of protein kinase signaling pathways. J Biol Chem 276(31):28719–28730
54. Huang L, Shen M, Morris GA, Xie J (2019) Sulfated polysaccharides: immunomodulation and signaling mechanisms. Trends Food Sci Technol 92:1–11
55. Hussain A, Wahab S, Ahmad MdP (2013) A systematic review of herbal immunomodulators in the Indian traditional health care system. Int J Inv Pharm Sci 1(3):261–266
56. Ilchmann A, Burgdorf S, Scheurer S, Waibler Z, Nagai R, Wellner A et al (2010) Glycation of a food allergen by the maillard reaction enhances its T-cell immunogenicity: role of macrophage scavenger receptor class A type I and II. J Allergy Clin Immun 125:175–183
57. Iwamoto Y, Xu X, Tamura T, Oda T, Muramatsu T (2003) Enzymatically depolymerized alginate oligomers that cause cytotoxic cytokine production in human mononuclear cells. Biosci Biotechnol Biochem 67(2):258–263
58. Jiang J, Wu C, Gao H, Song J, Li H (2010) Effects of Astragalus polysaccharides on immunologic function of erythrocyte in chickens infected with infectious bursa disease virus. Vaccine 28:5614–5616
59. Jiménez-Dalmaroni MJ, Gerswhin ME, Adamopoulos IE (2016) The critical role of toll-like receptors—from microbial recognition to autoimmunity: a comprehensive review. Autoimmun Rev 15(1):1–8
60. Jones SE, Paynich ML, Kearns DB, Knight KL (2014) Protection from intestinal inflammation by bacterial exopolysaccharides. J Immunol 192(10):4813–4820
61. Jung JY, Shin JS, Rhee YK, Cho CW, Lee MK, Hong HD, Lee KT (2015) In vitro and in vivo immunostimulatory activity of an exopolysaccharide-enriched fraction from *Bacillus subtilis*. J Appl Microbiol 118(3):739–752
62. Juvekar AR, Hule AK, Sakat SS, Chaughule VA (2009) In vitro and in vivo evaluation of immunomodulatory activity of methanol extract of *Momordica charantia* fruits. Drug Invent Today 1(2):89–94
63. Kalka-Moll WM, Tzianabos AO, Bryant PW, Niemeyer M, Ploegh HL., Kasper DL (2002) Zwitterionic polysaccharides stimulate T cells by MHC class II-dependent interactions. JImmunol 169(11):6149–6153
64. Kanmani P, Yuvaraj N, Paari KA, Pattukumar V, Arul V (2011) Production and purification of a novel exopolysaccharide from lactic acid bacterium *Streptococcus phocae* PI80 and its functional characteristics activity in vitro. Bioresour Technol 102(7):4827–4833

65. Kaur V, Bera, MB, Panesar PS. Kumar H, Kennedy JF (2014).Welan gum: microbial production, characterization, and applications. Int J Biomacromolecules, 65:454–461
66. Khan Z, Bhadouria P, Bisen PS (2005) Nutritional and therapeutic potential of Spirulina. Curr Pharm Biotechnol 6(5):373–379
67. Kim SJ, Park HJ, Shin HJ, Shon DH, Youn HS (2012) Suppression of TRIF-dependent signaling pathway of toll-like receptors by allyl isothiocyanate in RAW 264.7 macrophages. Int Immunopharmacol1 3(4):403–407
68. Kitazawa H, Harata T, Uemura J, Saito T, Kaneko T, Itoh T (1998) Phosphate group requirement for mitogenic activation of lymphocytes by an extracellular phosphopolysaccharide from *Lactobacillus delbrueckii* ssp. *bulgaricus*. Int J Food Microbiol 40:169–175
69. Kouakou K, Schepetkin IA, Yapi A, Kirpotina LN, Jutila MA, Quinn MT (2013) Immunomodulatory activity of polysaccharides isolated from *Alchornea cordifolia*. J Ethnopharmacol 146(1):232–242
195. Kouakou, K., Schepetkin, I. A., Yapi, A., Kirpotina, L. N., Jutila, M. A., & Quinn, M. T. (2013). Immunomodulatory activity of polysaccharides isolated from Alchornea cordifolia. Journal of ethnopharmacology, 146(1), 232-242.
71. Kumar AS, Mody K, Jha B (2007) Bacterial exopolysaccharides—a perception. J Basic Microbiol 47(2):103–117
72. Kumar A, Takada Y, Boriek AM, Aggarwal BB (2004) Nuclear factor-κB: its role in health and disease. J Mol Med 82(7):434–448
73. Kumar CG, Sujitha P (2014) Kocuran, an exopolysaccharide isolated from *Kocuria rosea* strain BS-1 and evaluation of its in vitro immunosuppression activities. Enzyme Microbial Technol 55:113–120
74. Kumar AS, Mody K (2009) Microbial exopolysaccharides: variety and potential applications. In: Microbial production of biopolymers and polymer precursors: applications and perspectives, pp 229–253
75. Laiño J, Villena J, Kanmani P, Kitazawa H (2016) Immunoregulatory effects triggered by lactic acid bacteria exopolysaccharides: new insights into molecular interactions with host cells. Microorganisms 4(3):27
76. Lake AC, Vassy R, Di Benedetto M, Lavigne D, Le Visage C, Perret GY, Letourneur D (2006) Low molecular weight fucoidan increases VEGF165-induced endothelial cell migration by enhancing VEGF165 binding to VEGFR-2 and NRP1. J Biol Chem 281:37844–37852
77. Lei N, Wang M, Zhang L, Xiao S, Fei C, Wang X, Zhang K, Zheng W, Wang C, Yang R, Xue F (2015) Effects of low molecular weight yeast β-glucan on antioxidant and immunological activities in mice. Int J Mol Sci 16(9):21575–90
78. Leung MYK, Liu C, Koon JCM, Fung KP (2006) Polysaccharide biological response modifiers. Immunol Lett 105(2):101–114
79. Li WJ, Tang XF, Shuai XX, Jiang CJ, Liu X, Wang LF, Xie MY (2017) Mannose receptor mediates the immune response to *Ganoderma atrum* polysaccharides in macrophages. J Agric Food Chem 65(2):348–357
80. Lin MH, Yang YL, Chen YP, Hua KF, Lu CP, Sheu F, Wu SH (2011) A novel exopolysaccharide from the biofilm of *Thermus aquaticus* YT-1 induces the immune response through Toll-like receptor 2. J Biol Chem 286(20):17736–17745
81. Liu M, Li N, Geng Y (2014) Influences of sulfated polysaccharide from Pine (*Pinus massoniana*) Pollen on the immunomodulatory effects of B lymphocytes in mice. Chinese J Cell Biol 36:461–469
82. Liu C, Lu J, Lu L, Liu Y, Wang F, Xiao M (2010) Isolation, structural characterization and immunological activity of an exopolysaccharide produced by *Bacillus licheniformis* 8-37-0-1. Bioresour Technol 101(14):5528–5533
83. Liu CF, Tseng KC, Chiang SS, Lee BH, Hsu WH, Pan TM (2011) Immunomodulatory and antioxidant potential of *Lactobacillus* exopolysaccharides. J Sci Food 91(12):2284–2289

84. Liu J, Yang S, Li X, Yan Q, Reaney MJ, Jiang Z (2019) Alginate oligosaccharides: production, biological activities, and potential applications. Compr Rev Food Sci Food Saf 18(6):1859–1881
85. Liu F, Zhang X, Ling P, Liao J, Zhao M, Mei L, Shao H, Jiang P, Song Z, Chen Q, Wang F (2017) Immunomodulatory effects of xanthan gum in LPS-stimulated RAW 264.7 macrophages. Carbohydr Polym 1;169:65–74
86. Llamas I, Amjres H, Mata JA, Quesada E, Béjar V (2012) The potential biotechnological applications of the exopolysaccharide produced by the halophilic bacterium *Halomonas almeriensis*. Molecules 17:7103–7120
87. Luo T, Qin J, Liu M, Luo J, Ding F, Wang M, Zheng L (2015) Astragalus polysaccharide attenuates lipopolysaccharide-induced inflammatory responses in microglial cells: regulation of protein kinase B and nuclear factor-κB signaling. Inflamm Res 64(3–4):205–212
88. Marth T, Kelsall BL (1997) Regulation of interleukin-12 by complement receptor 3 signaling. J Exp Med 185(11):1987–1995. https://doi.org/10.1084/jem.185.11.1987
89. Mata JA, Béjar V, Llamas I, Arias S, Bressollier P, Tallon R (2006) Exopolysaccharides produced by the recently described halophilic bacteria *Halomonas ventosae* and *Halomonas anticariensis*. Res Microbiol 157:827–835
90. Matsui MS, Muizzuddin N, Arad S et al (2003) Sulfated polysaccharides from red microalgae have antiinflammatory properties in vitro and in vivo. Appl Biochem Biotechnol 104:13–22. https://doi.org/10.1385/ABAB:104:1:13
91. Matsushita M, Fujita T (1995) Cleavage of the third component of complement (C3) by mannose-binding protein-associated serine protease (MASP) with subsequent complement activation. Immunobiology 194(4–5):443–448.78
92. Medzhitov R, Preston-Hurlburt P, Kopp E, Stadlen A, Chen C et al (1998) MyD88 is an adaptor protein in the hToll/IL-1 receptor family signaling pathways. Mol Cell 2(2):253–258
93. Meira DA, Pereira PCM, Marcondes-Machado J, Mendes RP, Barraviera B, Pellegrino J Jr, da Silva CL et al (1996) The use of glucan as immunostimulant in the treatment of paracoccidioidomycosis. Am J Trop Med Hyg 55(5):496–503
94. Ming H, Chen Y, Zhang F, Wang Q, Dong X, Gu J, Li Y (2015) Astragalus polysaccharides combined with cisplatin decreases the serum levels of CD44 and collagen type IV and hyaluronic acid in mice bearing Lewis lung cancer. Chinese J Cell Mol Immunol 31(7):909–913
95. Mizuno H, Tomotsune K, Islam M, Funabashi R, Albarracin L, Ikeda-Ohtsubo W, Sasaki Y (2020) Exopolysaccharides from *Streptococcus thermophilus* ST538 modulate the antiviral innate immune response in porcine intestinal epitheliocytes. Front Microbiol 11:894
96. Montoya S, Sanchez OJ, Levin L (2013) Polysaccharide production by submerged and solid-state cultures from several medicinal higher Basidiomycetes. Int J Med Mushrooms 15 (1):71–79
97. Mörk AC, Helmke RJ, Martinez JR, Michalek MT, Patchen ML, Zhang GH (1998) Effects of particulate and soluble (1–3)-β-glucans on Ca^{2+} influx in NR8383 alveolar macrophages. Immunopharmacology 40:77–89
98. Morris G, Harding S (2009) Polysaccharides, microbial. In: Encyclopedia of microbiology. Elsevier Inc., pp 482–494. https://doi.org/10.1016/B978-012373944-5.00135-8
99. Nakamura T, Suzuki H, Wada Y, Kodama T, Doi T (2006) Fucoidan induces nitric oxide production via p 38 mitogen-activated protein kinase and NF-kappaB-dependent signaling pathways through macrophage scavenger receptors. Biochem Biophys Res Commun 343:286–295
100. Netea MG, Schlitzer A, Placek K, Joosten LA, Schultze JL (2019) Innate and adaptive immune memory: an evolutionary continuum in the host's response to pathogens. Cell Host Microbe 25(1):13–26
101. Nowak B, Ciszek-Lenda M, Śróttek M et al (2012) *Lactobacillus rhamnosus* exopolysaccharide ameliorates arthritis induced by the systemic injection of collagen and lipopolysaccharide in DBA/1 mice. Arch Immunol Ther Exp 60:211–220. https://doi.org/10.1007/s00005-012-0170-5

102. Nutt SL, Hodgkin PD, Tarlinton DM, Corcoran LM (2015) The generation of antibody-secreting plasma cells. Nat Rev Immunol 15:160–171
103. Ogata H, Su I, Miyake K, Nagai Y, Akashi S et al (2000) The toll like receptor protein RP105 regulates lipopolysaccharide signaling in B cells. J Exp Med 192(1):23–29
104. Öner ET (2013) Microbial production of extracellular polysaccharides from biomass. In: Fang Z (ed) Pretreatment techniques for biofuels and biorefineries. Springer, Berlin Heidelberg, pp 35–56
105. Palomba S, Cavella S, Torrieri E, Piccolo A, Mazzei P, Blaiotta G, Ventorino V, Pepe O (2012) Wheat sourdough from Leuconostoc lactis and Lactobacillus curvatus exopolysaccharide-producing starter culture:polyphasic screening, homopolysaccharide composition and viscoelastic behavior. Appl Environ Microbiol
106. Pan D, Liu J, Zeng X, Liu L, Li H, Guo Y (2014) Immunomodulatory activity of selenium exopolysaccharide produced by *Lactococcus lactis subsp*, Lactis. Food Agr Immunol 26 (2):248–259. https://doi.org/10.1080/09540105.2014.894000
107. Pan D, Mei X (2010) Antioxidant activity of an exopolysaccharide purified from *Lactococcus lactis subsp. lactis12*. Carbohydr Polym 80(3):908–914
108. Park GT, Go RE, Lee HM, Lee GA, Kim CW, Seo JW, Hwang KA (2017) Potential anti-proliferative and immunomodulatory effects of marine microalgal exopolysaccharide on various human cancer cells and lymphocytes in vitro. Mar Biotechnol 19:136–146. https://doi.org/10.1007/s10126-017-9735-y
109. Patten DA, Collett A (2013) Exploring the immunomodulatory potential of microbial-associated molecular patterns derived from the enteric bacterial microbiota. Microbiology159(Pt_8):1535–1544
110. Paynich ML, Jones-Burrage SE, Knight KL (2017) Exopolysaccharide from *Bacillus subtilis* induces anti-inflammatory M2 macrophages that prevent T cell–mediated disease. J Immunol 198(7):2689–2698
111. Pelizon AC, Kaneno R, Soares AMVC, Meira DA, Sartori A (2005) Immunomodulatory activities associated with β-glucan derived from *Saccharomyces cerevisiae*. Physiol Res 54 (5):557–564
112. Pepe O, Ventorino V, Cavella S, Fagnano M, Brugno R (2013) Prebiotic content of bread prepared with flour from immature wheat grain and selected dextran-producing lactic acid bacteria. Appl Environ Microbiol 79(12):3779–3785
113. Perez Ramos A, Mohedano ML, Pardo MÁ, López P (2018) β-glucan-producing *Pediococcus parvulus* 2.6: test of probiotic and immunomodulatory properties in Zebrafish models. Front Microbiol 9:1684
114. Di Pippo F, Ellwood NT, Gismondi A, Bruno L, Rossi F, Magni P, De Philippis R (2013) Characterization of exopolysaccharides produced by seven biofilm-forming cyanobacterial strains for biotechnological applications. J Appl Phycol 25(6):1697–1708
115. Poli A, Anzelmo G, Tommonaro G, Pavlova K, Casaburi A, Nicolaus B (2010) Production and chemical characterization of an exopolysaccharide synthesized by psychrophilic yeast strain *Sporobolomyces salmonicolor* AL1 isolated from Livingston Island, Antarctica. Folia Microbiol 55(6):576–581
116. Poli A, Di Donato P, Abbamondi GR, Nicolaus B (2011) Synthesis, production, and biotechnological applications of exopolysaccharides and polyhydroxyalkanoates by archaea. Archaea 2011. https://doi.org/10.1155/2011/693253
117. Ponkshe CA, Indap MM (2002) In vivo and in vitro evaluation for immunomodulatory activity of three marine animal extracts with reference to phagocytosis. Indian J Exp Biol 40:1399–1402
118. PrabhuDas MR, Baldwin CL, Bollyky PL, Bowdish DM, Drickamer K, Febbraio M, McVicker B (2017) A consensus definitive classification of scavenger receptors and their roles in health and disease. J Immunol 198(10):3775–3789
119. Pérez Fernández ME, Quesada E, Gálvez J, Ruiz C (2000) Effect of exopolysaccharide V2-7, isolated from *Halomonas eurihalina*, on the proliferation in vitro of human peripheral blood lymphocytes. Immunopharm Immunot 22(1):131–141

120. Qi C, Cai Y, Gunn L, Ding C, Li B, Kloecker G, Qian K, Vasilakos J, Saijo S, Iwakura Y, Yannelli JR (2011) Differential pathways regulating innate and adaptive antitumor immune responses by particulate and soluble yeast-derived β-glucans. Am J Hematol 23;117 (25):6825–6836
121. Raposo MFDJ, De Morais RMSC, Bernardo de Morais AMM (2013) Bioactivity and applications of sulphated polysaccharides from marine microalgae. Mar Drugs 11(1):233–252
122. Raposo MFJ, Morais AMMB, Morais RMSC (2014) Bioactivity and applications of polysaccharides from marine microalgae. In: Merillon JM, Ramawat KG (eds) Polysaccharides: bioactivity and biotechnology. Springer, Switzerland, p 38
123. Ray A, Dittel BN (2010) Isolation of mouse peritoneal cavity cells. Jove J Vis Exp 35:e1488
124. Ren D, Li C, Qin Y, Yin R, Du S, Liu H, Jin N (2015) Evaluation of immunomodulatory activity of two potential probiotic Lactobacillus strains by in vivo tests. Anaerobe35:22–27
125. Roeder A, Kirschning CJ, Rupec RA, Schaller M, Weindl G, Korting HC (2004) Toll-like receptors as key mediators in innate antifungal immunity. Med Mycol J42(6):485–498
126. Ross GD (2000) Regulation of the adhesion versus cytotoxic functions of the Mac-1/CR3/alpha M beta2-integrin glycoprotein. Crit Rev Immunol 20:197–222
127. Rowe DC, McGettrick AF, Latz E, Monks BG, Gay NJ, Yamamoto M, Golenbock DT (2006) The myristoylation of TRIF-related adaptor molecule is essential for Toll-like receptor 4 signal transduction. Proc Natl Acad Sci USA 103:6299–6304
128. Ruiz-Bravo A, Jimenez-Valera M, Moreno E, Guerra V, Ramos-Cormenzana A (2001) Biological response modifier activity of an exopolysaccharide from *Paenibacillus jamilae* CP-7. Clin Diagn Lab Immunol 8(4):706–710
129. Rühmann B, Schmid J, Sieber V (2015) High throughput exopolysaccharide screening platform: from strain cultivation to monosaccharide composition and carbohydrate fingerprinting in one day. Carbohyd Polym 122:212–220
130. Saito Y, Watanabe K, Fujioka D, Nakamura T, Obata JE, Kawabata K, Shimizu T (2012) Disruption of group IVA cytosolic phospholipase A2 attenuates myocardial ischemia-reperfusion injury partly through inhibition of TNF-alpha-mediated pathway. Am J Physiol HeartCirc Physiol 302:2018–2030
131. Sajna KV, Kamat S (2020) Antibodies at work in the time of severe acute respiratory syndrome coronavirus 2. Cytotherapy (In press). https://doi.org/10.1016/j.jcyt.2020.08.009
132. Šandula J, Kogan G, Kačuráková M, Machová E (1999) Microbial (1 → 3)-β-d-glucans, their preparation, physico-chemical characterization and immunomodulatory activity. Carbohyd Polym 38(3):247–253. https://doi.org/10.1016/s0144-8617(98)00099-x
133. Schepetkin IA, Quinn MT (2006) Botanical polysaccharides: macrophage immunomodulation and therapeutic potential. Int Immunopharmacol 6(3):317–333
134. Schorey JS, Cooper AM (2003) Macrophage signalling upon mycobacterial infection: the MAP kinases lead the way. Cell Microbiol 5:133–142
135. Schuch RA, Oliveira TL, Collares TF, Monte LG, Inda GR, Dellagostin OA, Vendruscolo CT, Moreira AD, Hartwig DD (2017) The use of xanthan gum as vaccine adjuvant: an evaluation of immunostimulatory potential in balb/c mice and cytotoxicity in vitro. Biomed Res Int 7:2017
136. Selmi C, Leung PS, Fischer L et al (2011) The effects of Spirulina on anemia and immune function in senior citizens. Cell Mol Immunol 8(3):248–254
137. Shankar T, Vijayabaskar P, Sivasankara NS, Sivakumar T (2014) Screening of exopolysaccharide producing bacterium *Frateuria aurentia* from elephant dung. App Sci Report 1:105–109
138. Shao BM, Xu W, Dai H, Tu P, Li Z, Gao XM (2004) A study on the immune receptors for polysaccharides from the roots of *Astragalus membranaceus*, a Chinese medicinal herb. Biochem Biophys Res Commun 320(4):1103–1111
139. Shokri H, Khosravi A, Taghavi M (2014) Efficacy of *Spirulina platensis* on immune functions in cancer mice with systemic candidiasis. J Mycol Res 1(1):7–13

140. Shrestha PR, Handral MU (2017) Evaluation of immunomodulatory activity of extract from rind of *Nephelium lappaceum* fruit. Int J Pharm Pharm Sci 9(1):38–43
141. Smirnou D, Hrubošová D, Kulhánek J, Švík K, Bobková L, Moravcová V, Krčmář M, Franke L, Velebný V (2014) *Cryptococcus laurentii* extracellular biopolymer production for application in wound management. Appl Biochem Biotechnol 174:1344–1353. https://doi.org/10.1007/s12010-014-1105-x
142. Smith-Garvin JE, Koretzky GA, Jordan MS (2009) T cell activation. Annu Rev immunol 27:591–619
143. Sudha P, Asdaq SM, Dhamingi SS, Chandrakala GK (2010) Immunomodulatory activity of methanolic leaf extract of Moringa oleifera in animals. Indian J Physiol Pharmacol 54 (2):133–140
144. Sun L, Wang L, Zhou Y (2012) Immunomodulation and antitumor activities of different-molecular-weight polysaccharides from *Porphyridium cruentum*. Carbohyd Polym 87(2):1206–1210. https://doi.org/10.1016/j.carbpol.2011.08.097
145. Surayot U, Lee S, You S (2018) Effects of sulfated fucan from the sea cucumber *Stichopus japonicus* on natural killer cell activation and cytotoxicity. Int J Biol Macromol 1(108):177–184
146. Surayot U, Wang J, Seesuriyachan P, Kuntiya A, Tabarsa M, Lee Y, You S (2014) Exopolysaccharides from lactic acid bacteria: structural analysis, molecular weight effect on immunomodulation. Int J Biol Macromol 68:233–240
147. Surayot U, You S (2017) Structural effects of sulfated polysaccharides from *Codium fragile* on NK cell activation and cytotoxicity. Int J Biol Macromol 98:117–124
148. Surayot U, You S (2017) Structural effects of sulfated polysaccharides from Codium fragile on NK cell activation and cytotoxicity. Int J Biological Macromolecules 98:117–12
149. Surayot U, Lee JH, Park WJ, You SG (2016) Structural characteristics of polysaccharides extracted from *Cladophora glomerata* Kützing affecting nitric oxide releasing capacity of RAW 264.7 cells. Bioact Carbohydr 7(1):26–31
150. Surayot U, Wang J, Seesuriyachan P, Kuntiya A, Tabarsa M, Lee Y, You SG (2014) Exopolysaccharides from lactic acid bacteria: Structural analysis, molecular weight effect on immunomodulation. Int J Biological Macromol 68:233–240
151. Sutherland IW (1990) Biotechnology of microbial exopolysaccharides. Cambridge UniversityPress, Cambridge
152. Tamegai H, Takada Y, Okabe M, Asada Y, Kusano K, Katagiri YU, Nagahara Y (2013) *Aureobasidium* pullulans culture supernatant significantly stimulates R-848-activated phagocytosis of PMA-induced THP-1 macrophages. Immunopharmacol Immunotoxicol 35(4):455–461
153. Taylor ME (2001) Structure and function of the macrophage mannose receptor. Mammalian carbohydrate recognitionsystems. Springer, Berlin, Heidelberg, pp 105–121
154. Taylor ME, Conary JT, Lennartz MR, Stahl PD, Drickamer K (1990) Primary structure of the mannose receptor contains multiple motifs resembling carbohydrate recognition domains. Int J Biol Chem 265:12156–12162
155. Thirugnanasambandham K, Sivakumar V, Maran JP (2014) Process optimization and analysis of microwave assisted extraction of pectin from dragon fruit peel. Carbohyd Polym 112:622–626
156. Thoma-Uszynski S, Stenger S, Takeuchi O, Ochoa MT, Engele M, Sieling PA, Akira S (2001) Induction of direct antimicrobial activity through mammalian toll-like receptors. Science 291(5508):1544–1547
157. Thornton BP, Vetvicka V, Pitman M, Goldman RC, Ross GD (1996) Analysis of the sugar specificity and molecular location of the beta-glucan-binding lectin site of complement receptor type 3 (CD11b/CD18). J Immunol 156:1235–1246
158. Tzianabos AO (2000) Polysaccharide immunomodulators as therapeutic agents: structural aspects and biologic function. Clin Microbiol Rev 13(4):523–533

159. Tzianabos AO, Finberg RW, Wang Y, Chan M, Onderdonk AB, Jennings HJ, Kasper DL (2000) T cells activated by zwitterionic molecules prevent abscesses induced by pathogenic bacteria. J Biol Chem 275:6733
160. Ventorino V, Nicolaus B, Di Donato P, Pagliano G, Poli A, Robertiello A, Iavarone V, Pepe O (2019) Bioprospecting of exopolysaccharide-producing bacteria from different natural ecosystems for biopolymer synthesis from vinasse. Chem Biol Technol Agric 6 (1):18. https://doi.org/10.1186/s40538-019-0154-3
161. Wagner H (1990) Search for plant derived natural products with immunostimulatory activity: recent advances. Pure Appl Chem 62(7):1217–1222
162. Wang CL, Lu CY, Pi CC, Zhuang YJ, Chu CL, Liu WH et al (2012) Extracellular polysaccharides produced by *Ganoderma formosanum* stimulate macrophage activation via multiple pattern-recognition receptors. BMC Complement Altern Med 12:119. https://doi.org/10.1186/1472-6882-12-119
163. Wang F, Qiao L, Chen L, Zhang C, Wang Y, Wang Y, Liu Y, Zhang N (2016) The immunomodulatory activities of pullulan and its derivatives in human pDC-like CAL-1 cell line. Int J Biol 1(86):764–771
164. Wang Q, Zhao G, Lin J, Li C, Jiang N, Xu Q, Zhang J (2016) Role of the mannose receptor during *Aspergillus fumigatus* infection and interaction with Dectin-1 in corneal epithelial cells. Cornea 35(2):267–273
165. Wang G, Zhu L, Yu B, Chen K, Liu B, Liu, J, Chen K (2016) Exopolysaccharide from *Trichoderma pseudokoningii* induces macrophage activation. Carbohyd Polym 149:112–120
166. Weng BB, Lin YC, Hu CW, Kao MY, Wang SH, Lo DY, Lai TY, Kan LS, Chiou RY(2011) Toxicological and immunomodulatory assessments of *botryosphaeran* (β-glucan) produced by Botryosphaeria rhodina RCYU 30101. Food Chem Toxicol 1;49(4):910–916
167. Wesche H, Henzel WJ, Shillinglaw W, Li S, Cao Z (1997) MyD88: an adaptor that recruits IRAK to the IL-1 receptor complex. Immunity 7(6):837–847
168. Xaplanteri P, Lagoumintzis G, Dimitracopoulos G, Paliogianni F (2009) Synergistic regulation of *Pseudomonas aeruginosa*-induced cytokine production in human monocytes by mannose receptor and TLR2. Eur J Immunol 39(3):730–740
169. Xiao R, Zheng Y (2016) Overview of microalgal extracellular polymeric substances (EPS) and their applications. Biotechnol Adv 34(7):1225–1244
170. Xu CL, Wang YZ, Jin ML, Yang XQ (2009) Preparation, characterization and immunomodulatory activity of selenium-enriched exopolysaccharide produced by bacterium *Enterobacter cloacae* Z0206. Bioresour Technol 100(6):2095–2097
171. Xu, X, Wu, X, Wang, Q, Cai, N, Zhang, H, Jiang, Z, … & Oda, T. (2014) Immunomodulatory effects of alginate oligosaccharides on murine macrophage RAW264. 7 cells and their structure–activity relationships. J Agr Food Chem 62(14):3168–3176
172. Yamagishi T, Tsuboi T, Kikuchi K (2003) Potent natural immunomodulator, rice water-soluble polysaccharide fractions with anticomplementary activity. Cereal Chem 80 (1):5–8. https://doi.org/10.1094/cchem.2003.80.1.5
173. Yamamoto M, Sato S, Hemmi H, Uematsu S, Hoshino K, Kaisho T, Takeuchi O, Takeda K, Akira S (2003) TRAM is specifically involved in the Toll-like receptor 4—mediated MyD88-independent signaling pathway. Nat Immunol 4(11):1144–1150
174. Yang Y, Yin C, Zhang MW (2012) Immunomodulatory activities and mechanisms of polysaccharides on T/B lymphocytes. Chinese J Cell Biol 34:67–74
175. Yim JH, Son E, Pyo S, Lee HK (2005) Novel sulfated polysaccharide derived from red-tide microalga *Gyrodiniumim pudicum* strain KG03 with immunostimulating activity in vivo. Mar Biotechnol 7:331–338
176. Yin M, Zhang Y, Li H (2019) Advances in research on immunoregulation of macrophages by plant polysaccharides. Front Immunol 10:145
177. Yoon HS, Kim JW, Cho HR, Moon SB, Shin HD, Yang KJ, Lee HS, Kwon YS, Ku SK (2010) Immunomodulatory effects of *Aureobasidium* pullulans SM-2001 exopolymers on cyclophosphamide-treated mice. J Microbiol Biotechnol 20(2):438–445

178. You X, Li Z, Ma K, Zhang C, Chen X, Wang G, Li W (2020) Structural characterization and immunomodulatory activity of an exopolysaccharide produced by *Lactobacillus helveticus* LZ-R-5. Carbohyd Polym 235:
179. Yu W, Chen G, Zhang P, Chen K (2016) Purification, partial characterization and antitumor effect of an exopolysaccharide from *Rhizopus nigricans*. Int J Biol Macromol 82:299–307
180. Yu Q, Nie SP, Li WJ, Zheng WY, Yin PF, Gong DM, Xie MY (2013) Macrophage immunomodulatory activity of a purified polysaccharide isolated from *Ganodermaatrum*. Phytother Res 27(2):186–191
181. Yu L, Sun G, Wei J, Wang Y, Du C, Li J (2016) Activation of macrophages by an exopolysaccharide isolated from Antarctic Psychrobacter sp. B-3. Chin J Oceanol Limnol 34 (5):1064–1071
182. Zaidman BZ, Yassin M, Mahajna J, Wasser SP (2005) Medicinal mushroom modulators of molecular targets as cancer therapeutics. Appl Microbiol Biotechnol 67(4):453–468
183. Zamze S, Martinez-Pomares L, Jones H, Taylor PR, Stillion RJ, Gordon S, Wong SY (2002) Recognition of bacterial capsular polysaccharides and lipopolysaccharides by the macrophage mannose receptor. J Biol Chem 277(44):41613–41623
184. Zhang C, Huang K (2005) Characteristic immunostimulation by MAP, a polysaccharide isolated from the mucus of the loach, *Misgurnus anguillicaudatus*. Carbohyd Polym 59:75–82
185. Zhang FX, Kirschning CJ, Mancinelli R, Xu XP, Jin Y et al (1999) Bacterial lipopolysaccharide activates nuclear factor-kappa B through interleukin-1 signaling mediators in cultured human dermal endothelial cells and mononuclear phagocytes. J BiolChem 274(12):7611–7614
186. Zhang P, Liu W, Peng Y, Han B, Yang Y (2014) Toll like receptor 4 (TLR4) mediates the stimulating activities of chitosan oligosaccharide on macrophages. Int Immunopharmacol 23:254–261
187. Zhao T, Feng Y, Li J, Mao R, Zou Y, Feng W, Wu X (2014) Schisandra polysaccharide evokes immunomodulatory activity through TLR 4-mediated activation of macrophages. Int J Biol Macromol 65:33–40

Novel Insights of Microbial Exopolysaccharides as Bio-adsorbents for the Removal of Heavy Metals from Soil and Wastewater

Naga Raju Maddela, Laura Scalvenzi, and Matteo Radice

Abstract Exopolysaccharides (EPSs) are natural polymers of high molecular weight, are produced by both prokaryotic and eukaryotic microorganisms. However, the most extensively studied substances are bacterial EPSs, which have a wide range of chemical structures. Due to the unique properties they possess, EPSs are often used in food, pharmaceutical, and other industries. Above all else, EPSs are the preferable materials in the removal of heavy metal (HM) impurities in soils and wastewater, because, certain EPSs are acidic in nature and can tolerate HMs stress. Therefore, EPSs have high propensities to remove the HMs by forming organo-metal complexes. The present chapter is aimed to pay much attention to EPS-based bioremediation strategies for the removal of HMs from the contaminated sites, by setting up some successful case studies of the recent past. Besides these, this chapter will highlight the fundamentals of microbial EPSs and their characteristics.

Keywords Bacterial EPS · Heavy metals · Bioremediation · EPS functional groups · EPS-HMs interactions

1 Introduction

Human health and environmental security are under great threat due to extensive pollution of soils and other components of the environment by heavy metals (HMs) all over the world [95]. Evidently, in a recent survey, it has been found that >20 million hectares of land in China contaminated by HMs [35]. The principal reasons for the increased levels of HMs in the environment are industrial-

N. R. Maddela (✉)
Instituto de investigación, Universidad Técnica de Manabí, Portoviejo, Ecuador

Facultad la Ciencias de la Salud, Universidad Técnica de Manabí, Portoviejo 130105, Ecuador

L. Scalvenzi · M. Radice
Departamento de Ciencias de la Tierra, Universidad Estatal Amazónica, Puyo, Ecuador

A. K. Nadda et al. (eds.), *Microbial Exopolysaccharides as Novel and Significant Biomaterials*, Springer Series on Polymer and Composite Materials,
https://doi.org/10.1007/978-3-030-75289-7_10

ization [99], increased use of agrochemicals (e.g. fertilizers, [43]), etc. However, the actual global pollution by HMs could be more than these estimations due to the following facts: (i) consideration of small areas in the investigation, (ii) difficulty in conducting large-scale and long-term investigations in agroecosystems. For these reasons, potential HMs pollution level and their consequences on human health and ecosystem could be considered even worse. It should be remembered that HMs exhibit both carcinogenic and non-carcinogenic risks in humans [34]. Chronic exposure to HMs causes serious health disorders in humans, for instance, lung cancer, bone fractures, kidney dysfunction, these may further lead to several secondary disorders including but not limited to impairments in immune, nervous and endocrine systems. For these reasons, it is obligatory to reduce the environmental burden caused by HMs and the remediation of HMs-contaminated media.

There are different techniques available for the remediation of HMs-contaminated sites. Under one tag, i.e. physicochemical remediation techniques, examples include remediation approaches based on electrochemical and nanomaterials, encapsulation, extracted washing, landfill, solidification, soil flushing, surface capping, etc. It is noteworthy that though physicochemical remediation methods are rapid in the restoration of contaminated sites, these techniques are not economically effective and not eco-friendly as they do damage the ecosystem. On the other side, there is an eco-friendly method, in which remediation is achieved by using microorganisms (microbial remediation) and plants (phytoremediation), called 'bioremediation'. The following points deeply address the question "*Why bioremediation has become popular in the restoration of HMs-contaminated environmental media?*". Bioremediation offers undisputable advantages—(i) Transformation of contaminants (into either biomass and/or harmless products of metabolism) and become simmobilized or separated, (ii) highly efficient in the restoration of contaminated site, (iii) economically feasible, (iv) easily available technique, (v) eco-friendly. In several occasions, both plants [27] and microorganisms [41] have been successfully proved as potential tools for the remediation of HMs-polluted sites. However, regarding bacterial remediation contaminated sites, scientific communities are worried about the pathogenicity of microorganisms that are to be used. This problem can be overcome if remediation is designed with the microbial products rather than whole cells of microorganisms. Towards this, in the recent past, bacterial exopolymeric substances have become center of attraction in the remediation of HMs-contaminated sites as an adsorbent [6, 53].

Bacterial EPSs are the extensively studied byproducts that exhibit numerous applications in a wide range of industries such as food, pharmacy, nutraceutical, herbicides, and now in the field of bioremediation [53]. EPS molecules have high affinity towards HMS, as reported previously [44], and the principal mechanisms that mediate the binding between EPSs and HMs are complexation, ion exchange, surface micro-precipitation, etc. Two factors greatly affect the binding of EPSs with HMs are composition of EPS and the availability of binding sites on EPSs flocks. Interaction between EPSs and HMs leads to the formation of organo-metal complexes, where HMs are bound to different functional groups (e.g. hydroxyl, carboxyl, phosphoric amine groups, etc.) of EPSs. For example, hydrophilic interactions between HMs and carboxylic/phosphoric functional groups of EPSs are common during the complexation phenomenon [80]. On the other side, HMs are embedded on EPSs surfaces by hydrophobic interactions. Overall, pH of the environment plays an important role in the interactions of EPSs with HMs and in the formation of organo-metal complexes, because pH determines the distribution of charges on the EPS surfaces [98].

Keeping in view of the extension of environmental burden caused by HMs and the potential of EPSs in the restoration of HMs-contaminated environmental media, this chapter has been designed to provide in-depth insights over the bacterial EPSs and their implications in the field of bioremediation. In the beginning, there is detailed information on microbial EPS with special emphasis on the diversity in microbial EPSs and their commercial uses. Then, there is a detailed literature review on the structural characterization of EPS by Fourier transform infrared (FTIR) spectroscopy. Thereafter, special attention has been paid towards EPS-metal interactions, where the latest literature has been reviewed on several issues such as bacteria-heavy metal interactions, mechanism of interactions, EPS-mediated bioremediation of soil and water, and advantages of EPS-mediated bioremediation over conventional methods. Finally, knowledge gaps and future directions of research in the field of EPS-mediated bioremediation.

2 Microbial EPSs

Chemically, EPSs are highly heterogeneous and contain multiple functional groups which make them polyfunctional macromolecules of high molecular weight [30, 31]. Microbial polysaccharides are broadly classified into 3 categories based on their location and functions [97], they are capsular polysaccharides (CPSs), lipopolysaccharides (LPSs), and exopolysaccharides (EPSs). In general, CPSs [25] and LPSs [85] are associated with virulence functions, whereas EPSs offer several benefits to microorganisms such as protection from the extreme environment (e.g. cold, saline conditions), attachment to substrate, colony formation [2, 20, 73], and carbon and energy sources [87]. Microbial groups that do produce EPSs are bacteria and algae. In the following two subsections, the diversity of microbial EPSs and extraction methods are discussed.

2.1 Diversity in Microbial EPSs

Chemically, microbial EPSs are made up of both carbohydrate and non-carbohydrate (e.g. acetate, phosphate, pyruvate, and succinate) compounds, and broadly can be divided into homopolysaccharides (e.g. cellulose, dextran, pullulan) and heteropolysaccharides (e.g. hyaluronic acid, xanthan) [38]. Microbial EPSs exhibit diverse rheological properties, thus they readily produce viscous pseudoplastic liquids [97]. In one of our very recent investigations, we characterized EPSs of 23 sludge-bacteria and found that EPS physicochemical properties are highly strain-dependent [69]. EPS dry weight of 23 sludge bacteria was in the range of 0.58–1.28 g L^{-1}. There was a higher carbohydrate content (52–530 mg L^{-1}) than the protein content (5.2–53.0 mg L^{-1}), therefore, protein-to-polysaccharides ratios of all 23 EPSs were <1.0. Apparent viscosities at a shear rate of 112.5 s^{-1} was in the range of 1.92–2.37 mPa s^{-1}. Furthermore, viscosity values have been increased by 24–59% when shear rate was increased from 112.5 to 244.9 s^{-1}. It is believed that EPS with higher viscosity values show much higher adherence to the surfaces than the EPSs having lower viscosities. We also determined the zeta potential of all 23 EPSs, and the values were in the broad ranges such as −1.61 to −17.40 mV.

The chemical linkage that is found between the monomers in EPS is highly varied among different microorganisms. For instance, EPS of *Gluconacetobater, Agrobacterium Rhizobium* and *Sarcina* is made up of cellulose contains β-(1,4)-d-glucan linked glucose monomers [10, 24]. Similarly, β-(1,3) glycosidic linked glucose or sucrose contained curdlan is found in the EPS of *Agrobacterium* sp., *Bacillus* sp., *Cellulomonas* sp., and *Rhizobium* sp. [61], α-(1,6)-glycosidic linked glucose contained dextran is found in the EPS of *Leuconostoc, Streptococcus, Weissella, Pediococcus* and *Lactobacillus* [78]. In case of *Sphingomonas paucimobilis*, EPS is made up of gellan contains two types of monomers (e.g., glucose and rhamnose) with multiple chemical linkages such as β-D-glucose, β-D-glucuronic acid, β-D-glucose, and α-L-rhamnose [60]. Likewise, microbial EPS are known to contain different types of chemicals such as hyaluronic acid, levan, pullulan, mutan, xanthan gum, etc., in different bacteria (e.g. *Acetobacter, Aureobasidium pullulans, Bacillus, Brenneria, Cryphonectria parasitica, Cytaria* spp., *Geobacillus, Halomonas, Lactobacillus, Rhodototula bacarum, Saccharomyces, Streptococcus equi, Streptococcus equisimilis, Streptococcus mutans, Streptococcus pyogenes, Streptococcus thermophilus, Teloschistes flavicans, Xanthomonas campestris* and *Zymomonas*), where the EPS monomers (e.g. glucuronic acid, N-acetylglucosamine, fructose, glucose, glucuronic acid, and pyruvic acid) are linked by different chemical linkages such as β-1,3 N-acetylglucosamine and β-1,4 glucuronic acid [40], β-(2,6) glycosidic bonds β-(1,2)-linked side chains [47], α-(1,6)-linked α-(1,4)-D-triglucoside maltotriose units [92], α-(1,3)—linked glucose α-(1,6) bonds side chains [54], and 1,4-linked β-D-glucose residues [50]. These insights clearly imply that EPS is a highly diversified biopolymer in microorganisms, in terms of chemical composition and linkage. This is the reason why the biofouling formation potentials (on the membrane surfaces) [69] and biofilm formation potentials (in microtiter plates) [65] by different environmental isolates (in this case, bacteria) are purely strain-dependent.

2.2 Commercial Uses

Microbial EPSs have widest applications in different areas as viscosifier, stabilizers, emulsifiers. Nevertheless, microbial cells are the richest source for the recovery of different biopolymers (so-called EPSs), such as cellulose, curdlan, dextran, emulsan, gellan, hyaluronic acid, levan, pullulan, and xanthan gum [97]. In food industries, the majority of the above biopolymers (except emulsan and levan) are used for various purposes. For example, as a natural non-digestible fibers, cellulose is used in the food industries [24]. Curdlan has several functions, such as it improves the viscous nature, stability, creaminess of food, besides being used in the preparation of gels and as immobilization matrix [15]. Dextran has multiple functions in food industries, such as it improves the softness or moisture retention, viscosity, rheology, texture, and volume, and it prevents the crystallization of food substances [78]. Dextran is also used in cosmetics as a moisturizer and thickener, besides using dextran as a microcarrier in the culturing of tissues and cells. Gellan is widely used as a food additive, attributed to its stabilizing and thickening functions [4]. Also, lactic acid bacteria are immobilized by gellan microcapsules. Pullulan has applications both in food and pharmaceutical industries. Especially in the pharmacy, pullulan acts as a potential carrier for the delivery of genes and proteins embedded in the pullulan film which are physiologically acceptable and orally consumable [84]. Whereas xanthan gum is familiar as an emulsifier, stabilizer, suspending agent and viscosifier, thus, xanthan gum has wide implications in food industry and others (e.g. agriculture, cosmetic, oil recovery, pharmaceutical, textile, etc.) [49]. Similarly, another biopolymer emulsan is also widely used in different areas, such as in the recovery of crude oil, as a cleansing agent in the preparation of several personal care products (e.g., creams, lotion, soap, shampoo, and toothpaste) [16]. The other biopolymers, like hyaluronic acid [14] is familiar in food and cosmetic industries, and levan [76] is widely used in the preparation of confectionaries (food items that are rich in sugar and carbohydrates) due to viscous and stabilizing properties.

3 Characterization of EPS by Fourier Transform Infrared (FTIR) Spectroscopy

EPS molecules are characterized by several spectroscopic methods, such as UV-Visible absorption spectroscopy, differential absorbance spectroscopy (DAS), fluorescence spectroscopy (FS), fluorescence excitation-emission matrix (EEM), Fourier-transform infrared spectroscopy (FTIR), attenuated total reflection (ATR) FTIR and Raman spectroscopy [13]. However, vibrational spectroscopic methods do provide precise structural information than the fluorescence methods, therefore, in this section, characterization of EPS molecules by FTIR is presented. In general, spectral ranges from 4000 to 400 cm^{-1} is used to reveal the structural features (basic vibrations and associated rotational-vibrational) of EPSs. There are two regions in the IR spectra, one is a high wavelength region (4000–1350 cm^{-1}) used to study the functional groups, and another is a low wavelength region (1350–400 cm^{-1}) called fingerprint region. Both high and low wavelength regions are useful in the identification and discrimination of the chemical structure of mixed compounds. Chemical bonds that are present in the EPS molecules can be identified by different spectral tools such as FTIR and attenuated total reflection (ATR) FTIR spectra. According to these spectral analyses, any peaks at 1650 and 1560 cm^{-1} are confined to amide I and amide II bonds, respectively [71]. Amide I and amide II are the two unique bonds of protein secondary structure. Presence of a broad band at 1600–1650 cm^{-1} indicates quinone and ketone C = O bonds or aromatic rings [12]. A peak at 1400 cm^{-1} belongs to $-COO^{-}$ bond or phenolic C–O bond, and at 1100 cm^{-1} is related to C–O–C bond of polysaccharide structure [11]. Whereas peaks at 1720 cm^{-1} and 1275 cm^{-1} are attributed to C = O (carboxylic group) and C–N bonds, respectively [94]. Presence of peaks at ~1020 cm^{-1} and 920 cm^{-1} are belonging to uronic acid [7] and α-1,4-glycosidic bond, respectively. In one of our recent investigations, EPS has been isolated from 23 sludge-bacteria, and characterization of EPS by FTIR revealed the following functional groups [69]: amide I (1640 cm^{-1}), amide II (1550 cm^{-1}), $-COO^{-}$ (1400 cm^{-1}), polysaccharides region (~1100 cm^{-1}), uronic acid (~1020 cm^{-1}), α-1,4-glycosidic bond (920 cm^{-1}). Nevertheless, three functional groups predominantly observed in each region of protein (1400, 1550, and 1640 cm^{-1}) and polysaccharides (920, 1020, and 1100 cm^{-1}) in the EPSs of 23 sludge-bacteria. Furthermore, there are certain software programs that does help in the process of I R spectral data in order to have in-depth information about the hidden peaks. One of the examples for the peak separation and analysis program is PeakFit, that is an automated nonlinear software package which can find and fit a maximum number of peaks (e.g. 100) to a given data set at a time. It is possible to obtain clearly distinguished curve-fitted peaks by different applications of PeakFit program, and some of these applications are as follows—second derivative revolution enhancement, Fourier deconvolution of Gaussian instrument response function, AutoFit Peaks II second derivative function, AutoFit Peaks III deconvolution function, etc. By using PeakFit and its functions, curve-fitted spectra were obtained for the EPSs of three bacteria species (*Vagococcus* sp. JSB21, *Proteus* sp. JSB21 and *Bacillus* sp. JSB10), and able to distinguish the differences at the spectra at ~920 (α-1,4-glycosidic bond), 1600–1700 (amide I) and 1500–1600 cm^{-1} (amide II) of EPS spectra of among these species [69]. As per the PeakFit analysis, the IR absorbance intensity of α-1,4-glycosidic bond (a) and amide II (b) of the EPSs of *Vagococcus* sp. JSB21, *Proteus* sp. JSB20 and *Bacillus* sp. JSB10 were as follows (a) = 0.05, 0.12, 0.18 and (b) = 0.06, 0.23 and 0.26, respectively [69]. Thus, PeakFit is the automatic choice that automatically places the peaks in three different ways such as (i) the Residual procedure, (ii) the Second Derivative procedure, and (iii) the Deconvolution procedure (https://systatsoftware.com/products/peakfit/).

Also, combined use of FTIR and two-dimensional correlation spectroscopy (2D-COS) analysis helps in revealing the several subtle structural changes, interaction mechanisms in response to external perturbations [93]. The impact of Ca^{2+} ions on EPS functional groups and the order of reactivity of functional groups with Ca^{2+} has been studied in detail by using 2D-FTIR-COS analysis [69]. In this study, it was found that the order of reactivity of EPS functional groups with Ca^{2+} ions were significantly different among the bacteria. For instance, *Bacillus* sp. JSB10, the order of reactivity of functions groups with Ca^{2+} was α-1,4-glycosidic bond (920 cm^{-1}) > uronic acids (1020 cm^{-1}) > 1100 cm^{-1}) > amide I (1640 cm^{-1}) > amide II (1550 cm^{-1}) > C = O (1400 cm^{-1}). However, this order was seemingly different in *Proteus* sp. JSB20 (amide II > uronic acids > C = O > α-1,4-glycosidic bond > amide I) and *Vagococcus* sp. JSB21 (1100 > C = O > amide I > amide II > uronic acids > α-1,4-glycosidic bond).

4 EPS-Metal Interactions

4.1 Bacteria EPS-Heavy Metal Interactions

Certain EPS molecules may be acidic in nature, which is attributed to specific functional groups, e.g., uronic acids, pyruvate, and inorganics (e.g. SO_4^{2-} and PO_4^{2-}) [18]. Furthermore, it has been reported that uronic acids are the important function group in the binding of EPS with metals [22], and high content of uronic acid in the EPS is directly linked with the high anionicity of cell surface (cell wall) of *Marinobacter* sp. with a high propensity of metal adsorption specific pH conditions (e.g. neutral pH) [5] which is a direct evidence that EPS functional groups have high influence on the surface properties of microbial cells. Especially, in an environment with heavy metal pollution, microbial EPS is not only responsible for the sequestration of pollutants (here it is heavy metal) but also helps the bacterium by preventing the entry of pollutants into the cell [72]. A study focused on EPS-metal interactions revealed that secreted form of EPS has higher metal removal ability than the biomass that secretes the EPS which clearly suggest that cell-free EPS has significant importance in the metal removal [17].

There are several experimental proofs to unravel the relation between microbial EPS and heavy metals. There was an increased production of EPS by bacterial colonies in the medium containing heavy metals [18]. Similarly, the accumulation of copper is linearly correlated with the amount of EPS produced by a culture [46]. It is noteworthy that both gram-positive and gram-negative bacteria equally bind with the heavy metals, which is attributed to the presence of cell wall-associated EPS substances, which act as a good biosorbent matrix for heavy metals [91]. It is surprising that bacterial EPS (e.g. *Methylobacterium organophilum*) is able to adsorb the heavy metals (e.g., Cu and Pb) with in 30 min of incubation [51], which implies the efficiency bacterial EPS to bind heavy metals. Furthermore, bacterial EPS is able to bind with the heavy metals in the wide range of pH, which implies that they are good absorbents for the removal of heavy metals in the environments of different pH. In general, heavy metal resistance and heavy metal absorbing propensities are quite common in microorganisms that are living in the petroleum hydrocarbon contaminated sites. In our previous investigations, we isolated two bacteria (were belonging to *Bacillus cereus, Bacillus thuringiensis*) and two fungi (were belonging to *Geomyces* sp., *Geomyces pannorum*) [64], and these strains found degrade diesel fuel, crude oil, and spent lubricant oil in different conditions such as laboratory medium and soil microcosms (in vitro and in vivo) [63, 67, 68, 77]. However, interactions of these four strains with heavy metals are interesting. Two bacterial strains showed 100% resistance to 100 mM of $MgCl_2$, $FeCl_3$ and $Al_2(SO_4)_3$ and less tolerance to $ZnCl_2$ and $CuSO_4.5H_2O$ present in the nutrient medium [66]. On the other side, the biomass of two filamentous fungi (*G. pannorum* and *Geomyces* sp.) was found to absorb 97–100% of copper (5 mg L^{-1}) from the growth medium within 7 days. These results imply that crude oil-polluted sites are the richest source for the isolation of heavy metal absorbing or resistant microorganisms. Though in the above studies we did not characterize the EPS of four microbial strains, their heavy metal resistance and absorbing propensities could be attributed to the presence of EPS molecules.

Likewise, there are several EPS-producing bacteria are known to remove the heavy metals (e.g. Pb, Hg, Cr, Cu, Cd, Zn, and Sb) from the contaminated sites at a remediation efficiency of 30–98% (refer a review [70]). EPS producing bacteria that have been involved in the metal removal including *Alcaligenes faecalis, Alteromonas macleodii* subsp. *fijiensis, Azotobacter chroococcum* XU1, *Bacillus cereus* KMS3-1, *Bacillus licheniformes, Bacillus* sp. S3, *Burkholderia cenocepalia, Enterobacter cloacae, Ochrobactrum* sp. HG16, *Paenibacillus jamilae, Paenibacillus polymyxa*, and *Serratia marcescens* HG19 [70]. List of microbial EPSs that can remove the HMs are presented in Table 1.

4.2 Mechanism

There are multiple mechanisms behind the sorption and immobilization of heavy metals by microbial EPSs, they include ion exchange, complexation or chelation, precipitation, etc. (Fig. 1).

Binding of EPS with the heavy metal is purely a surface phenomenon, and this property is not linked with any metabolic reaction, instead there is an interaction between the charged surfaces (i.e. negatively charged EPS and positively charged metal ions). Negative charge of EPS is attributed to the presence of different EPS functional groups such as amine, carbonyl, hydroxyl, phosphate, sulfhydral, and other molecules (e.g., techoic acid, peptidoglycan layer) [100]. Since the metal-EPS binding is regulated by different functional groups and ionic surfaces, pH of the environment greatly influences the metal adsorption by EPS [86]. Also, temperature has a significant impact on the EPS-metal interactions [3].

Table 1 Role of microbial EPS in the removal of heavy metals

No.	EPS source (name of microorganism)	Heavy metal	Reference
1	*Arthrobacter* ps-5	Copper, Lead, Chromium	[82]
2	*Athelia rolfsii*	Cadmium, Copper, Zinc	[58]
3	*Azobacter chroococcum*	Lead, Mercury	[55]
4	*Bacillus* sp. CIK-516	Nickel	[55]
5	*Bacillus firmus*	Lead, Cobalt, Zinc	[55]
6	*Bacillus vietnamensis* AB403	Arsenic	[55]
7	*Cellulosimicrobium funkei* AR8 and AR6	Chromium	[55]
8	*Chlorella* sp.	Lead	[1]
9	*Ensifer meliloti*	Lead, Nickel, Zinc	[55]
10	*Enterobacter cloaceae*	Chromium	[39]
11	*Enterococcus faecalis*	Zinc	[90]
12	*Euglena mutabilis*	Lead	[1]
13	*Gordonia alkanivorans* (SMV185.1, SMV185.5, SMV207.37)	Arsenic, Mercury	[55]
14	*Kocuria flava AB402*	Arsenic	[55]
15	*Lactobacillus plantarum*	Lead	[55]
16	*Lysinibacillus macrolides*	Arsenic, Mercury	[55]
17	*Macrococcos caseolysticus*	Arsenic, Mercury	[55]
18	*Ochrobactanthropi*	Chromium, Cadmium, Copper	[75]
19	*Ochrobactrum intermedium* AM7	Thorium	[83]
20	*Paenibacillus jamilae*	Lead, Cadmium	[55]
21	*Pseudomonas aureofaciens*	Zinc	[23]
22	*Pseudomonas* sp.	Zinc	[55]
23	*Stenotrophomonas* sp. CIK-517Y	Nickel	[55]
24	*Phanerochaete chrysosporium*	Lead	[59]
25	*Aspergillus fumigatus*	Copper	[26]
26	*Cordyceps militaris*	Lead	[96]

4.3 Bioremediation of HM-Contaminated Soil

Microbial EPS has a strong potential to adsorb different heavy metals in the soil system. EPS extracted from *Rhizobium radiobacter* VBCK1062 strain showed strong uptake propensity of arsenate [19], it is not surprising that one gram of biomass adsorbed 0.068 mg of arsenate. A free-living nitrogen-fixing bacterium i.e. *Azotobacter* is known to influence plant growth positively, however, this bacteria has a capacity to remove mercury from the soil where EPS formation plays a dominant role [33]. Biofilms of *Achyranthus aspera* strain have a capacity to purify

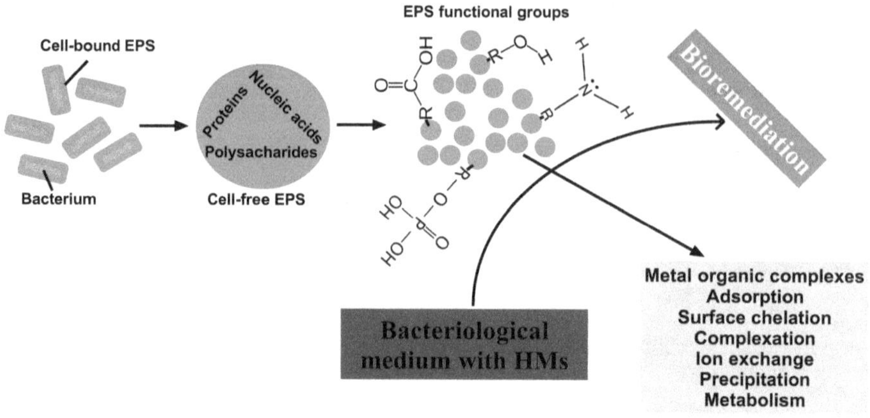

Fig. 1 Mechanism of HMs removal by bacterial EPSs

Ni (II) contaminated soils [45]. It should be remembered that biofilm matrix in different biofilm-forming microorganisms is made up of EPSs. The direct link between EPS and HM adsorption has been revealed at gene level in *Pseudomonas* Psd strain [89]. EPS of strain Psd has a strong capacity of zinc biosorption, and the EPS synthesis in this strain is regulated by *algeS* gene, which encodes a key component of EPS i.e. alginate polymerase subunit. Interestingly, knockout of *algeS* gene resulted in the decreased zing adsorption in the Psd strain. Crude EPS of *Pseudomonas stutzeri* AS22 has a capacity to adsorb 460 mg of lead per gram of EPS [62], which seems that this bacterium is a good tool for the removal of lead from the contaminated sites. As of now, EPS of several bacteria were known to be involved in the bioremediation of different HMs-contaminated soils [53], following are the names of bacteria from which EPS was formed and HM that underwent bioremediation: *Bacillus* sp. AS2—Pb(II); *Eromonas diversa*—Cr(III); *Azotobacter chroococcum* XU1—Pb(II) and Hg; *Pseudomonas stutzeri* AS22—Cu(II), Co(II), Pb(II), Cd(II), Fe(II); *Rhizobacteria*—Cd(II); *Anthyllis vulneraria*—Zn(II); *Enterobacter ludwigii*—Cd(II); *Rhizobium radiobacter* strain VBCK1062—As(V), Cu(II), Pb(II), Cr(II), Ni(II), Cd(II); *Bacillus cereus* BDBC01—Cr(VI), Ni(II), Cd (II); *Bacillus thuringiensis* UEAB3—Cr(III) etc. These insights clearly imply that bioremediation of HMs-contaminated sites is highly feasible by bacterial EPS.

4.4 Bioremediation of HM-Contaminated Water

Microbial biopolymers have also the capacity to adsorb HMs in the contaminated water. For example, alginate of both Gramnegative and Gram-positive bacteria have a strong propensity to adsorb copper from synthetic wastewater [9]. Furthermore, mixing of alginate with zeolite and clinoptilolite could adsorb ~ 130 mg g^{-1} at pH

4.5, indicates that EPS-based bioremediation is highly suitable to remediate the HMs-polluted acidic wastewater [52]. In another study, it has been found that the cadmium and lead removal efficiency by *Enterococci* MC1 was 46% and 43%, respectively, where as *Bacillus acidiproducens* SM2 exhibited 43% of removal efficiency of chromium [37]. Furthermore, in the EPS-based bioremediation, recovery of HMs is highly possible after the removal of HMs from the contaminated media. There was a successful recovery of chromium from *Spirulina platensis*, and the % of recovery was increased by three folds in an acid treatment, which is attributed to the decrease in the negative charges on the surface of EPSs. It has been suggested that the bioadsorption of HMs by microbial cells occurred in two stages, first there is surface adsorption, then there is a transportation of metal ion into the cytoplasm [32].

In other occasions also, bacterial EPS has proved their efficiency in the removal of HMs from the wastewater. EPS from *Bacillus* sp. JH7 is known to remove calcium carbonate successfully, and this removal efficiency has been improved in the addition of glycerol, attributed to the formation of ionic bonds between bacterial EPS and Ca(II) ions [48]. Key environmental factors that do affect the HM adsorption ability of EPS are pH, temperature, concentration of HMs, and contact time (EPS and HMs) of solution. EPSs (EPS-605) isolated from lactic acid bacteria strain (LCC-605) showed wide range of HMs adsorption, for example, maximum adsorption capacity in 24 h were as follows, cadmium (2097 mg g^{-1}), copper (2987 mg g^{-1}), lead (2327 mg g^{-1}) and molybdenum (3029 mg g^{-1}) [57]. According to a FTIR analysis, key functional groups that were responsible for the attachment of entophytic bacteria *Kocuria rhizophila* KF875448 with Cd and Cr were alkyne, methyl, phosphate, amide, and hydroxyl groups [29]. Sulphated rhamnoglecan is a uroic acid containing EPS that has a strong ability to adsorb or eliminate HMs such as Cd, Cu, and Pb [88]. It is noteworthy that the efficiency of EPSs to adsorb HMs can be improved by chemical pretreatments. For example, acid pretreatment (A) and base pretreatment (B) of EPSs of *Cyanotheca* sp. CCY0119 resulted in the adsorption efficiencies of Cu, Cd and Pb were continuously A = 86%, 43%, 22%, and B = 318%, 26%, 92%, respectively [74]. Similarly, *Nostoc sphaeroides* is a potential agent in removing various HMs (e.g. Ni, Pb, Cr, Cd, and Cu) from water. When *N. sphaeroides* treated with a synthetic water containing single metal ions such as Pb, Cd, Cu, Mn, Cr, and Ni, the adsorption capacities were continuously 53, 47, 30, 24, 93 and 28%, respectively [42]. There is an in-depth study on the kinetics of HMs adsorption by EPS of selected Grampositive bacterial species (*Rhodococcus opacus* and *Rhodococcus rhodochrous*) in the agrochemicals-contaminated water [21]. *R. opacus* and *R. rhodochrous* showed sorption equilibrium after 30 and 1 min in the water containing Cd, Pb, Ni, Co, and Cr. EPS of these two bacterial strains showed highest adsorption of HMs in the pH range of 2–7.5, and most HMs (e.g. Cd, Pb, Ni, Co, Cr) were adsorbed effectively at 25 °C, whereas Ni was adsorbed effectively at 35 °C.

4.5 *Advantages Over Conventional Methods*

There are several undisputable advantages offered by EPS [70]. Though EPS is an easily bioavailable substance, is a non-living bio-sorbent. Bioremediation with EPS doesn't raise the issue of pathogenicity, which is quite common when bioremediation involves living microorganisms. EPS production is only depending on the source of microorganisms, but not on the colony size of the source microorganism. It should be remembered that microbial EPS is highly stable to environmental stresses, in addition to having a higher surface to volume ratio, such features are hard to find in the EPS of other microorganisms [81]. EPSs are the good bioemulsifiers, and thus increase the solubility, viscosity, and degradation of contaminants (especially hydrocarbons), in addition to bioflocculating and bioadsorbing HMs in the different environmental sectors [8, 28, 36, 82]. Emulsifiers of EPSs of microbial origin (microbial emulsifiers) have great advantages over the synthetic emulsifiers. Hallmarks of microbial emulsifiers are less toxicity, high degradability, good compatibility, and selectivity, resistant to extreme environmental conditions (e.g. pH, temperature, salinity). Another noteworthy about EPS-mediated bioremediation is that EPS of microbial cell forms a biological barrier during landfill [70], this is extremely important in avoiding the groundwater contamination. Furthermore, this EPS matrix with its adhesive properties, can efficiently reduce the transport of contaminants (e.g. solids and HMs) in runoff and stabilize soil.

Microbial EPSs play additional roles in the agricultural soils besides contaminant removal. For example, EPS produced by *Rhizobium leguminosarum* helps in the adhesion of soil particles and stable aggregates which subsequently improve the vegetative growth [79], and all these are attributed to the adhesive and gel formation properties of EPS of *R. leguminosarum* [56]. Furthermore, EPSs also increase the root structure, which ultimately results in the high node densities and increased biomass. It is not surprising that EPSs also create enough micropores that facilitate the easy movement of water in silt soils [79]. These insights clearly suggest that rhizobial EPS has a wide range of functions in agricultural soils which greatly helps in the plant growth and development.

5 Knowledge Gaps and Future Directions

Undoubtedly, EPS-based bioremediation of HMs-contaminated environments is being received considerable attention as this type of bioremediation proved undisputable advantages while restoring the contaminated media at lab- and small-scale experimentation. However, for the full-scale implications, several research gaps are there in the area of EPS-based bioremediation. Key areas that need to be focused on this area are as follows:

- Field-scale studies using EPS-mediated bioremediation are obligatory in order to understand the specificity and selectivity of HMs and their removal by EPS.
- Physico-chemical properties of EPS which affect the HMs biosorption by EPS.
- Desorption studies for the successful recovery or recycling of HMs from the EPS, these insights are useful in the circular economy.
- There is a need for characterization of enzymes released by bacteria during EPS production. Such enzymes have prominent importance in the HMs removal.

It is highly believable that the results of above studies will not only fill the research gaps but also have a significant impact on the successful implication of EPS-assisted bioremediation of HMs-contaminated sites in various environments.

6 Conclusions

Main conclusions regarding microbial EPSs and their relevance in the restoration of HMs-contaminated sites are as follows:

- There are three main microbial groups that produce EPSs are bacteria, fungi, and algae, and microbial EPSs have widest applications in different areas as viscosifier, stabilizers, emulsifiers.
- Several spectroscopic methods can be used to characterize EPS molecules, such as UV-Visible absorption spectroscopy, differential absorbance spectroscopy (DAS), fluorescence spectroscopy (FS), fluorescence excitation-emission matrix (EEM), fourier-transform infrared spectroscopy (FTIR), attenuated total reflection (ATR) FTIR and Raman spectroscopy.
- EPS molecules are usually acidic in nature due to specific functional groups (e.g., uronic acids, pyruvate, and inorganics (e.g. SO_4^{2-} and PO_4^{2-}). Such functional groups enhance the anionicity of microbial cell surface (cell wall), which is responsible for a high propensity of metal adsorption.
- There are multiple mechanisms behind the sorption and immobilization of heavy metals by microbial EPSs, they include metal-organic complexes, adsorption, surface chelation, complexation, ion exchange, precipitation, and metabolism.
- There is much evidence for the successful removal of HMs by microbial EPS in different environmental media such as soil and wastewater systems.
- Bioemulsification, bioflocculation, and bioadsorption are highly possible by microbial EPSs, and such events are key in the remediation of contaminated environments.
- Microbial EPSs are also responsible for the adhesion of soil particles, formation of stable aggregates which subsequently improve the vegetative growth, root structure, high node densities, increased biomass, easy movement of water in silt soils, and decrease the evapotranspiration.

Above conclusions provide great hope that it is not far for the emergence of microbial EPSs as sustainable agents for the restoration of HMs-contaminated sites in an eco-friendly and economic way.

References

1. Aguilera A, Souza-Egipsy V, San Martín-Úriz P, Amils R (2008) Extracellular matrix assembly in extreme acidic eukaryotic biofilms and their possible implications in heavy metal adsorption. Aquat Toxicol 88(4):257–266
2. Aluwihare LI, Repeta DJ (1999) A comparison of the chemical characteristics of oceanic DOM and extracellular DOM produced by marine algae. Mar Ecol Prog Ser 186:105–117. https://doi.org/10.3354/meps186105
3. Arjoon A, Olaniran AO, Pillay B (2013) Co-contamination of water with chlorinated hydrocarbons and heavy metals: challenges and current bioremediation strategies. Int J Environ Sci Technol 10(2):395–412. https://doi.org/10.1007/s13762-012-0122-y
4. Bajaj IB, Survase SA, Saudagar PS, Singhal RS (2007) Gellan gum: fermentative production, downstream processing and applications. Food Technol Biotechnol 45(4):341–354. https://www.scopus.com/inward/record.uri?eid=2-s2.0-40749150054&partnerID=40&md5=aa0c148cae1b9d6a5189ff0772dbfa48
5. Bhaskar PV, Bhosle NB (2006) Bacterial extracellular polymeric substance (EPS): a carrier of heavy metals in the marine food-chain. Environ Int 32(2):191–198. https://doi.org/10.1016/j.envint.2005.08.010
6. Bhunia B, Prasad Uday US, Oinam G, Mondal A, Bandyopadhyay TK, Tiwari ON (2018) Characterization, genetic regulation and production of cyanobacterial exopolysaccharides and its applicability for heavy metal removal. Carbohydr Polym 179:228–243. https://doi.org/10.1016/j.carbpol.2017.09.091
7. Bramhachari PV, Dubey SK (2006) Isolation and characterization of exopolysaccharide produced by Vibrio harveyi strain VB23. Lett Appl Microbiol 43(5):571–577. https://doi.org/10.1111/j.1472-765X.2006.01967.x
8. Calvo C, Martinez-Checa F, Mota A, Bejar V, Quesada E (1998) Effect of cations, pH and sulfate content on the viscosity and emulsifying activity of the Halomonas eurihalina exopolysaccharide. J Ind Microbiol Biotechnol 20(3–4):205–209. https://doi.org/10.1038/sj.jim.2900513
9. Chauhan M, Solanki M, Nehra K (2017) Putative Mechanism of cadmium bioremediation employed by resistant bacteria. Jordan J Biol Sci 10(2)
10. Chawla PR, Bajaj IB, Survase SA, Singhal RS (2009) Microbial cellulose: fermentative production and applications. Food Technol Biotechnol 47(2):107–124. https://www.scopus.com/inward/record.uri?eid=2-s2.0-70349397464&partnerID=40&md5=9faae7692d0a4d5fe25693dba380b026
11. Chen W, Habibul N, Liu XY, Sheng GP, Yu HQ (2015) FTIR and synchronous fluorescence heterospectral two-dimensional correlation analyses on the binding characteristics of copper onto dissolved organic matter. Environ Sci Technol 49(4):2052–2058. https://doi.org/10.1021/es5049495
12. Chen W, Qian C, Liu XY, Yu HQ (2014) Two-dimensional correlation spectroscopic analysis on the interaction between humic acids and TiO_2 nanoparticles. Environ Sci Technol 48(19):11119–11126. https://doi.org/10.1021/es502502n
13. Chen W, Qian C, Zhou K-G, Yu H-Q (2018) Molecular spectroscopic characterization of membrane fouling: a critical review. Chem 4(7):1492–1509. https://doi.org/10.1016/j.chempr.2018.03.011

14. Chen X, Siu KC, Cheung YC, Wu JY (2014) Structure and properties of a (1 → 3)-β-d-glucan from ultrasound-degraded exopolysaccharides of a medicinal fungus. Carbohyd Polym 106(1):270–275. https://doi.org/10.1016/j.carbpol.2014.02.040
15. Chien CY, Enomoto-Rogers Y, Takemura A, Iwata T (2017) Synthesis and characterization of regioselectively substituted curdlan hetero esters via an unexpected acyl migration. Carbohyd Polym 155:440–447. https://doi.org/10.1016/j.carbpol.2016.08.067
16. Choi JW, Choi HG, Lee WH (1996) Effects of ethanol and phosphate on emulsan production by Acinetobacter calcoaceticus RAG-1. J Biotechnol 45(3):217–225. https://doi.org/10.1016/0168-1656(95)00175-1
17. De Philippis R, Colica G, Micheletti E (2011) Exopolysaccharide-producing cyanobacteria in heavy metal removal from water: molecular basis and practical applicability of the biosorption process. Appl Microbiol Biotechnol 92(4):697–708. https://doi.org/10.1007/s00253-011-3601-z
18. De Philippis R, Paperi R, Sili C (2007) Heavy metal sorption by released polysaccharides and whole cultures of two exopolysaccharide-producing cyanobacteria. Biodegradation 18 (2):181–187. https://doi.org/10.1007/s10532-006-9053-y
19. Deepika KV, Raghuram M, Kariali E, Bramhachari PV (2016) Biological responses of symbiotic Rhizobium radiobacter strain VBCK1062 to the arsenic contaminated rhizosphere soils of mung bean. Ecotoxicol Environ Saf 134:1–10
20. Delattre C, Pierre G, Laroche C, Michaud P (2016) Production, extraction and characterization of microalgal and cyanobacterial exopolysaccharides. Biotechnol Adv 34(7):1159–1179. https://doi.org/10.1016/j.biotechadv.2016.08.001
21. Dobrowolski R, Szcześ A, Czemierska M, Jarosz-Wikołazka A (2017) Studies of cadmium (II), lead (II), nickel (II), cobalt (II) and chromium (VI) sorption on extracellular polymeric substances produced by Rhodococcus opacus and Rhodococcus rhodochrous. Biores Technol 225:113–120
22. Dogan NM, Doganli GA, Dogan G, Bozkaya O (2015) Characterization of extracellular polysaccharides (EPS) produced by thermal bacillus and determination of environmental conditions affecting exopolysaccharide production. Int J Environ Res 9(3):1107–1116. https://www.scopus.com/inward/record.uri?eid=2-s2.0-84939523584&partnerID=40&md5=766950e102918b507348d2f71a164575
23. Drozdova OY, Pokrovsky OS, Lapitskiy SA, Shirokova LS, González AG, Demin VV (2014) Decrease in zinc adsorption onto soil in the presence of EPS-rich and EPS-poor Pseudomonas aureofaciens. J Colloid Interface Sci 435:59–66
24. Fang L, Catchmark JM (2015) Characterization of cellulose and other exopolysaccharides produced from Gluconacetobacter strains. Carbohyd Polym 115:663–669. https://doi.org/10.1016/j.carbpol.2014.09.028
25. Fattom A, Li X, Cho YH, Burns A, Hawwari A, Shepherd SE et al (1995) Effect of conjugation methodology, carrier protein, and adjuvants on the immune response to Staphylococcus aureus capsular polysaccharides. Vaccine 13(14):1288–1293. https://doi.org/10.1016/0264-410X(95)00052-3
26. Ghaed S, Shirazi EK, Marandi R (2013) Biosorption of copper ions by Bacillus and Aspergillus species. Adsorpt Sci Technol 31(10):869–890
27. Ghosh M, Singh SP (2005) A review on phytoremediation of heavy metals and utilization of it's by products. Asian J Energy Environ 6(4):18
28. Han PP, Sun Y, Wu XY, Yuan YJ, Dai YJ, Jia SR (2014) Emulsifying, flocculating, and physicochemical properties of exopolysaccharide produced by cyanobacterium Nostoc flagelliforme. Appl Biochem Biotechnol 172(1):36–49. https://doi.org/10.1007/s12010-013-0505-7
29. Haq F, Butt M, Ali H, Chaudhary HJ (2016) Biosorption of cadmium and chromium from water by endophytic Kocuria rhizophila: equilibrium and kinetic studies. Desalin Water Treat 57(42):19946–19958
30. Hassler CS, Alasonati E, Mancuso Nichols CA, Slaveykova VI (2011) Exopolysaccharides produced by bacteria isolated from the pelagic Southern Ocean—role in Fe binding,

chemical reactivity, and bioavailability. Mar Chem 123(1–4):88–98. https://doi.org/10.1016/j.marchem.2010.10.003
31. Hassler CS, Schoemann V, Nichols CM, Butler ECV, Boyd PW (2011) Saccharides enhance iron bioavailability to southern ocean phytoplankton. Proc Natl Acad Sci USA 108(3):1076–1081. https://doi.org/10.1073/pnas.1010963108
32. Hegde SM, Babu RL, Vijayalakshmi E, Patil RH, Naveen Kumar M, Kiran Kumar KM et al (2016) Biosorption of hexavalent chromium from aqueous solution using chemically modified Spirulina platensis algal biomass: an ecofriendly approach. Desalin Water Treat 57 (18):8504–8513
33. Hindersah R, Mulyani O, Osok R (2017) Proliferation and exopolysaccharide production of Azotobacter in the presence of mercury. Biodivers J 8(1):21–26
34. Hu B, Shao S, Ni H, Fu Z, Hu L, Zhou Y et al (2020) Current status, spatial features, health risks, and potential driving factors of soil heavy metal pollution in China at province level. Environ Pollut 266: https://doi.org/10.1016/j.envpol.2020.114961
35. Hu H, Jin Q, Kavan P (2014) A study of heavy metal pollution in China: current status, pollution-control policies and countermeasures. Sustainability 6(9):5820–5838
36. Huang KH, Chen BY, Shen FT, Young CC (2012) Optimization of exopolysaccharide production and diesel oil emulsifying properties in root nodulating bacteria. World J Microbiol Biotechnol 28(4):1367–1373. https://doi.org/10.1007/s11274-011-0936-7
37. Huët MAL, Puchooa D (2017) Bioremediation of heavy metals from aquatic environment through microbial processes: a potential role for probiotics. J Appl Biol Biotechnol 5(6):14–23
38. Hussain A, Zia KM, Tabasum S, Noreen A, Ali M, Iqbal R et al (2017) Blends and composites of exopolysaccharides; properties and applications: a review. Int J Biol Macromol 94:10–27
39. Iyer A, Mody K, Jha B (2004) Accumulation of hexavalent chromium by an exopolysaccharide producing marine Enterobacter cloaceae. Mar Pollut Bull 49(11–12):974–977
40. Izawa N, Hanamizu T, Iizuka R, Sone T, Mizukoshi H, Kimura K et al (2009) Streptococcus thermophilus produces exopolysaccharides including hyaluronic acid. J Biosci Bioeng 107 (2):119–123. https://doi.org/10.1016/j.jbiosc.2008.11.007
41. Jan AT, Azam M, Ali A, Haq QMR (2014) Prospects for exploiting bacteria for bioremediation of metal pollution. Crit Rev Environ Sci Technol 44(5):519–560
42. Jiang J, Zhang N, Yang X, Song L, Yang S (2016) Toxic metal biosorption by macrocolonies of cyanobacterium Nostoc sphaeroides Kützing. J Appl Phycol 28(4):2265–2277
43. Jing F, Chen X, Yang Z, Guo B (2018) Heavy metals status, transport mechanisms, sources, and factors affecting their mobility in Chinese agricultural soils. Environ Earth Sci 77(3):104
44. Kantar C, Demiray H, Dogan NM, Dodge CJ (2011) Role of microbial exopolymeric substances (EPS) on chromium sorption and transport in heterogeneous subsurface soils: I. Cr (III) complexation with EPS in aqueous solution. Chemosphere 82(10):1489–1495
45. Karthik C, Oves M, Thangabalu R, Sharma R, Santhosh SB, Arulselvi PI (2016) Cellulosimicrobium funkei-like enhances the growth of Phaseolus vulgaris by modulating oxidative damage under Chromium (VI) toxicity. J Adv Res 7(6):839–850
46. Kazy SK, Sar P, Singh SP, Sen AK, D'Souza SF (2002) Extracellular polysaccharides of a copper-sensitive and a copper-resistant Pseudomonas aeruginosa strain: synthesis, chemical nature and copper binding. World J Microbiol Biotechnol 18(6):583–588. https://doi.org/10.1023/A:1016354713289
47. Kekez B, Gojgić-Cvijović G, Jakovljević D, Pavlović V, Beškoski V, Popović A et al (2016) Synthesis and characterization of a new type of levan-graft-polystyrene copolymer. Carbohyd Polym 154:20–29. https://doi.org/10.1016/j.carbpol.2016.08.001
48. Kim HJ, Shin B, Lee YS, Park W (2017a) Modulation of calcium carbonate precipitation by exopolysaccharide in Bacillus sp. JH7. Appl Microbiol Biotechnol 101(16):6551–6561
49. Kim J, Hwang J, Seo Y, Jo Y, Son J, Choi J (2017) Engineered chitosan–xanthan gum biopolymers effectively adhere to cells and readily release incorporated antiseptic molecules in a sustained manner. J Ind Eng Chem 46:68–79. https://doi.org/10.1016/j.jiec.2016.10.017

50. Kim JY, Kim HJ, Choi K, Nam B (2017) Mental health conditions among North Korean female refugee victims of sexual violence. Int Migr 55(2):68–79. https://doi.org/10.1111/imig.12300
51. Kim SY, Kim JH, Kim CJ, Oh DK (1996) Metal adsorption of the polysaccharide produced from Methylobacterium organophilum. Biotech Lett 18(10):1161–1164. https://doi.org/10.1007/BF00128585
52. Kıvılcımdan Moral Ç, Yıldız M (2016) Alginate production from alternative carbon sources and use of polymer based adsorbent in heavy metal removal. Int J Polym Sci 2016
53. Kranthi Raj K, Sardar UR, Bhargavi E, Devi I, Bhunia B, Tiwari ON (2018) Advances in exopolysaccharides based bioremediation of heavy metals in soil and water: a critical review. Carbohyd Polym 199:353–364. https://doi.org/10.1016/j.carbpol.2018.07.037
54. Kwon HJ, Kim JM, Han KI, Jung EG, Kim YH, Patnaik BB et al (2016) Mutan: a mixed linkage α-[(1,3)- and (1,6)]-d-glucan from Streptococcus mutans, that induces osteoclast differentiation and promotes alveolar bone loss. Carbohyd Polym 137:561–569. https://doi.org/10.1016/j.carbpol.2015.11.013
55. Lal S, Ratna S, Said OB, Kumar R (2018) Biosurfactant and exopolysaccharide-assisted rhizobacterial technique for the remediation of heavy metal contaminated soil: an advancement in metal phytoremediation technology. Environ Technol Innovation 10:243–263
56. Laus MC, Van Brussel AAN, Kijne JW (2005) Role of cellulose fibrils and exopolysaccharides of Rhizobium leguminosarum in attachment to and infection of Vicia sativa root hairs. Mol Plant Microbe Interact 18(6):533–538. https://doi.org/10.1094/MPMI-18-0533
57. Li C, Zhou L, Yang H, Lv R, Tian P, Li X et al (2017) Self-assembled exopolysaccharide nanoparticles for bioremediation and green synthesis of noble metal nanoparticles. ACS Appl Mater Interfaces 9(27):22808–22818
58. Li H, Wei M, Min W, Gao Y, Liu X, Liu J (2016) Removal of heavy metal Ions in aqueous solution by Exopolysaccharides from Athelia rolfsii. Biocatal Agric Biotechnol 6:28–32
59. Li N, Liu J, Yang R, Wu L (2020) Distribution, characteristics of extracellular polymeric substances of Phanerochaete chrysosporium under lead ion stress and the influence on Pb removal. Sci Rep 10(1):17633. https://doi.org/10.1038/s41598-020-74983-0
60. Liang C, Hu X, Ni Y, Wu J, Chen F, Liao X (2006) Effect of hydrocolloids on pulp sediment, white sediment, turbidity and viscosity of reconstituted carrot juice. Food Hydrocolloids 20(8):1190–1197. https://doi.org/10.1016/j.foodhyd.2006.01.010
61. Liu Y, Gu Q, Ofosu FK, Yu X (2015) Isolation and characterization of curdlan produced by Agrobacterium HX1126 using α-lactose as substrate. Int J Biol Macromol 81:498–503. https://doi.org/10.1016/j.ijbiomac.2015.08.045
62. Maalej H, Hmidet N, Boisset C, Buon L, Heyraud A, Nasri M (2015) Optimization of exopolysaccharide production from P seudomonas stutzeri AS 22 and examination of its metal-binding abilities. J Appl Microbiol 118(2):356–367
63. Maddela NR, Burgos R, Kadiyala V, Carrion AR, Bangeppagari M (2016) Removal of petroleum hydrocarbons from crude oil in solid and slurry phase by mixed soil microorganisms isolated from Ecuadorian oil fields. Int Biodeterior Biodegradation 108:85–90. https://doi.org/10.1016/j.ibiod.2015.12.015
64. Maddela NR, Masabanda M, Leiva-Mora M (2015) Novel diesel-oil-degrading bacteria and fungi from the Ecuadorian Amazon rainforest. Water Sci Technol 71(10):1554–1561. https://doi.org/10.2166/wst.2015.142
65. Maddela NR, Meng F (2020) Discrepant roles of a quorum quenching bacterium (Rhodococcus sp. BH4) in growing dual-species biofilms. Sci Total Environ 713:136402. https://doi.org/10.1016/j.scitotenv.2019.136402
66. Maddela NR, Reyes JJM, Viafara D, Gooty JM (2015) Biosorption of copper (II) by the microorganisms Isolated from the crude-oil-contaminated soil. Soil Sed Contam 24(8):898–908. https://doi.org/10.1080/15320383.2015.1064089
67. Maddela NR, Scalvenzi L, Kadiyala V (2016b) Microbial degradation of total petroleum hydrocarbons in crude oil: a field-scale study at the low-land rainforest of Ecuador. Environ Technol 1–24. https://doi.org/10.1080/09593330.2016.1270356

68. Maddela NR, Scalvenzi L, Pérez M, Montero C, Gooty JM (2015) Efficiency of indigenous filamentous fungi for biodegradation of petroleum hydrocarbons in medium and soil: laboratory study from Ecuador. Bull Environ Contam Toxicol 95(3):385–394
69. Maddela NR, Zhou Z, Yu Z, Zhao S, Meng F (2018) Functional determinants of extracellular polymeric substances in membrane biofouling: experimental evidence from pure-cultured sludge bacteria. Appl Environ Microbiol 84(15):e00756–e00818. https://doi.org/10.1128/AEM.00756-18
70. Mahapatra B, Dhal NK, Pradhan A, Panda BP (2020) Application of bacterial extracellular polymeric substances for detoxification of heavy metals from contaminated environment: a mini-review. Mater Today: Proc https://doi.org/10.1016/j.matpr.2020.01.490
71. Meng F, Liao B, Liang S, Yang F, Zhang H, Song L (2010) Morphological visualization, componential characterization and microbiological identification of membrane fouling in membrane bioreactors (MBRs). J Membr Sci 361(1–2):1–14. https://doi.org/10.1016/j.memsci.2010.06.006
72. More TT, Yadav JSS, Yan S, Tyagi RD, Surampalli RY (2014) Extracellular polymeric substances of bacteria and their potential environmental applications. J Environ Manage 144:1–25. https://doi.org/10.1016/j.jenvman.2014.05.010
73. Moscovici M (2015) Present and future medical applications of microbial exopolysaccharides. Front Microbiol 6:1012
74. Mota R, Rossi F, Andrenelli L, Pereira SB, De Philippis R, Tamagnini P (2016) Released polysaccharides (RPS) from Cyanothece sp. CCY 0110 as biosorbent for heavy metals bioremediation: interactions between metals and RPS binding sites. Appl Microbiol Biotechnol 100(17):7765–7775
75. Ozdemir G, Ozturk T, Ceyhan N, Isler R, Cosar T (2003) Heavy metal biosorption by biomass of Ochrobactrum anthropi producing exopolysaccharide in activated sludge. Biores Technol 90(1):71–74
76. Patel A, Prajapati JB (2013) Food and health applications of exopolysaccharides produced by lactic acid bacteria. Adv Dairy Res 1(2):1–7
77. Raju MN, Leo R, Herminia SS, Moran RE, Venkateswarlu K, Laura S (2017) Biodegradation of diesel, crude oil and spent lubricating oil by soil isolates of Bacillus spp. Bull Environ Contam Toxicol 98(5):698–705. https://doi.org/10.1007/s00128-017-2039-0
78. Robyt JF, Yoon SH, Mukerjea R (2008) Dextransucrase and the mechanism for dextran biosynthesis. Carbohyd Res 343(18):3039–3048. https://doi.org/10.1016/j.carres.2008.09.012
79. Rossi F, Potrafka RM, Pichel FG, De Philippis R (2012) The role of the exopolysaccharides in enhancing hydraulic conductivity of biological soil crusts. Soil Biol Biochem 46:33–40. https://doi.org/10.1016/j.soilbio.2011.10.016
80. Sheng GP, Xu J, Li WH, Yu HQ (2013) Quantification of the interactions between Ca_2+, Hg_2+ and extracellular polymeric substances (EPS) of sludge. Chemosphere 93(7):1436–1441. https://doi.org/10.1016/j.chemosphere.2013.07.076
81. Sheng GP, Yu HQ, Yue ZB (2005) Production of extracellular polymeric substances from Rhodopseudomonas acidophila in the presence of toxic substances. Appl Microbiol Biotechnol 69(2):216–222. https://doi.org/10.1007/s00253-005-1990-6
82. Shuhong Y, Meiping Z, Hong Y, Han W, Shan X, Yan L et al (2014) Biosorption of Cu_2+, Pb_2+ and Cr_6+ by a novel exopolysaccharide from Arthrobacter ps-5. Carbohyd Polym 101(1):50–56. https://doi.org/10.1016/j.carbpol.2013.09.021
83. Shukla A, Parmar P, Goswami D, Patel B, Saraf M (2020) Characterization of novel thorium tolerant Ochrobactrum intermedium AM7 in consort with assessing its EPS-Thorium binding. J Hazard Mater 388:
84. Singh RS, Kaur N, Rana V, Kennedy JF (2017) Pullulan: a novel molecule for biomedical applications. Carbohyd Polym 171:102–121. https://doi.org/10.1016/j.carbpol.2017.04.089
85. Stromberg LR, Stromberg ZR, Banisadr A, Graves SW, Moxley RA, Mukundan H (2015) Purification and characterization of lipopolysaccharides from six strains of non-O157 Shiga

toxin-producing Escherichia coli. J Microbiol Methods 116:1–7. https://doi.org/10.1016/j.mimet.2015.06.008
86. Su C, Jiang L, Zhang W (2014) A review on heavy metal contamination in the soil worldwide: situation, impact and remediation techniques. Environ Skeptics Critics 3(2):24–38
87. Surayot U, Wang J, Seesuriyachan P, Kuntiya A, Tabarsa M, Lee Y et al (2014) Exopolysaccharides from lactic acid bacteria: structural analysis, molecular weight effect on immunomodulation. Int J Biol Macromol 68:233–240. https://doi.org/10.1016/j.ijbiomac.2014.05.005
88. Tran HT, Vu ND, Matsukawa M, Okajima M, Kaneko T, Ohki K et al (2016) Heavy metal biosorption from aqueous solutions by algae inhabiting rice paddies in Vietnam. J Environ Chem Eng 4(2):2529–2535
89. Upadhyay A, Kochar M, Rajam MV, Srivastava S (2017) Players over the surface: unraveling the role of exopolysaccharides in zinc biosorption by fluorescent Pseudomonas strain psd. Front Microbiol 8:284
90. Venkatesh P, Balraj M, Ayyanna R, Ankaiah D, Arul V (2016) Physicochemical and biosorption properties of novel exopolysaccharide produced by Enterococcus faecalis. LWT-Food Sci Technol 68:606–614
91. Vijayaraghavan K, Yun YS (2008) Bacterial biosorbents and biosorption. Biotechnol Adv 26(3):266–291. https://doi.org/10.1016/j.biotechadv.2008.02.002
92. Vuddanda PR, Montenegro-Nicolini M, Morales JO, Velaga S (2017) Effect of plasticizers on the physico-mechanical properties of pullulan based pharmaceutical oral films. Eur J Pharm Sci 96:290–298. https://doi.org/10.1016/j.ejps.2016.09.011
93. Wang L, Chen W, Song X, Li Y, Zhang W, Zhang H et al (2020) Cultivation substrata differentiate the properties of river biofilm EPS and their binding of heavy metals: a spectroscopic insight. Environ Res 182: https://doi.org/10.1016/j.envres.2019.109052
94. Wang LL, Wang LF, Ren XM, Ye XD, Li WW, Yuan SJ et al (2012) PH dependence of structure and surface properties of microbial EPS. Environ Sci Technol 46(2):737–744. https://doi.org/10.1021/es203540w
95. Wang S, Cai L-M, Wen H-H, Luo J, Wang Q-S, Liu X (2019) Spatial distribution and source apportionment of heavy metals in soil from a typical county-level city of Guangdong Province, China. Sci Total Environ 655:92–101
96. Yang H, Wu Z, He D, Zhou H, Yang H (2017) Enzyme-assisted extraction and Pb_2+ biosorption of polysaccharide from cordyceps militaris. J Polym Environ 25(4):1033–1043. https://doi.org/10.1007/s10924-016-0882-4
97. Yildiz H, Karatas N (2018) Microbial exopolysaccharides: resources and bioactive properties. Process Biochem 72:41–46. https://doi.org/10.1016/j.procbio.2018.06.009
98. Zhang Z, Wang P, Zhang J, Xia S (2014) Removal and mechanism of Cu (II) and Cd (II) from aqueous single-metal solutions by a novel biosorbent from waste-activated sludge. Environ Sci Pollut Res 21(18):10823–10829
99. Zhou X-Y, Wang X-R (2019) Impact of industrial activities on heavy metal contamination in soils in three major urban agglomerations of China. J Clean Prod 230:1–10
100. Zouboulis AI, Loukidou MX, Matis KA (2004) Biosorption of toxic metals from aqueous solutions by bacteria strains isolated from metal-polluted soils. Process Biochem 39(8):909–916. https://doi.org/10.1016/S0032-9592(03)00200-0

Applications of EPS in Environmental Bioremediations

Tarun Kumar Kumawat, Varsha Kumawat, Swati Sharma, Nirat Kandwani, and Manish Biyani

Abstract Micro-organisms (several fungi, bacteria, and algae) provide several biopolymers from renewable sources in which exopolysaccharides gain importance over other biopolymers. EPSs are useful for a vast variety of industrially important biomaterials. EPS accrues on the cell superficial of microbes and keeps safe against tough environmental conditions. In this book chapter, first, an introduction about EPS is given and then several current applications, potentials of exopolysaccharide which have been applied in bioremediation of heavy metals (HMs), colorant-containing residual, and toxic chemicals are deliberated.

Keywords Bioremediation · Exopolysaccharides (EPS) · Heavy metals · Environment · Biosorption · Dyes · Detoxification · Biofilm

Abbreviations

EPS	Exopolysaccharide
CPS	Capsular polysaccharide
HMs	Heavy metals
WHO	World Health Organization
PAHs	Polycyclic aromatic hydrocarbons

T. K. Kumawat (✉) · N. Kandwani
Department of Biotechnology, Biyani Girls College, Jaipur, Rajasthan, India

V. Kumawat
Naturilk Organic & Dairy Foods Pvt. Ltd., Jaipur, Rajasthan, India

S. Sharma
Chandigarh University, Mohali, Punjab, India

M. Biyani
Department of Bioscience and Biotechnology, Japan Advanced Institute of Science and Technology, 1-1 Asahidai, Nomi City, Ishikawa, Japan

A. K. Nadda et al. (eds.), *Microbial Exopolysaccharides as Novel and Significant Biomaterials*, Springer Series on Polymer and Composite Materials,
https://doi.org/10.1007/978-3-030-75289-7_11

1 Introduction

Exopolysaccharides (EPS) are complex biosynthetic polymers with microbial origins [10]. The polysaccharides from microbial origin are classified into two different forms (a) Capsular Polysaccharide (CPS) and (b) Exopolysaccharide (EPS) [88]. Depending on their location, EPS exhibit two typologies (a) Cell bound exopolysaccharides and (b) Released exopolysaccharide. Microbes secrete exopolysaccharide as carbohydrate polymers which form a layer on cell surface [103]. Exopolysaccharides (EPS) are organic macromolecule and water-soluble biopolymers attributed to various applications [118, 135]. Exopolysaccharides are divided into homopolysaccharides and heteropolysaccharides (HePS) (Molecular weight 10–30 kDa) [73]. Exopolysaccharides (EPSs) can be formed by prokaryotes (archaebacteria and eubacteria) and eukaryotes (Fungi, phytoplankton, and algae) have received more attention from scientists [145]. Different carbon sources, C/N ratio, pH, and temperature influence the composition and production of EPS [102]. EPSs are pondered as ubiquitous and plentiful bio-products [144].

The microbes are protected by exopolysaccharide against toxicity of heavy metal, drought, and salinity [26]. Exopolysaccharide has active and ionizable functional groups and substituents of non-carbohydrates which are responsible for the polymer's negative charge. Due to this property, ion exchange, complexation, and trapping, such as mechanisms, can bind different heavy metals to EPS and resulting in choice for the procedure of bioremediation [31].

Many microbial EPSs demonstrate the activity of metal ion biosorbent. Recently, heavy metal contamination has aroused in the environment and possess high risk to the health [81]. Heavy metal accumulation in environment due to manufacturing of electrical machinery, metal mining operations, tannery factories, municipal sewage disposal pesticides, and chemical fertilizers [119]. Biomineralization, biosorption, phytostabilization, mycoremediation, hyperaccumulation, rhizoremediation, cyano-remediation, dendroremediation, biostimulation, and genoremediation are among recent biotechnological tactics to bioremediation [87].

2 Environment Pollutants

Heavy metals are naturally occurring elements with relatively high density compared to H_2O and causes serious environmental issues [142]. Rapid industrialization is the main reason for the exposure of toxic heavy metals from soil colloids and water supplies. Heavy metal is persistent inorganic contaminants that are toxic to plants, animals, and humans due to food chain accumulation [152]. Heavy metal contamination is caused by waste from factories, and refineries, as well as indirectly by pollutants into the water stream [39]. Lead (Pb), copper (Cu), arsenic (As), mercury (Hg), cadmium (Cd) are considered to be the extreme common pollution causing heavy metals [61, 89].

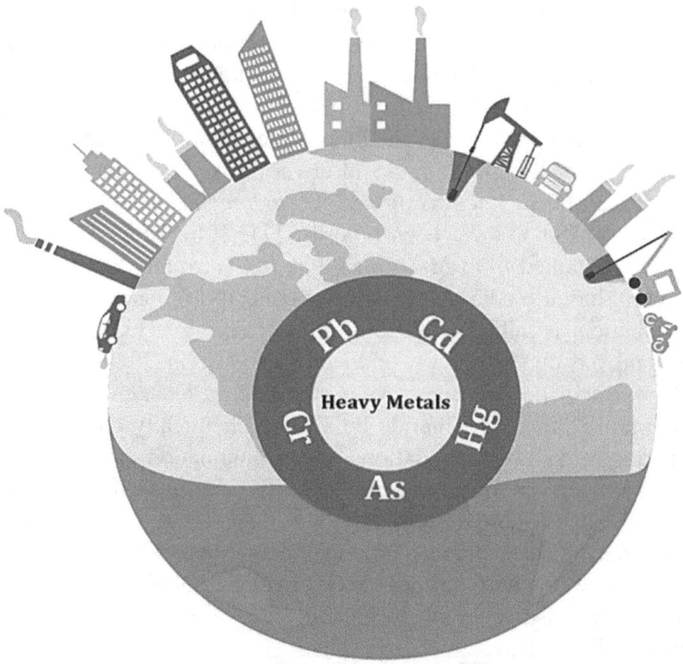

Fig. 1 Heavy metal contamination on earth

Textile industry is mainly accountable for the water pollution that releases untreated sewage water into river, water bodies and poses risks to environment [65]. Dyes (synthetic) are broadly used in textile, paper, pulp industries, color photography, chemical, and cosmetics industries (Fig. 1) [12].

3 Sustainable Bioremediation of Toxic Heavy Metals

The application of EPS biopolymer started in 1960s, and their marketable uses have increased considerably since then [115]. Microbes can absorb and remove pollutants from the atmosphere through different pathways [33]. Initially, EPS research focused primarily on the treatment and elimination of toxic heavy metal from manufacturing as well as domestic wastewater [112]. The bioremediation of hazardous metals by microbes can be helpful in comparison of traditional physico-chemical procedures [95]. Conventional remediation approaches are either costly or produce dangerous by-products that adversely affect the atmosphere [49]. Due to physical, structural, and chemical diversity, scientists show the research curiosity in microorganisms' exopolysaccharides, and their uses in environmental bioremediation have increased [9, 72]. The bioremediation mechanism efficiently minimizes the hazard of contamination by removal of toxic metals from soil and groundwater [136].

In recent times, use of biomass (bacteria both living and dead conditions) for bioremediation has arisen as safe, low waste production, environment-friendly, cost-effective, low energy demand, and self-sustainability alternative for HMs removal (Fig. 2) [45]. Biosorption from microbes is also an innovative procedure for the management of waste [67]. The innovative exopolysaccharide might be used as novel source for heavy metals removal in various sectors. Adsorption of various HMs by bacterial EPS (*Bacillus firmus, Pseudomonas pachastrellae* KMS2-2, *Bacillus cereus* KMS3-1) have been reported [71, 126]. EPS-M816 (extracellular polysaccharide) obtained from *Mesorhizobium loti* (formerly known as *Rhizobium loti*) on glycerol-based media and is used as emulsifying agent [32].

EPS production is allied with biofilm processing which is important for biosorption and bio-mineralization of metal ion [41]. In 2020, Rusinova-Videva et al. [123] reported the proficiency of *Cryptococcus laurentii* (AL65) for heavy metal biosorption. Exopolysaccharide be able to chemically changed via acetylation, methylation, carboxymethylation, and sulphonation, that amends natural activities of EPS, thus refining the applicability of the polymer [104]. In harsh conditions such as geothermal springs, saline lakes, and deep-sea hydrothermal vents, EPSs often make a coating around the microbe cell [101]. Hydrothermal deep-sea vents may be a virtuous source of novel exopolysaccharide [47]. Various microorganisms are also reported for EPS production and used for environmental bioremediation (Table 1).

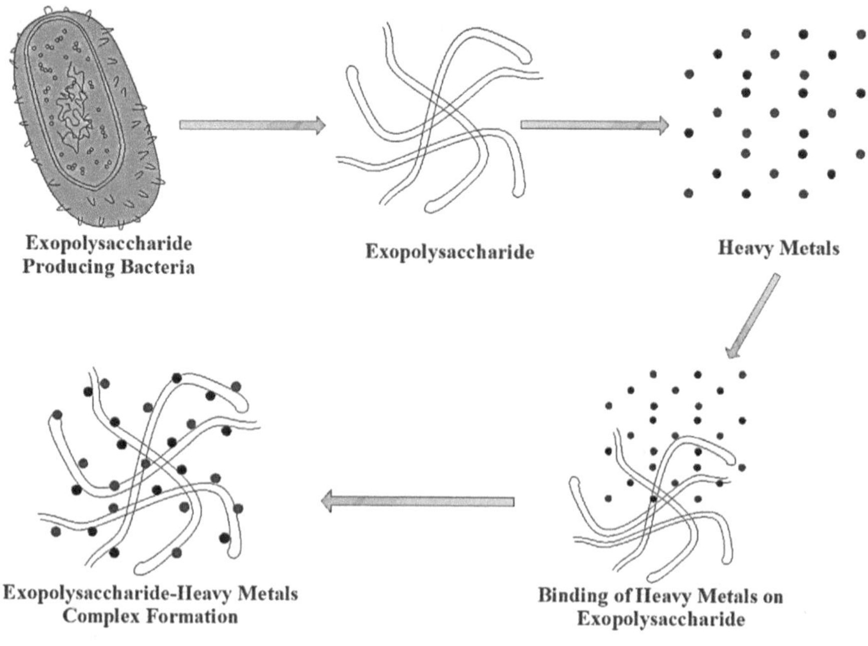

Fig. 2 Mechanism of exopolysaccharide interaction with heavy metals

Table 1 Production of EPS from microbes and their applications

EPS producing microbes	Applications	References
Bacillus licheniformis MS3	Excellent emulsifying properties against hydrocarbons and oils	[10]
Sphingomonas sp. MKIV	Ionic liquids (ILs) bioremediation	[70]
Klebsiella oxytoca strain DSM 29614	As(III) and As(V) removal from contaminated water	[16]
Bacillus strain TCL	Bioremediation of contaminated leather industry effluent and chromium reduction	[11]
Pantoea agglomerans	Metal bioremediation	[91]
Bacillus sonorensis MJM60135	Emulsification of toluene and bioremediation of hydrocarbons	[110]
Escherichia coli ATCC 25922 (ECBK) and *Staphylococcus epidermidis* RP62A (SEBK)	Removal of Cr(VI) and Zn(II)	[116]
Sphingomonas pseudosanguinis and *Sphingomonas yabuuchiae*	Bioremediation of biodiesel-derived waste glycerol (WG)	[117]
Bacillus cereus VK1	Bioremediation of Hg^{2+} polluted eco-systems	[64]
Alteromonas sp. JL2810	Removal of heavy metals Cu^{2+}, Ni^{2+}, and Cr^{6+}	[153]
Leptothrix cholodnii SP-6SL	Metal adsorption	[100]
Achromobacter xylosoxidans strain TERI L1	Produce EPS bioflocculant	[138]
Arthrobacter sp. SUK 1205	Bioremediation of environmental Cr pollution	[36]
Shewanella oneidensis	Adsorption of Cd(II)	[151]
Pseudomonas sp. Lk9	Removal of heavy metal ions (Cd^{2+} and Cu^{2+})	[84]
Synechocystis sp. BASO671	Metal removal	[108]
Arthrobacter ps-5	Biosorption capacity on metal ions (Cu^{2+}, Pb^{2+} and Cr^{6+})	[134]
Azotobacter chroococcum XU1	Biosorption of Pb and Hg	[119]
Paracoccus sp., *Alteromonas* sp., *Vibrio* sp., One unidentified	Cu(II) and Ag(I) bioremediation	[35]
Enterobacter cloacae strain P2B	Bioremediation of Lead	[98]
Rhizobium miluonense CC-B-L1, *Burkholderia seminalis* CC-IDD2w and *Ensifer adhaerens* CC-GSB4	Diesel oil emulsification	[56]
Pseudoalteromonas sp. A28 *Vibrio proteolyticus*	Biodegradation of a Bioemulsificant Exopolysaccharide (EPS_{2003})	[15]
Aspergillus fumigatus	Sorption efficiency copper and lead	[150]
Gordonia polyisoprenivorans CCT 7137	Removal of contaminating oil	[43]

(continued)

Table 1 (continued)

EPS producing microbes	Applications	References
Synechocystis sp.	Cr(VI) removal	[107]
Penicillium simplicissimum	Cd(II), Zn(II), and Pb(II) biosorption	[40]
Gordonia alkanivorans CC-JG39	Diesel biodegradation	[139]
Psuedoalteromonas sp. strain TG12	Emulsifying activity against oil substrate and metal binding activity	[51]
Paenibacillus jamilae	Production of a metal-binding EPS for adsorption	[93]
Pestalotiopsis sp. KCTC 8637	Wastewater treatment	[92]
Chryseomonas luteola TEM05	Cadmium and cobalt ions adsorption	[106]
Pseudomonas putida CZ1	Copper(II) and Zinc(II) biosorption	[21]
Bacillus subtilis	Adsorption of neutral Hg(II)	[30]
Cyanothece CE 4 and *Cyanospira capsulata*	Remove Cu^{2+} from aqueous solution	[34]

Bioremediation of Lead (Pb): Lead released from processing factories, recycling plants, and automotive industries [63]. In the environment, lead is a non-bioessential, hazardous, and pervasive metal that affects all living beings [97, 143]. The adsorption of lead by EPS has been investigated in several studies. Concordio-Reis et al. [24] reported that polysaccharide FucoPol (fucose-containing exopolysaccharide) secreted by *Enterobacter* A47 have good potential as biodegradable and safe biosorbent for the management of Pb^{2+} contaminated water. *Paenibacillus jamilae*'s EPS adsorbs Pb, Cd, Co, Ni, Zn, and Cu. The EPS showed good interaction with Pb with binding capacity of 303.03 mg/g [94].

Bioremediation of Cadmium (Cd): Cadmium is a persuasive pollutant and environmental contaminant that can unfavorably affect anthropoid health [20]. At extremely low concentrations of around 0.001–0.1 mg/L, this heavy metal is dangerous to individuals [140]. One of the greatest tactics of microbes to combat against toxicity of HMs is the exopolysaccharide (EPS) production [90]. Bacterial exopolysaccharide from *Arthrobacter viscosus* can significantly increase biosorption of Cd [132]. *Halomonas* sp., is an exopolysaccharide-producing halophilic bacterium which uptake more than 50% cadmium and might be used for bioremediation of Cd in polluted saline soils [7].

Bioremediation of Arsenic (As): According to World Health Organization (WHO), the concentration of As in drinking water is 10 μg/L [146]. As is toxic metalloid released into the atmosphere by natural and anthropogenic activities. The major activities include i.e., mining, weathering, combustion of fossil fuels and volcanic happenings Arsenic causes several severe diseases in human and animals [14, 82, 131, 141]. The polyanionic EPS produced biofilm is pondered as admirable biosorbent material for the bioremediation of arsenic and other metals [124].

Halomonas sp. Exo1 was proficient producer of exopolysaccharide and bio-remediate arsenic. Apart from As, EPS has also role in biosorption of Cr, Hg, and Mn [96].

Bioremediation of Mercury (Hg): Mercury is significantly dispersed in environment and found in the water, soil, and air. This metal is being highly poisonous to human beings, animals, and plants at low concentrations, so its remediation is incredibly essential [137]. Metal mining, fossil combustion, acetaldehyde industries, chloralkali plants, amalgamation industries, and paint industries are primary source of mercury pollution [80]. Mercury toxicity depends primarily on the speciation of Hg [4]. Mercury enters the environment mainly as Hg^{2+} from industrial sources [121]. Exopolysaccharide shows substantial role in mercury detoxification by heavy metal bioremediation method. *Vibrio fluvialis* was evaluated at four separate concentrations (100, 150, 200, and 250 μg/ml) for Hg removal potential and effective bioremediation was observed at 250 μg/ml (60% Hg removal) [129]. *Bacillus thuringiensis* strain RGN1.2 is used for bioremediation of mercury pollution and remove 96.72%, 90.67%, and 90.10% of mercury at 10, 25, and 50 ppm, respectively [114].

Bioremediation of Chromium (Cr): Chromium is an imperative metal for living beings, but it is also harmful at extremely low concentrations due to its high toxicity, and carcinogenicity [22]. Industrial development, such as electroplating, metallurgy, dyeing, paper, and pulp making, and tanning, leads chromium contamination in the environment [53]. *Enterobacter cloaceae* produced exopolysaccharide with Cr (VI) at 25, 50, and 100 ppm, and approx. 60–70% Cr was accumulated by this bacterium [58]. Hexavalent chromium [Cr(VI)] is a possible mutagen and carcinogen, which is introduced in the atmosphere by the anthropogenic activities [27]. In India, around Sukinda valley, pollution of hexavalent chromium was due to Chrome mining activity. Exopolysaccharide from *Enterobacter cloacae* SUKCr1D has been able to minimize Cr(VI) concentration by 31.7% at 10 mg/l [52].

Apart from the bacterial polysaccharide, some fungal EPS also produced by *Aspergillus terreus, Aspergillus niger, Lentinula edodes, Fusarium solani, Fusarium oxyporum* species [19, 48, 133]. Scientists have research focus on fungal polysaccharides in several biotechnological fields [105]. Basidiomycota and Ascomycota fungal strains, and yeast species are recognized for their capability to produce EPSs under different cultural environments [6]. The significant biological properties of exopolysaccharides produced by fungi were recognized only 20 years ago [147]. Since several studies have shown various applications in environmental bioremediation, fungal exopolysaccharide has also gained significance from last few decades [113, 120]. Mushrooms are known to be good producer of EPS and have applications in various bio-productive fields as well as play part in bioremediation of various heavy metals [23, 111]. Exopolysaccharides produced from fungi have also antioxidants, anti-cholesterol, and anti-cancer properties [85]. Algal extracellular polymeric substances (EPSs) are also used as bioflocculant in the bioremediation process [74]. The EPS from *Anabaena spiroides* (cynobacterium) was efficient in the binding capacity of Mn(II), Pb(II), Hg(II), and Cu(II) [42].

4 Applications of EPS in Bioremediation of Colorant-Containing Residual from Textile Effluent

On earth, water is the most vital resource for life. Textile industries are imperative for economic development for nations but it not only consumes enormous quantities of water but also pollute water bodies via discharge of dye effluents and are responsible for major environmental pollution [19, 76, 149]. Bioremediation is a biological process for the elimination of colorant-containing residual by microorganisms (Table 2) [109]. Azo dyes i.e., Methyl orange (MO) and Congo red (CR) are degraded by exopolysaccharide-stabilized silver nanoparticles (AgNPs) produced by *Leuconostoc lactis* KC117496. EPS-AgNPs are synthesized through reduction of Ag^{+} into Ag^{o} from $AgNO_3$ which are effective, inexpensive, and environmentally safe candidate for biodegradation of hazardous textile dyes [130]. Li et al. [79] reported first time EPS-605 (a non-glucan EPS) that showed greatest biosorption capability for Pb^{2+}, Cu^{2+}, Cd^{2+}, and methylene blue. Halophilic *Exiguobacterium* sp. VK1 (indigenous salt pan isolate) removed carcinogenic Malachite Green with biosorption capacity of 79.1% at pH 6.0 and 40 °C temperature [65].

Pseudomonas aeruginosa and *Ochrobactrum* sp. produced EPS 456.4 mg l^{-1} and 404.6 mg l^{-1} after incubation for 48 h (40 °C) and 72 h (30 °C) respectively. EPS produced by *P. aeruginosa* and *Ochrobactrum* sp. removed Remazol Blue (RB) with a maximum yield of 12.5% and 89.4% respectively [68]. Novel hetero-exopolysaccharide-R040 (EPS-R040) prepared from *Lactobacillus plantarum* used as adsorbent for the removal of Methylene Blue [77].

5 EPS in Bioremediation of Oil Spills and Petroleum-Contaminated Sites

With the continual population growth, industrialization, and dependence on petroleum goods, environment pollution has increased globally. Petroleum oil-contaminated environment is not safe for both human and animals [5, 69, 148]. Oil biodegradation was measured respirometrically and based on variations in the composition of oil [122]. Ibrahim et al. [57] isolated halophilic bacterium *Halobacillus* sp. strain EG1HP4QL from Lake Qarun, Fayoum Province, Egypt that produced EPS and showed bioremediation prospective to consume crude oil (35.3%) as the solitary carbon source.

Vibrio harveyi (VB23) exopolysaccharide is hetero-polysaccharide and has strong activity of emulsification [13]. EPS produced by *Enterobacter cloaceae* (EPS 71a) emulsified benzene, hexane, kerosene, xylene, paraffin oil [59]. EPS produced from species of *Halomonas* have amphiphilic properties and contributed to the ultimate oil removal and oil aggregate formation [50]. *Sporosacina halophila* produced EPS, biosurfactant, and laccase for toluene biodegradation. It may also use for bioremediation of hydrocarbon adulterated sea and waste-water [1].

Table 2 Bioremediation of various toxic dyes by using exopolysaccharides

Name of dye	EPS producing microorganism	Application	References
Acridine orange and crystal violet	*Serratia* sp. ISTD04	Removal of dye	[75]
Azo dyes	*Rhizopus arrhizus*	Removal of azo dyes from aqueous dye solution	[127]
Methylene blue	*Rhodotorula mucilaginosa* strain UANL-001L	Methylene blue (MB) adsorption using EPS	[44]
Reactive dyes	*A. niger*, *A. oryzae* and *R. arrhizus*	Biosorption of textile dyes	[128]
Amaranth dye	*Bacillus* sp. AK1, *Lysinibacillus* sp. AK2, and *Kerstersia* sp. VKY1 strains biofilm	Decolorization of Amaranth dye	[8]
Congo red and indigo carmine	*Dietzia* sp. PD1	Biosorption of acid dye from aqueous solution	[125]
Remazol blue (RB)	*Pseudomonas aeruginosa* and *Ochrobactrum* sp.	Dye removal	[68]
Malachite green (MG)	*Pseudomonas aeruginosa* NCIM 2074	Decolorization process	[66]
Acid black 172	*Pseudomonas* sp. strain DY1	Biosorption of dyes from wastewater	[38]
Methylene blue	*Lactobacillus plantarum* JNULCC001	Excellent biosorption ability toward methylene blue	[78]
Reactive dyes	*Aeromonas* sp., *Pseudomonas luteola, Escherichia coli, Bacillus subtilis* and *Staphylococcus aureus*	Adsorption of reactive dyes	[54]

6 EPS in Bioremediation of Toxic Chemicals (PCBs, PAHs, CP)

Polycyclic Aromatic Hydrocarbons (PAHs) are environmental pollutants emitted from coal mining, power plants, and chemical industries. It is well-thought-out as a highly toxic pollutant because of their toxic, and carcinogenic properties [2, 46]. Exopolysaccharide (EPS) is produced by several species of rhizobia (soil bacterium) and play roles as bioemulsifier with possible applications for hydrocarbon degradation. *Rhizobium tropici* is member of the Rhizobiaceae and produced exopolysaccharides [17, 18]. *Rhizobium* sp. produced exopolysaccharide that showed flocculating and metal sorption properties [28]. Exopolysaccharide (MW $\sim 2 \times 10(5)$ Da) produced from *Klebsiella* sp. PB12 showed emulsifying activity with toluene, *n*-hexadecane, olive oil, and kerosene of 66.6%, 65%, 63.3% and 50% respectively [86]. Kachlany et al. [62] observed that hydrocarbon-degrading bacterium (*Pseudomonas putida* G7) produced EPS and show great metal binding capacity. EPSs of *Aspergillus niger* and *Zoogloea* sp. degrade more than 80% pyrene (polycyclic aromatic hydrocarbon) in polluted soils [60]. *Paenibacillus jamila*'s EPS contributes in bioremediation of wastewater generated during olive oil process [3]. Naseem and Bano [99] observed that the soil moisture, plant biomass was improved from *Alcaligenes faecalis* (AF3), *Proteus penneri* (Pp1), and *P. aeruginosa* (Pa2) (EPS producing bacteria) by seed bacterization of maize. *Enterobacter cloacae* strain TU (Gram-negative rod-shaped bacterium) release extracellular polymeric substance and exhibiting bio-emulsifying activity. The bacterium also utilizes alkanes and polycyclic aromatic hydrocarbons (PAHs) [55].

7 Concluding Remarks

Microbial exopolysaccharides are vastly diverse, sustainable, and environmentally safe alternatives in various applications especially in bioremediation. EPS producing microbes remediate the colorant residues, toxic chemicals, and hazardous heavy metals from environmental effluents. Biofilm-mediated remediation has been establishing as safer alternative to bioremediation. EPSs received significant research consideration from scientists in recent years because of their biodegradability and biocompatibility. According to existing information on exopolysaccharides, these biopolymers can cover a broad array of research in greener technologies.

Acknowledgements We thank the Director, Research and Development, Biyani Group of Colleges, Jaipur for support and encouragement.

Conflict of Interest The authors declare that there is no conflict of interest.

References

1. Abari AH, Emtiazi G, Ghasemi SM (2012) The role of exopolysaccharide, biosurfactant and peroxidase enzymes on toluene degradation by bacteria isolated from marine and wastewater environments. Jundishapur J Microbiol 5:479–485
2. Abdel-Shafy HI, Mansour MSM (2016) A review on polycyclic aromatic hydrocarbons: source, environmental impact, effect on human health and remediation. Egypt J Pet 25:107–123

3. Aguilera M, Quesada MT, del Águila VG et al (2008) Characterisation of *Paenibacillus jamilae* strains that produce exopolysaccharide during growth on and detoxification of olive mill wastewaters. Bioresour Technol 99:5640–5644
4. Al-Manhel AJA (2017) Production of exopolysaccharide from local fungal isolate. Curr Res Nutr Food Sci 5:338–346
5. Ali N, Dashti N, Khanafer M, Al-Awadhi H, Radwan S (2020) Bioremediation of soils saturated with spilled crude oil. Sci Rep 10:1116
6. Ali H, Khan E, Ilahi I (2019) Environmental chemistry and ecotoxicology of hazardous heavy metals: environmental persistence, toxicity, and bioaccumulation. J Chem 6730305
7. Amoozegar MA, Ghazanfari N, Didari M (2012) Lead and cadmium bioremoval by *Halomonas* sp., an exopolysaccharide-producing halophilic bacterium. Prog Biol Sci 2:1–11
8. Anjaneya O, Shrishailnath SS, Guruprasad K, Nayak AS, Mashetty SB, Karegoudar TB (2013) Decolourization of Amaranth dye by bacterial biofilm in batch and continuous packed bed bioreactor. Int Biodeterior Biodegradation 79:64–72
9. Annadurai ST, Arivalagan P, Sundaram R, Mariappan R, Pudukadu Munusamy A (2019) Batch and column approach on biosorption of fluoride from aqueous medium using live, dead and various pretreated *Aspergillus niger* (FS18) biomass. Surf Interfaces 15:60–69
10. Asgher M, Urooj Y, Qamar SA, Khalid N (2020) Improved exopolysaccharide production from *Bacillus licheniformis* MS3: optimization and structural/functional characterization. Int J Biol Macromol 151:984–992
11. Banerjee S, Misra A, Chaudhury S, Dam B (2019) A *Bacillus* strain TCL isolated from Jharia coalmine with remarkable stress responses, chromium reduction capability and bioremediation potential. J Hazard Mater 367:215–223
12. Bayoumi MN, Al-wasify RS, Hamed SR (2014) Bioremediation of textile wastewater dyes using local bacterial isolates. Int J Curr miccrobiology Appl Sci 3:962–970
13. Bramhachari PV, Dubey SK (2006) Isolation and characterization of exopolysaccharide produced by *Vibrio harveyi* strain VB23. Lett Appl Microbiol 43:571–577
14. Butt AS, Rehman A (2011) Isolation of arsenite-oxidizing bacteria from industrial effluents and their potential use in wastewater treatment. World J Microbiol Biotechnol 27:2435–2441
15. Cappello S, Crisari A, Denaro R, Crescenzi F, Porcelli F, Yakimov MM (2011) Biodegradation of a bioemulsificant exopolysaccharide (EPS_{2003}) by marine bacteria. Water Air Soil Pollut 214:645–652
16. Casentini B, Gallo M, Baldi F (2019) Arsenate and arsenite removal from contaminated water by iron oxides nanoparticles formed inside a bacterial exopolysaccharide. J Environ Chem Eng 7:
17. Castellane TCL, Campanharo JC, Colnago LA et al (2017) Characterization of new exopolysaccharide production by *Rhizobium tropici* during growth on hydrocarbon substrate. Int J Biol Macromol 96:361–369
18. Castellane TCL, Lemos MVF, Lemos EGDM (2014) Evaluation of the biotechnological potential of I strains for exopolysaccharide production. Carbohydr Polym 111:191–197
19. Castro AM, Nogueira V, Lopes I, Rocha-Santos T, Pereira R (2019) Evaluation of the potential toxicity of effluents from the textile industry before and after treatment. Appl Sci 9:3804
20. Chellaiah ER (2018) Cadmium (heavy metals) bioremediation by *Pseudomonas aeruginosa*: a minireview. Appl Water Sci 8:154
21. Chen XC, Wang YP, Lin Q, Shi JY, Wu WX, Chen YX (2005) Biosorption of copper(II) and zinc(II) from aqueous solution by *Pseudomonas putida* CZ1. Colloids Surfaces B Biointerfaces 46:101–107
22. Cheung KH, Gu JD (2003) Reduction of chromate (CrO^{42-}) by an enrichment consortium and an isolate of marine sulfate-reducing bacteria. Chemosphere 52:1523–1529
23. Choudhary S (2020) Characterization and applications of mushroom exopolysaccharides. In: Gehlot P, Singh J (eds) New and Future Developments in Microbial Biotechnology and Bioengineering. Elsevier, Netherlands, pp 171–181

24. Concordio-Reis P, Reis MAM, Freitas F (2020) Biosorption of heavy metals by the bacterial exopolysaccharide fucopol. Appl Sci 10:6708
25. Costa OYA, Raaijmakers JM, Kuramae EE (2018) Microbial extracellular polymeric substances: Ecological function and impact on soil aggregation. Front Microbiol 9:1636
26. Das A, Mishra S (2010) Biodegradation of the metallic carcinogen hexavalent chromium Cr (VI) by an indigenously isolated bacterial strain. J Carcinog 9:6
27. Das S, Sen IK, Kati A et al (2019) Flocculating, emulsification and metal sorption properties of a partial characterized novel exopolysaccharide produced by *Rhizobium tropici* SRA1 isolated from *Psophocarpus tetragonolobus* (L) D.C. Int Microbiol 22:91–101
28. Daughney CJ, Siciliano SD, Rencz AN, Lean D, Fortin D (2002) Hg(II) adsorption by bacteria: a surface complexation model and its application to shallow acidic lakes and wetlands in Kejimkujik National Park, Nova Scotia, Canada. Environ Sci Technol 36:1546–1553
29. Dave SR, Upadhyay KH, Vaishnav AM, Tipre DR (2020) Exopolysaccharides from marine bacteria: production, recovery and applications. Environ Sustain 3:139–154
30. Deschatre M, Ghillebaert F, Guezennec J, Colin CS (2013) Sorption of copper(II) and silver (I) by four bacterial exopolysaccharides. Appl Biochem Biotechnol 171:1313–1327
31. Dey S, Paul AK (2015) Hexavalent chromate reduction during growth and by immobilized cells of *Arthrobacter* sp. SUK 1205. Sci Technol Dev 34:158–168
32. Du LN, Wang B, Li G, Wang S, Crowley DE, Zhao YH (2012) Biosorption of the metal-complex dye acid black 172 by live and heat-treated biomass of *Pseudomonas* sp. strain DY1: kinetics and sorption mechanisms. J Hazard Mater 205–206:47–54
33. Edward Raja C, Selvam GS (2009) Plasmid profile and curing analysis of *Pseudomonas aeruginosa* as metal resistant. Int J Environ Sci Technol 62:259–266
34. Fan T, Liu Y, Feng B et al (2008) Biosorption of cadmium(II), zinc(II) and lead(II) by *Penicillium simplicissimum*: isotherms, kinetics and thermodynamics. J Hazard Mater 160:655–661
35. Francois F, Lombard C, Guigner JM et al (2012) Isolation and characterization of environmental bacteria capable of extracellular biosorption of mercury. Appl Environ Microbiol 78:1097–1106
36. Freire-Nordi CS, Vieira AAH, Nascimento OR (2005) The metal binding capacity of *Anabaena spiroides* extracellular polysaccharide: an EPR study. Process Biochem 40:2215–2224
37. Fusconi R, Maria Nascimento Assuncao R, de Moura Guimaraes R, Rodrigues Filho G, Eduardo da Hora Machado A (2010) Exopolysaccharide produced by *Gordonia polyisoprenivorans* CCT 7137 in GYM commercial medium and sugarcane molasses alternative medium: FT-IR study and emulsifying activity. Carbohydr Polym 79:403–408
38. Garza MTG, Perez DB, Rodriguez AV et al (2016) Metal-induced production of a novel bioadsorbent exopolysaccharide in a native *Rhodotorula mucilaginosa* from the mexican northeastern region. PLoS ONE 11:
39. Ghoniem AA, El-Naggar NEA, Saber WEIA, El-Hersh MS, El-khateeb AY (2020) Statistical modeling-approach for optimization of Cu^+ biosorption by *Azotobacter nigricans* NEWG-1; characterization and application of immobilized cells for metal removal. Sci Rep 10:9491
40. Ghosal D, Ghosh S, Dutta TK, Ahn Y (2016) Current state of knowledge in microbial degradation of polycyclic aromatic hydrocarbons (PAHs): a review. Front Microbiol 7:1369
41. Guezennec J (2002) Deep-sea hydrothermal vents: A new source of innovative bacterial exopolysaccharides of biotechnological interest? J Ind Microbiol Biotechnol 29:204–208
42. Guo S, Mao W, Li Y, Tian J, Xu J (2012) Structural elucidation of the exopolysaccharide produced by fungus *Fusarium oxysporum* Y24-2. Carbohydr Res 365:9–13
43. Gupta P, Diwan B (2017) Bacterial exopolysaccharide mediated heavy metal removal: a review on biosynthesis, mechanism and remediation strategies. Biotechnol Rep 13:58–71
44. Gutierrez T, Berry D, Yang T et al (2013) Role of bacterial exopolysaccharides (EPS) in the fate of the oil released during the deepwater horizon oil spill. PLoS ONE 8:

45. Gutierrez T, Shimmield T, Haidon C, Black K, Green DH (2008) Emulsifying and metal ion binding activity of a glycoprotein exopolymer produced by *Pseudoalteromonas* sp. strain TG12. Appl Environ Microbiol 74:4867–4876
46. Harish R, Samuel J, Mishra R, Chandrasekaran N, Mukherjee A (2012) Bio-reduction of Cr (VI) by exopolysaccharides (EPS) from indigenous bacterial species of Sukinda chromite mine, India. Biodegradation 23:487–496
47. He C, Gu L, Xu Z et al (2020) Cleaning chromium pollution in aquatic environments by bioremediation, photocatalytic remediation, electrochemical remediation and coupled remediation systems. Environ Chem Lett 18:561–576
48. Hu TL (1996) Removal of reactive dyes from aqueous solution by different bacterial genera. Water Sci Technol 34:89–95
49. Hua X, Wu Z, Zhang H et al (2010) Degradation of hexadecane by *Enterobacter cloacae* strain TU that secretes an exopolysaccharide as a bioemulsifier. Chemosphere 80:951–956
50. Huang KH, Chen BY, Shen FT, Young CC (2012) Optimization of exopolysaccharide production and diesel oil emulsifying properties in root nodulating bacteria. World J Microbiol Biotechnol 28:1367–1373
51. Ibrahim IM, Konnova SA, Sigida EN et al (2020) Bioremediation potential of a halophilic *Halobacillus* sp. strain, EG1HP4QL: exopolysaccharide production, crude oil degradation, and heavy metal tolerance. Extremophiles 24:157–166
52. Iyer A, Mody K, Jha B (2004) Accumulation of hexavalent chromium by an exopolysaccharide producing marine *Enterobacter cloaceae*. Mar Pollut Bull 49:974–977
53. Iyer A, Mody K, Jha B (2006) Emulsifying properties of a marine bacterial exopolysaccharide. Enzyme Microb Technol 38:220–222
54. Jia C, Li P, Li X, Tai P, Liu W, Gong Z (2011) Degradation of pyrene in soils by extracellular polymeric substances (EPS) extracted from liquid cultures. Process Biochem 46:1627–1631
55. Jia L, Wang W, Li Y, Yang L (2010) Heavy metals in soil and crops of an intensively farmed area: a case study in Yucheng City, Shandong Province, China. Int J Environ Res Public Health 7:395–412
56. Kachlany SC, Levery SB, Kim JS, Reuhs BL, Lion LW, Ghiorse WC (2001) Structure and carbohydrate analysis of the exopolysaccharide capsule of *Pseudomonas putida* G7. Environ Microbiol 3:774–784
57. Kalita D, Joshi SR (2017) Study on bioremediation of Lead by exopolysaccharide producing metallophilic bacterium isolated from extreme habitat. Biotechnol Reports 16:48–57
58. Kalpana R, Angelaalincy MJ, Kamatchirajan BV et al (2018) Exopolysaccharide from *Bacillus cereus* VK1: Enhancement, characterization and its potential application in heavy metal removal. Colloids Surf B Biointerfaces 171:327–334
59. Kalpana R, Maheshwaran M, Vimali E et al (2020) Decolorization of textile dye by halophilic *Exiguobacterium* sp. VK1: biomass and exopolysaccharide (EPS) enhancement for bioremediation of malachite green. Chem Select 5:8787
60. Kalyani DC, Telke AA, Surwase SN, Jadhav SB, Lee JK, Jadhav JP (2012) Effectual decolorization and detoxification of triphenylmethane dye malachite green (MG) by *Pseudomonas aeruginosa* NCIM 2074 and its enzyme system. Clean Technol Environ Policy 14:989–1001
61. Kamala K, Sivaperumal P, Thilagaraj R, Natarajan E (2020) Bioremediation of Sr^{2+} ion radionuclide by using marine *Streptomyces* sp. CuOff24 extracellular polymeric substances. J Chem Technol Biotechnol 95:893–903
62. Kilic NK, Donmez G (2012) Remazol blue removal and EPS production by *Pseudomonas aeruginosa* and *Ochrobactrum* sp. Polish J Environ Stud 21:123–128
63. Koshlaf ES, Ball A (2017) Soil bioremediation approaches for petroleum hydrocarbon polluted environments. AIMS Microbiol 3:25–49
64. Koutinas M, Vasquez MI, Nicolaou E et al (2019) Biodegradation and toxicity of emerging contaminants: Isolation of an exopolysaccharide-producing *Sphingomonas* sp. for ionic liquids bioremediation. J Hazard Mater 365:88–96

65. Krishnamurthy M, Jayaraman Uthaya C, Thangavel M, Annadurai V, Rajendran R, Gurusamy A (2020) Optimization, compositional analysis, and characterization of exopolysaccharides produced by multi-metal resistant *Bacillus cereus* KMS3-1. Carbohydr Polym 227:
66. Kumar D, Kastanek P, Adhikary SP (2018) Exopolysaccharides from cyanobacteria and microalgae and their commercial application. Curr Sci 115:234
67. Kumar M, Kumar M, Pandey A, Thakur IS (2019) Genomic analysis of carbon dioxide sequestering bacterium for exopolysaccharides production. Sci Rep 9:4270
68. Kumar AS, Mody K, Jha B (2007) Bacterial exopolysaccharides—a perception. J Basic Microbiol 47:103–117
69. Kumar AS, Mody K (2009) Microbial exopolysaccharides: variety and potential applications. In: Rehm BHA (ed) Microbial production biopolymers and polymer precursors—application and perspectives. Caister Academic Press, Norfolk, pp 229–253
70. Lellis B, Fávaro-Polonio CZ, Pamphile JA, Polonio JC (2019) Effects of textile dyes on health and the environment and bioremediation potential of living organisms. Biotechnol Res Innov 3:275–290
71. Li C, Chen D, Ding J, Shi Z (2020) A novel hetero-exopolysaccharide for the adsorption of methylene blue from aqueous solutions: Isotherm, kinetic, and mechanism studies. J Clean Prod 265:
72. Li C, Ding J, Chen D, Shi Z, Wang L (2020) Bioconversion of cheese whey into a hetero-exopolysaccharide via a one-step bioprocess and its applications. Biochem Eng J 161:
73. Li P, Feng XB, Qiu GL, Shang LH, Li ZG (2009) Mercury pollution in Asia: a review of the contaminated sites. J Hazard Mater 168:591–601
74. Li WW, Yu HQ (2014) Insight into the roles of microbial extracellular polymer substances in metal biosorption. Bioresour Technol 160:15–23
75. Li C, Zhou L, Yang H et al (2017) Self-assembled exopolysaccharide nanoparticles for bioremediation and green synthesis of noble metal nanoparticles. ACS Appl Mater Interfaces 9:22808–22818
76. Lim KT, Shukor MY, Wasoh H (2014) Physical, chemical, and biological methods for the removal of arsenic compounds. Biomed Res Int 2014:1–9
77. Luo S, Li X, Chen L, Chen J, Wan Y, Liu C (2014) Layer-by-layer strategy for adsorption capacity fattening of endophytic bacterial biomass for highly effective removal of heavy metals. Chem Eng J 239:312–321
78. Mahapatra S, Banerjee D (2013) Fungal exopolysaccharide: production, composition and applications. Microbiol Insights 29:1–16
79. Mandal AK, Yadav KK, Sen IK et al (2013) Partial characterization and flocculating behavior of an exopolysaccharide produced in nutrient-poor medium by a facultative oligotroph *Klebsiella* sp. PB12. J Biosci Bioeng 115:76–81
80. Mani D, Kumar C (2014) Biotechnological advances in bioremediation of heavy metals contaminated ecosystems: an overview with special reference to phytoremediation. Int J Environ Sci Technol 11:843–872
81. Mishra A, Jha B (2013) Microbial exopolysaccharides. In: Rosenberg E, DeLong EF, Lory S, Stackebrandt E, Thompson F (eds) The prokaryotes. Springer, Berlin, Heidelberg
82. Mishra RK, Sharma V (2017) Biotic strategies for toxic heavy metal decontamination. Recent Pat Biotechnol 11:218–228
83. Mohite BV, Koli SH, Narkhede CP, Patil SN, Patil SV (2017) Prospective of microbial exopolysaccharide for heavy metal exclusion. Appl Biochem Biotechnol 183:582–600
84. Mohite BV, Koli SH, Patil SV (2018) Heavy metal stress and its consequences on exopolysaccharide (EPS)-producing *Pantoea agglomerans*. Appl Biochem Biotechnol 186:199–216
85. Moon SH, Park CS, Kim YJ, Il Park Y (2006) Biosorption isotherms of Pb (II) and Zn (II) on pestan, an extracellular polysaccharide, of *Pestalotiopsis* sp. KCTC 8637P. Process Biochem 41:312–316

86. Morillo JA, Aguilera M, Ramos-Cormenzana A, Monteoliva-Sánchez M (2006) Production of a metal-binding exopolysaccharide by *Paenibacillus jamilae* using two-phase olive-mill waste as fermentation substrate. Curr Microbiol 53:189–193
87. Morillo Perez JA, Garcia-Ribera R, Quesada T, Aguilera M, Ramos-Cormenzana A, Monteoliva-Sanchez M (2008) Biosorption of heavy metals by the exopolysaccharide produced by *Paenibacillus jamilae*. World J Microbiol Biotechnol 24:2699
88. Mota R, Rossi F, Andrenelli L, Pereira SB, De Philippis R, Tamagnini P (2016) Released polysaccharides (RPS) from *Cyanothece* sp. CCY 0110 as biosorbent for heavy metals bioremediation: interactions between metals and RPS binding sites. Appl Microbiol Biotechnol 100:7765–7775
89. Mukherjee P, Mitra A, Roy M (2019) *Halomonas* Rhizobacteria of *Avicennia marina* of Indian sundarbans promote rice growth under saline and heavy metal stresses through exopolysaccharide production. Front Microbiol 10:1207
90. Naik MM, Dubey SK (2013) Lead resistant bacteria: lead resistance mechanisms, their applications in lead bioremediation and biomonitoring. Ecotoxicol Environ Saf 98:1–7
91. Naik MM, Pandey A, Dubey SK (2012) Biological characterization of lead-enhanced exopolysaccharide produced by a lead resistant *Enterobacter cloacae* strain P2B. Biodegradation 23:775–783
92. Naseem H, Bano A (2014) Role of plant growth-promoting rhizobacteria and their exopolysaccharide in drought tolerance of maize. J Plant Interact 9:689–701
93. Nell RM, Szymanowski JES, Fein JB (2016) Divalent metal cation adsorption onto *Leptothrix cholodnii* SP-6SL bacterial cells. Chem Geol 439:132–138
94. Nicolaus B, Kambourova M, Oner ET (2010) Exopolysaccharides from extremophiles: from fundamentals to biotechnology. Environ Technol 31:1145–1158
95. Noghabi KA, Zahiri HS, Yoon SC (2007) The production of a cold-induced extracellular biopolymer by *Pseudomonas fluorescens* BM07 under various growth conditions and its role in heavy metals absorption. Process Biochem 42:847–855
96. Nouha K, Yan S, Tyagi RD, Surampalli RY (2015) EPS producing microorganisms from municipal wastewater activated sludge. J Pet Environ Biotechnol 7:
97. Ojuederie OB, Babalola OO (2017) Microbial and plant-assisted bioremediation of heavy metal polluted environments: a review. Int J Environ Res Public Health 14:1504
98. Osinska-Jaroszuk M, Jarosz-Wilkołazka A, Jaroszuk-Sciseł J et al (2015) Extracellular polysaccharides from ascomycota and basidiomycota: production conditions, biochemical characteristics, and biological properties. World J Microbiol Biotechnol 31:1823–1844
99. Ozdemir G, Ceyhan N, Manav E (2005) Utilization of an exopolysaccharide produced by *Chryseomonas luteola* TEM05 in alginate beads for adsorption of cadmium and cobalt ions. Bioresour Technol 96:1677–1682
100. Ozturk S, Aslim B, Suludere Z (2009) Evaluation of chromium(VI) removal behaviour by two isolates of *Synechocystis* sp. in terms of exopolysaccharide (EPS) production and monomer composition. Bioresour Technol 100:5588–5593
101. Ozturk S, Aslim B, Suludere Z, Tan S (2014) Metal removal of cyanobacterial exopolysaccharides by uronic acid content and monosaccharide composition. Carbohydr Polym 101:265–271
102. Pal AK, Singh J, Soni R et al (2020) The role of microorganism in bioremediation for sustainable environment management. In: Pandey VC, Singh V (eds) Bioremediation of pollutants, pp 227–249
103. Palaniyandi SA, Damodharan K, Suh JW, Yang SH (2018) Functional characterization of an exopolysaccharide produced by *Bacillus sonorensis* MJM60135 isolated from Ganjang. J Microbiol Biotechnol 28:663–670
104. Patel M, Patel U, Gupte S (2014) Production of exopolysaccharide (EPS) and its application by new fungal isolates SGMP 1 and SGMP 2. Int J Agric Environ Biotechnol 7:511–523
105. Pereira SB, Sousa A, Santos M et al (2019) Strategies to obtain designer polymers based on cyanobacterial extracellular polymeric substances (EPS). Int J Mol Sci 20:5693

106. De Philippis R, Colica G, Micheletti E (2011) Exopolysaccharide-producing cyanobacteria in heavy metal removal from water: molecular basis and practical applicability of the biosorption process. Appl Microbiol Biotechnol 92:697–708
107. De Philippis R, Sili C, Paperi R, Vincenzini M (2001) Exopolysaccharide-producing cyanobacteria and their possible exploitation: a review. J Appl Phycol 13:293–299
108. Prathyusha AMVN, Mohana Sheela G, Bramhachari PV (2018) Chemical characterization and antioxidant properties of exopolysaccharides from mangrove filamentous fungi *Fusarium equiseti* ANP2. Biotechnol Rep 19:
109. Pushkar B, Sevak P, Sounderajan S (2019) Assessment of the bioremediation efficacy of the mercury resistant bacterium isolated from the Mithi river. Water Sci Technol Water Supply 19:191–199
110. Quesada E, Bejar V, Calvo C (1993) Exopolysaccharide production by *Volcaniella eurihalina*. Experientia 49:1037–1041
111. Quiton KG, Doma B, Futalan CM, Wan MW (2018) Removal of chromium(VI) and zinc(II) from aqueous solution using kaolin-supported bacterial biofilms of Gram-negative *E. coli* and gram-positive *Staphylococcus epidermidis*. Sustain Environ Res 28:206–213
112. Raghunandan K, Kumar A, Kumar S, Permaul K, Singh S (2018) Production of gellan gum, an exopolysaccharide, from biodiesel-derived waste glycerol by *Sphingomonas* spp. 3 Biotech 8:71
113. Raj K, Sardar UR, Bhargavi E, Devi I, Bhunia B, Tiwari ON (2018) Advances in exopolysaccharides based bioremediation of heavy metals in soil and water: a critical review. Carbohydr Polym 199:353–364
114. Rasulov BA, Yili A, Aisa HA (2013) Biosorption of metal ions by exopolysaccharide produced by *Azotobacter chroococcum* XU1. J Environ Prot (Irvine, Calif) 4:989–993
115. Rodrigues ML, Nimrichter L, Cordero RJB, Casadevall A (2011) Fungal polysaccharides: biological activity beyond the usual structural properties. Front Microbiol 2:171
116. Rojas LA, Yanez C, González M, Lobos S, Smalla K, Seeger M (2011) Characterization of the metabolically modified heavy metal-resistant *Cupriavidus metallidurans* strain MSR33 generated for mercury bioremediation. PLoS ONE 6:
117. Roling WFM, Milner MG, Jones DM et al (2002) Robust hydrocarbon degradation and dynamics of bacterial communities during nutrient-enhanced oil spill bioremediation. Appl Environ Microbiol 68:5537–5548
118. Rusinova-Videva S, Nachkova S, Adamov A, Dimitrova-Dyulgerova I (2020) Antarctic yeast *Cryptococcus laurentii* (AL_{65}): biomass and exopolysaccharide production and biosorption of metals. J Chem Technol Biotechnol 95:1372–1379
119. Saba Rehman Y, Ahmed M, Sabri AN (2019) Potential role of bacterial extracellular polymeric substances as biosorbent material for arsenic bioremediation. Bioremediat J 23:72–81
120. Saha PD, Bhattacharya P, Sinha K, Chowdhury S (2013) Biosorption of Congo red and Indigo carmine by nonviable biomass of a new *Dietzia* strain isolated from the effluent of a textile industry. Desalin Water Treat 51:5840–5847
121. Salehizadeh H, Shojaosadati SA (2003) Removal of metal ions from aqueous solution by polysaccharide produced from *Bacillus frmus*. Water Res 37:4231–4235
122. Salvi NA, Chattopadhyay S (2017) Biosorption of Azo dyes by spent *Rhizopus arrhizus* biomass. Appl Water Sci 7:3041–3054
123. Saraf S, Vaidya VK (2015) Comparative study of biosorption of textile dyes using fungal biosorbents. Int J Curr Microbiol App Sci 2:357–365
124. Saranya K, Sundaramanickam A, Shekhar S, Swaminathan S, Balasubramanian T (2017) Bioremediation of mercury by *Vibrio fluvialis* screened from industrial effluents. Biomed Res Int 6509648:6
125. Saravanan C, Rajesh R, Kaviarasan T, Muthukumar K, Kavitake D, Shetty PH (2017) Synthesis of silver nanoparticles using bacterial exopolysaccharide and its application for degradation of azo-dyes. Biotechnol Rep 15:33–40

126. Satyapal GK, Mishra SK, Srivastava A et al (2018) Possible bioremediation of arsenic toxicity by isolating indigenous bacteria from the middle Gangetic plain of Bihar, India. Biotechnol Rep 17:117–125
127. Scott JA, Palmer SJ (1988) Cadmium bio-sorption by bacterial exopolysaccharide. Biotechnol Lett 10:21–24
128. Sharma S, Khanna PK, Kapoor S (2016) Optimised isolation of polysaccharides from *Lentinula edodes* strain NCBI JX915793 using response surface methodology and their antibacterial activities. Nat Prod Res 30:616–21
129. Shuhong Y, Meiping Z, Hong Y et al (2014) Biosorption of Cu^{2+}, Pb^{2+} and Cr^{6+} by a novel exopolysaccharide from *Arthrobacter* ps-5. Carbohydr Polym 101:50–56
130. Shukla A, Mehta K, Parmar J, Pandya J, Saraf M (2019) Depicting the exemplary knowledge of microbial exopolysaccharides in a nutshell. Eur Polym J 119:298–310
131. Siebielec S, Siebielec G, Stuczynski T, Sugier P, Grzeda E, Grzadziel J (2018) Long term insight into biodiversity of a smelter wasteland reclaimed with biosolids and by-product lime. Sci Total Environ 636:1048–1057
132. Singh S, Kumar V (2020) Mercury detoxification by absorption, mercuric ion reductase, and exopolysaccharides: a comprehensive study. Environ Sci Pollut Res 27:27181–27201
133. Subudhi S, Bisht V, Batta N, Pathak M, Devi A, Lal B (2016) Purification and characterization of exopolysaccharide bioflocculant produced by heavy metal resistant *Achromobacter xylosoxidans*. Carbohydr Polym 137:441–451
134. Ta-Chen L, Chang JS, Young CC (2008) Exopolysaccharides produced by *Gordonia alkanivorans* enhance bacterial degradation activity for diesel. Biotechnol Lett 30:1201–1206
135. Tang XY, Zhu YG, Cui YS, Duan J, Tang L (2006) The effect of ageing on the bioaccessibility and fractionation of cadmium in some typical soils of China. Environ Int 32:682–689
136. Taran M, Fateh R, Rezaei S, Gholi MK (2019) Isolation of arsenic accumulating bacteria from garbage leachates for possible application in bioremediation. Iran J Microbiol 11:60–66
137. Tchounwou PB, Yedjou CG, Patlolla AK, Sutton DJ (2012) Heavy metal toxicity and the environment. Exp Suppl 101:133–164
138. Tiquia-Arashiro SM (2018) Lead absorption mechanisms in bacteria as strategies for lead bioremediation. Appl Microbiol Biotechnol 102:5437–5444
139. Tiwari ON, Muthuraj M, Bhunia B et al (2020) Biosynthesis, purification and structure-property relationships of new cyanobacterial exopolysaccharides. Polym Test 89:
140. Trivedi R (2020) Exopolysaccharides: production and application in industrial wastewater treatment. In: Shah M, Banerjee A (eds) Combined application of physico-chemical & microbiological processes for industrial effluent treatment plant. Springer, Singapore
141. Wang S, Zhao X (2009) On the potential of biological treatment for arsenic contaminated soils and groundwater. J Environ Manage 90:2367–2376
142. Wingender J, Neu TR, Flemming HC (1999) What are bacterial extracellular polymeric substances? In: Wingender J, Neu TR, Flemming HC (eds) Microbial extracellular polymeric substances. Springer, Berlin, Heidelberg
143. Xu X, Liu W, Tian S et al (2018) Petroleum hydrocarbon-degrading bacteria for the remediation of oil pollution under aerobic conditions: a perspective analysis. Front Microbiol 9:2885
144. Yaseen DA, Scholz M (2019) Textile dye wastewater characteristics and constituents of synthetic effluents: a critical review. Int J Environ Sci Technol 16:1193–1226
145. Yin Y, Hu Y, Xiong F (2011) Sorption of Cu(II) and Cd(II) by extracellular polymeric substances (EPS) from *Aspergillus fumigatus*. Int Biodeterior Biodegrad 65:1012–1018
146. Yu Q, Fein JB (2015) The effect of metal loading on Cd adsorption onto *Shewanella oneidensis* bacterial cell envelopes: the role of sulfhydryl sites. Geochim Cosmochim Acta 167:1–10

147. Zainab N, Amna Din BU et al (2020) Deciphering metal toxicity responses of flax (*Linum usitatissimum* L.) with exopolysaccharide and ACC-deaminase producing bacteria in industrially contaminated soils. Plant Physiol Biochem 152:90–99
148. Zhang Z, Cai R, Zhang W, Fu Y, Jiao N (2017) A novel exopolysaccharide with metal adsorption capacity produced by a marine bacterium *Alteromonas* sp. JL2810. Mar Drugs 15:175
149. de Oliveira JM, Amaral SA, Burkert CAV (2018) Rheological, textural and emulsifying properties of an exopolysaccharide produced by *Mesorhizobium loti* grown on a crude glycerol-based medium. Int J Biol Macromol 120(Part B):2180–2187

Cost-Benefit Analysis and Industrial Potential of Exopolysaccharides

Kenji Fukuda and Hiroichi Kono

Abstract A large number of researches has recently been reported on finding novel microbial exopolysaccharides (EPSs) with multifunctional characteristics. Although all such microbial EPSs are highly attractive, their application use in several industries is in general not as easy as expected. One of the reasons is presumably that an economic point of view is usually lacking or even ignored at the initial stage of exploring microbial EPSs that are potential for industrial application. In fact, it is rather complicated or even impossible to perform appropriate cost-benefit analysis (CBA) at the beginning of such researches due to lack of precise information necessary to do that. In this chapter, general requirements for CBA are first summarized in brief. Further, recently found microbial EPSs are overviewed to understand the trend in this research area, and then technological aspects required to achieve better cost-benefit ratio in implementation of microbial EPSs will be discussed.

Keywords Accuracy in market survey · Central composite design · Cost-benefit ratio · Costs in preparation of EPS · Mass production of EPS · Metabolic engineering · Nanoparticles

---

K. Fukuda (✉)
Research Center for Global Agromedicine, Obihiro University of Agriculture and Veterinary Medicine, Inada-cho, Obihiro, Hokkaido 080-8555, Japan
e-mail: fuku@obihiro.ac.jp

H. Kono
Department of Agro-environmental Science, Obihiro University of Agriculture and Veterinary Medicine, Obihiro, Japan
e-mail: kono@obihiro.ac.jp

A. K. Nadda et al. (eds.), *Microbial Exopolysaccharides as Novel and Significant Biomaterials*, Springer Series on Polymer and Composite Materials,
https://doi.org/10.1007/978-3-030-75289-7_12

1 General Introduction to Cost-Benefit Analysis of Microbial Exopolysaccharides (EPSs)

Cost-benefit analysis (CBA) assesses the effects of countermeasures on costs and benefits in monetary units and serves as important reference information for companies in their decision-making. In the case of EPS, companies compare the necessary research and development (R&D) cost (C) for technological innovation, which is expected to increase EPS productivity, and the benefit (B) when the developed product is marketed. If the company determines that the benefit is greater than the cost (B > C), then EPS R&D investment will proceed and EPS-based products or foods will be marketed.

Benefits are influenced by various consumer factors, which differ owing to societies that consumers belong to. For example, there has been growing interest among Japanese consumers in the consumption of health functional foods due to increasing health consciousness presumably along with elevating population of elderly, and the market is expected to grow further in the future. Consumer decision depends on not only willingness to pay but also social demands, e.g., reduction of antimicrobial resistance (AMR) and achievement of sustainable development goals (SDGs), in some cases.

Balance between demand and supply can be expressed representatively as shown in Fig. 1. In general, price (P) should be set at the equilibrium point of demand curve (D) and supply curve (S). These curves can be depicted with significance by accurate estimation of individual cost and benefit involved with product manufacturing. Boardman [29] suggested an instruction to perform appropriate CBA as follows: the concept of product should be defined clearly first. Next, a list of stakeholders should be prepared. Most importantly, a precise evaluation of all cost and benefit elements should be performed to predict resulting costs and benefits during the relevant time. And then, all the costs and benefits should be converted into a prevalent currency by applying discount rate. Finally, net present value (NPV) of the production process should be calculated, and then sensitivity analysis should be performed to predict future risks.

For companies developing a new product, early R&D investment may increase manufacturing costs (supply curves are shifted to S_1), product price increases to P_1, and product demand (or supply) may decrease to Q_1. If the technological innovation enables mass production, the supply curve shifts from S_1 to S_2 and the price is expected to drop to P_2. Consumers will respond to lower prices (P_2) and consume more health functional foods (change from Q_1 to Q_2). Goto et al. [64] clarified the prioritized factors when selecting functional foods in four countries: Japan, the USA, the UK, and Italy. According to the results, the first- and second-ranking factors shared among the four countries were “low price” and “made from natural ingredients”, in that order, and for Japan, the third-ranking factor was “easy intake.” By means of enabling mass production of differentiated products which reflect these consumer preferences (shifting from S_1 to S_2), product demand is expected to shift from D to D_1 in the context of increased health awareness, especially in developed

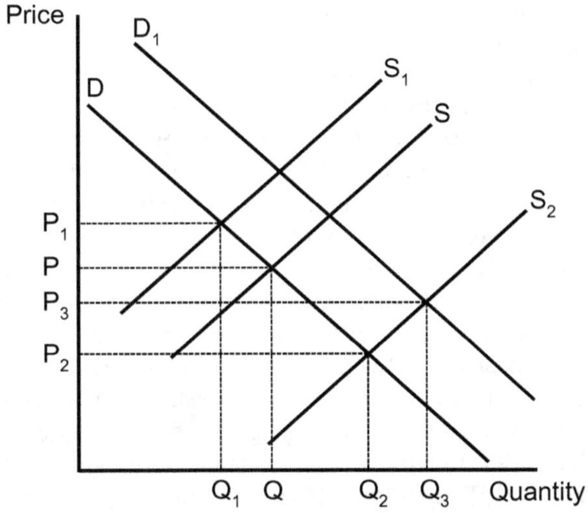

Fig. 1 Demand and supply in the functional foods market

countries. As a result, product demand (sales) will grow further (change from Q_2 to Q_3), suggesting that the cost-benefit ratio (B/C) for the enterprise may significantly exceed 1.0.

Microbial EPSs are recognized to have various physicochemical properties such as viscosity, emulsifying, gelling, and flocculation abilities and health functionalities, including blood pressure-lowering, antibacterial, anti-allergic and hypocholesterolemic activities. Meanwhile, it is apparent that technological innovations toward mass production are primarily needed to promote the industrial use of microbial EPSs. Characteristics of recently reported microbial EPSs and technical advances expected to improve B/C are reviewed, and further factors affecting B/C in industrial application of microbial EPSs will be clarified in the following sections.

2 Microbial EPSs Potential for Industrial Application

Chemical synthesis of polysaccharide is still rather challenging, primarily because of difficulty in control of regioselectivity and stereospecificity of glycosidic bond formation during polymerization process [224]. In contrast, microbial EPSs are advantageous for the low cost of the mass production in industrial scale and for the availability of different types of complicated chemical structures that one can expect a wide variety of physicochemical and/or biological properties. In fact, major industrial polysaccharides are produced mainly by enzymes secreted from EPS-producing microbes. Briefly, such enzymes, namely glucansucrases, produce homo-EPSs, e.g. dextran and levan, using sucrose as a substrate. By different biosynthetic pathways, microbial cells produce many homo- and hetero-EPSs such as gellan gum by *Pseudomonas elodea* [85], β-glucan from various fungi [171], hyaluronic acid by *Streptococcus* sp. [111], pullulan by *Aureobasidium pullulans* [186], succinoglycan by *Agrobacterium*, *Rhizobium* or *Pseudomonas*, etc. [69], and xanthan gum by *Xanthomonas campestris* [151]. These EPSs have already been commercialized, owing to their relatively high production yields. In general, at least a few tens of grams of EPS in one litter of culture medium is ideally required for their practical applicability at the industrial level.

An enormous variety of microbial EPSs exist in nature, and only a very small part of them have been studied to date. Moreover, the number of microbial EPSs that have been successfully applied in the industry is further less. Therefore, microbial EPSs are still considered to be a frontier for exploring noble EPSs, which are valuable in a wide variety of industries, e.g., bioremediation [68], food processing [165], pharmaceutical [132], waste water treatment [154]. In addition, microbial EPSs are emerging interests in agriculture, cosmetic, and livestock industries. In terms of health claims on bacterial EPSs, anticoagulant [89], antitumor [214], antiulcer [169], antiviral [11], cholesterol-lowering activities [44], treatment of rheumatoid arthritis [143] have been suggested. Excellent reviews have already been present on well-established microbial EPSs, e.g., dextran [138], gellan gum [20], β-glucan [171], hyaluronic acid [95], levan [146], pullulan [186], succinoglycan [69], and xanthan gum [151], which have already been relevant to their significance in the industrial application. Apart from such already established microbial EPSs in terms of their industrial use, a great number of potentially functional microbial EPSs are emerging in the last decade. Those are originated from bacteria, fungi, yeast, archaea, and alga, which are going to be focused in this section. Due to the limited space, only the EPSs, whose chemical structures and biological and/or physicochemical characteristics have been clarified to some extent very recently, are mainly dealt with in view of their potential for the future industrial application. By the way, many chemical structures of EPSs including mannose residue as a monosaccharide constituent have been recently reported. Needless to say, but mannan derived from yeast cell wall can be contaminated in the EPS fractions unless thorough purification process is applied. Therefore, it should be pointed out that a careful validation on the purity of EPSs is prerequisite when yeast extract is used as culture constituent or when EPSs are purified from yeast.

Characteristics and potential functionality of EPSs seem to be somewhat in relation to the environment that the producers inhabit. In this context, EPSs has been explored for isolation mainly in bacteria, including food-borne and environmental lactic acid bacteria, extremophiles, soil bacteria, marine bacteria, and pathogenic bacteria for long time and it is still a trend as shown in Table 1. In addition, fungi, yeast, microalgae, and archaea have also been explored to find novel functional EPSs as summarized in Table 1.

Table 1 Recently characterized microbial EPSs with apparent monosaccharide composition and beneficial properties

Strain	Monosaccharide composition or chemical structure	Molecular weight (Da)	Bioactivity	Physicochemical characteristic	Production yield (mg/L)	References
Bacteria						
Bacillus licheniformis MS3	Mannose:Glucose:Fructose (20.6:46.8:32.58)	Unknown	Unknown	Emulsification	1560 (crude)	[12]
Bacillus mycoides BS4	Galactose, Mannose, Glucose and Glucuronic acid	1.9×10^4	Antitumor	Unknown	8020	[51]
Cyanobacterium CCC 746	Glucose, Xylose and Glucuronic acid (the backbone is → 4)-Glc*p*-(1 →)	CPS, 1.3×10^4 EPS, 0.9×10^4	Antioxidant	Viscosity	CPS, 184 EPS, 341	[205]
Deinococcus radiodurans R1	Xylose:Galactose:Fucose:Glucose: Arabinose:Fructose (10.6:6.1:4.2:3.8:2.6:1.0)	$0.8 \sim 1.0 \times 10^5$	Antioxidant	Unknown	Unknown	[107]
Enterococcus faecium MS79	Arabinose:Mannose:Glucose (0.8:1.7:11.3)	8.3×10^5	Antioxidant Antitumor ANtupathogenic	Viscosity	Unknown	[17]
Glutamicibacter halophytocola KLBMP 5180	Rhamnose:Galacturonic acid:Glucose: Glucuronic acid:Xylose:Arabinose (5180EPS-1, 17.2:34.6:24.7:23.5:0:0) (5180EPS-2, 11.2:35.3:13.8:0:16.5:23.2)	5180EPS-1, 5.9×10^4 5180EPS-2, 1.1×10^4	Antioxidant	Unknown	2770 (crude)	[227]
Gracilibacillus sp. SCU50	Mannose:Galactose:Glucose:Fucose (90.81:5.76:2.22:1.21)	5.9×10^4	Unknown	Emulsification	Unknown	[54]
Halomonas elongata S6	Glucosamine:Mannose:Rhamnose: Glucose (1:0.9:0.7:0.3)	2.7×10^5	Antioxidant DNA protection Anti-biofilm formation	Flocculation Emulsification	96	[90]
Kosakonia sp. CCTCC M2018092	L-Fucose:D-Glucose:D-Galactose: D-Glucuronic acid:Pyruvic acid (2.03:1.00:1.18:0.64:0.67)	3.7×10^5	Antibacterial	Unknown	13,500	[105]

(continued)

Table 1 (continued)

Strain	Monosaccharide composition or chemical structure	Molecular weight (Da)	Bioactivity	Physicochemical characteristic	Production yield (mg/L)	References
Lactobacillus coryniformis NA-3	α-Rhamnose:α-Mannose:α-Galactose:α-Glucose (2.6:1.0:5.0:3.3)	8.6×10^6	Antioxidant Antibiofilm Antibacterial	Unknown	Unknown	[230]
Lactobacillus fermentum YL-11	Galactose:Glucose:Mannose:Arabinose (48.0:30.3:11.8:6.0)	Unknown	Antitumor	Unknown	84.5	[221]
Lactobacillus helveticus LZ-R-5	→6)-β-D-Gal*p*-(1 → 3)-β-D-Glc*p*-(1 → 3)-β-D-Glc*p*-(1 → 3)-β-D-Glc*p*-(1 → 3)-β-D-Glc*p*-(1→	5.4×10^5	Immunomodulatory	Unknown	128	[233]
Lactobacillus pentosus LZ-R-17	→2)-α-D-Gal*p*-(1 → 4)-β-D-Glc*p*-(1 → 4)-β-D-Glc*p*-(1 → 4)-β-D-Glc*p*-(1→	1.2×10^6	Immunomodulatory	Unknown	185	[234]
Lactobacillus plantarum C70	Arabinose:Mannose:Glucose:Galactose (13.3:7.1:74.6:5.0)	3.8×10^5	Antitumor Antioxidant Antipathogenic	Viscosity	Unknown	[19]
Lactobacillus plantarum JLAU103	Arabinose:Rhamnose:Fucose:Xylose:Mannose:Fructose:Galactose:Glucose (4.05: 6.04: 6.29: 5.22: 1.47: 5.21: 2.24: 1.83)	1.2×10^4	Antioxidant Immunomodulatory	Ferrous ions chelating	75 (crude)	[129, 212]
Lactobacillus plantarum PFC311	Glucose, Galactose, and Fructose	Unknown	Unknown	Viscosity	120–400	[238]
Lactobacillus plantarum WLPL09	Mannose:Glucose:Galactose (89.69:8.65:1.66)	3.3×10^4	Antitumor	Unknown	Unknown	[87, 88]

(continued)

Table 1 (continued)

Strain	Monosaccharide composition or chemical structure	Molecular weight (Da)	Bioactivity	Physicochemical characteristic	Production yield (mg/L)	References
Lactococcus garvieae C47	→3)[α-D-Glc(1 → 6)β-D-Glc(1 → 2)] [α-D-Xyl(1 → 4)β-D-Glc (1 → 3)α-D-Glc(1 → 2)β-D-Glc (1 → 4)]β-D-Xyl(1→ (with Arabinose at the terminals)	7.3×10^6	Antioxidant Antidiabetic Antitumor Antipathogenic	Viscosity	Unknown	[16]
Lactococcus lactis LL-2A	→5)-β-D-Gal*f*-(1 → 3)-β-D-Glc*p*-(1→ Side chain: β-D-Gal*p*-(1 → 4)-β-D-Glc*p*-(1→ (branched at C3 of β-D-Gal*f* in the mainchain)	3.3×10^6–1.3×10^6	Unknown	Viscosity	267	[137]
Nostoc sp.	(1 → 4)-linked α-L-Arabinopyranose, β-D-Glucopyranose, β-D-Xylopyranose and (1 → 3)-linked β-D-Mannopyranose, two different Uronic acids, and a lactyl group, with (1 → 4,6)-linked β-D-Glucopyranose as the only branch point	2.1×10^5	Immunomodulatory	Unknown	2400	[208]
Pantoea sp. YU16-S3	Glucose:Galactose:N-Acetyl galactosamine:Glucosamine (1.9:1.0:0.4:0.02)	1.8×10^5	Wound healing	Unknown	2600–2800	[173]
Pediococcus pentosaceus M41	→3)α-D-Glc(1 → 2)β-D-Man (1 → 2)α-D-Glc(1 → 6)α-D-Glc (1 → 4)α-D-Glc(1 → 4)α-D-Gal(1→ with Arabinose linked at the terminals	6.8×10^5	Antioxidant Antitumor α-Amylase inhibition α-Glucosidase inhibition	Viscosity	Unknown	[18]
Pseudomonas sp. BGI-2	Glucose, Galactose and Glucosamine	Unknown	Unknown	Cryoprotective	2010	[6]

(continued)

Table 1 (continued)

Strain	Monosaccharide composition or chemical structure	Molecular weight (Da)	Bioactivity	Physicochemical characteristic	Production yield (mg/L)	References
Rhodococcus erythropolis HX-2	Glucose:Galactose:Fucose:Mannose: Glucuronic acid (27.29:24.83:4.79:26.66:15.84)	1.0×10^6	Anticancer	Unknown	3736	[78]
Streptococcus thermophilus DGCC7919	→4)-β-D-Gal*p*-(1 → 4)-α-D-Glc*p*NAc-(1 → 2)-α-L-Rha*p*-(1 → 4)-β-D-Glc*p*-(1 → 6)-α-D-Glc*p*-(1 → 3)-β-L-Rha*p*-(1→	2.2×10^6–1.0×10^6	Unknown	Viscosity	351	[137]
Virgibacillus salarius BM02	Mannose:Arabinose:Glucose (1.0:0.26:0.08)	Unknown	Unknown	Antioxidant Emulsification Viscosity	5870 (crude)	[62]
Weissella confusa KR780676	Galactose	Unknown	Aflatoxin binding	Emulsifying	Unknown	[91, 92]
Fungi						
Mortierella alpina	→4)GlcNAc(β1→	4.9×10^5	Antioxidant Antitumor	Gelation	1510	[66]
Monascus purpureus	Mannose:Glucose:Galactose (EPS-1, 8:1:11)	7.0×10^4	Immunomodulatory	Unknown	8.6 wt% of freeze-dried powder from filtrated medium of liquid fermentation	[213]
Floccularia luteovirens	Man:Glc:Fuc:Gal:GlcA:Ara (ALF1, 1.00:0.65:0.39:0.47:0.29:0.01) Glc:Man:Rha:Gal (ALF2, 1.00:0.40:0.34:0.54)	ALF1, 2.8×10^4 ALF2, 1.9×10^4	Antioxidant Antitumor	Unknown	Unknown	[109]

(continued)

Table 1 (continued)

Strain	Monosaccharide composition or chemical structure	Molecular weight (Da)	Bioactivity	Physicochemical characteristic	Production yield (mg/L)	References
Tremella fuciformis	Rhamnose, Arabinose, Mannose, Galactose and Glucose	TFP-1, 5.2×10^6 TFP-2, 1.7×10^5 TFP-3, 6.6×10^5	Antioxidant	Unknown	Unknown	[244]
Yeasts						
Rhodotorula mucilaginosa sp. GUMS16	Glucose:Mannose (85:15)	8.4×10^4	Antioxidant	Unknown	28,500	[70]
Lipomyces starkeyi VIT-MN03	Glucose, Mannose, and Rhamnose	Unknown	Antioxidant Biosurfactant Cholesterol-removal Mutagen-binding	Viscosity	4870	[162]
Microalgae						
Chlorella vulgaris (BEIJ. 1890, strain P13/1998)	Backbone repeating unit; →2)-α-L-Rha (1 → 3)-α-L-Rha(1→ Branched by long 1,6-linked α-D-Gal*p* side chains Further branched at C2, C3, or C4 by α-L-Ara*f*, α-D-Gal*f* and β-D-Gal*f* residues (α-L-Ara*f* form longer 1,2-linked chains branched at C3, C4, or C5; Gal*f* residues are localized as terminal units predominantly in the β configuration, while α-D-Gal*p* and α-L-Ara*f* may be partially O-methylated)	8.4×10^5	Anti-inflammatory Anti-remodelling	Unknown	~3000	[33]
Porphyridium sordidum	Xylose, Galactose, Glucose, and Glucuronic acid	Unknown	Unknown	Viscosity	160	[123, 124]

(continued)

Table 1 (continued)

Strain	Monosaccharide composition or chemical structure	Molecular weight (Da)	Bioactivity	Physicochemical characteristic	Production yield (mg/L)	References
Porphyridium purpureum	Xylose, Galactose, Glucose and Glucuronic acid	Unknown	Unknown	Viscosity	240	[123, 124]
Chlorella vulgaris	β-D-Galactan (1,3-Gal, 1,6-Gal, and 1,3,6-Gal residues)	Unknown	Immunostimulatory	Unknown	Unknown	[53]
Archaea						
Haloferax mucosum DSM 27191	Unknown	$7.6 \times 10^4 1.5 \times 10^5$	Unknown	Viscosity Emulsification	7150 (crude)	[117]
Haloterrigena turkmenica DSM-5511	Glucose, Galactose, Glucosamine, Galactosamine, and Glucuronic acid	8.0×10^5 2.1×10^5	Antioxidant	Emulsification Moisture retention	206.8	[195]

2.1 Bacterial EPSs

Dairy products are the ideal source of EPS-producing bacteria. Ayyash et al. [16] isolated *Lactococcus garvieae* C47 as an EPS-producer from fermented camel milk. The strain produces a neutral EPS composed of 82.51% glucose (Glc), 5.32% arabinose (Ara), and 12.17% xylose (Xyl) with several bioactivities such as α-amylase inhibitory, antioxidant, and antitumor activities. Solution of the EPS showed shear-thinning behavior, which was influenced by pH and presence of salts. The same author reported isolation of another EPS-producer, *Lactobacillus plantarum* C70, isolated from camel milk [17]. EPS-C70, an EPS produced by the strain, was composed by 13.3% Ara, 7.1% mannose (Man), 74.6% Glc, and 5.0% galactose (Gal) as the major monosaccharide constituents. It showed antioxidant, antitumor, and viscous properties. *Lactobacillus helveticus* LZ-R-5 was isolated from fermented milk and produced an EPS (R-5-EPS), containing linear repeating units of → 6)-β-D-Gal*p*-(1 → 3)-β-D-Glc*p*-(1 → 3)-β-D-Glc*p*-(1 → 3)-β-D-Glc*p*-(1 → 3)-β-D-Glc*p*-(1 → . R-5-EPS stimulated the growth of RAW264.7 macrophages in vitro with elevated levels of acid phosphatase activity, cytokines production, nitric oxide production, and phagocytosis [233]. You et al. [234] also reported that *Lactobacillus pentosus* LZ-R-17 isolated from Tibetan kefir grains produced an EPS (R-17-EPS) composed of linear repeating units of → 2)-α-D-Gal*p*-(1 → 4)-β-D-Glc*p*-(1 → 4)-β-D-Glc*p*-(1 → 4)-β-D-Glc*p*-(1 → . R-17-EPS also promoted the proliferation, cytokines, and nitric oxide productions and phagocytosis of RAW264.7 macrophages. *Lactobacillus plantarum* JLAU103 has been isolated from a Chinese traditional fermented dairy food, *Hurood.* The strain produced an acidic exopolysaccharide (EPS103) that was consisted of Ara, rhamnose (Rha), fucose (Fuc), Xyl, Man, fructose (Fru), Gal, and Glc in an approximate molar ratio of 4.05: 6.04: 6.29: 5.22: 1.47: 5.21: 2.24: 1.83, showing strong antioxidant activity in vitro [128]. Kavitake et al. [91] reported that *Weissella confusa* KR780676 was isolated from an Indian traditional fermented food, *idli* butter, as a producer of galactan. Besides the good emulsifying activity of the galactan, it showed a unique aflatoxin B1 binding ability dose-dependently, up to 100 mg/mL. Therefore, the galactan is potential to remove mycotoxins in feed and food industries [92]. *Lactobacillus fermentum* YL-11, capable of producing an EPS, was isolated from fermented milk by Wei et al. [221]. The EPS mainly comprised 48.0% Gal, 30.3% Glc, 11.8% Man, and 6.0% Ara with antitumor activity. Highly precise structural analysis has been performed by Nachtigall et al. [137] on viscous EPSs secreted by *Lactococcus lactis* LL-2A and *Streptococcus thermophilus* DGCC7919, which are commercial dairy starters, clarifying a significant contribution of side -hain structure of EPS molecules on the viscosity.

Human milk is a fascinating source of probiotics due to their possible transition into infant's gastrointestinal tract during lactation period [135]. *Lactobacillus plantarum* WLPL09 was isolated from human breast milk. It was found that the strain secreted two EPSs, namely NPS and APS, mainly composed of Man and Glc in molar ratio of 85.35:14.65 and of Man, Glc, and Gal in molar ratio of

89.69:8.65:1.66, respectively. Among them, APS displayed strong antiproliferative effect and induction of apoptosis in vitro against hepatocellular carcinoma cells, HepG2, and colon adenocarcinoma cells, HCT-8, by up-regulating expression levels of mRNAs of apoptosis-related genes [9].

Several other types of fermented and non-fermented foods are also good sources of probiotics as commonly recognized. *Lactobacillus plantarum* PFC311 was isolated from *Tarhana*, which is Turkish dried food in which grain and fermented milk mixed. The strain secreted a viscous EPS composed of Glc, Gal, and Fru [238]. Xu et al. [230] isolated *Lactobacillus coryniformis* NA-3 from northeast Chinese sauerkraut, *Suan Cai*, as a producer of an EPS (EPS-NA3). EPS-NA3 was composed of α-Rha, α-Man, α-Gal, and α-Glc in a molar ratio of 2.6:1.0:5.0:3.3. It showed bioactivities of radical-scavenging and attenuation the formation of biofilms produced by *Bacillus cereus* and *Salmonella* Typhimurium. *Enterococcus faecium* MS79 was isolated from seafoods with low-water activity and identified to be a probiotic candidate [19]. The strain can produce a viscous EPS composed of Ara, Man and Glc in a molar ratio of 0.8:1.7:11.3 with antioxidant, antitumor, and antipathogenic activities.

As shown above, EPSs derived from food-borne bacteria are mainly oriented to apply in food and pharmaceutical industries. In food industry, EPS purification process is not necessarily because the EPS-producer is used as fermentation starter, and hence the B/C primarily depends on the health claim, consumer's preference, and production cost. Although many health beneficial activities were assigned to EPSs, the mild effectiveness and purification cost hamper the application use of most of natural EPSs in pharmaceutical industry. To overcome this bottleneck, a technique to conjugate EPSs with metal nanoparticle seems to be one of promising solutions as described in Sect. 4.1.2.

Extremophiles are organisms living in extreme environment such as high and low temperatures, acid and alkaline, ionizing and ultraviolet radiation, high pressure, high salt concentration, desiccation, and the Earth's crust. Bacteria inhabits in such extreme environments produce EPSs that exhibit unique characteristics, probably as a result of adaptation to such harsh conditions. From this point of view, the number of EPSs isolated from extremophilic bacteria have been increasing. For example, an extremely halotolerant *Halomonas elongata* strain S6 was isolated and its EPS (EPS-S6) was characterized by Joulak et al. [90]. The authors reported that EPS-S6 was mostly composed of glucosamine (GlcN), Man, Rha, and Glc in a molar ratio of 1:0.9:0.7:0.3 with highly negatively charged characteristic. It was found that EPS-S6 showed several activities such as antioxidant, DNA protection, inhibition and disruption of biofilms produced by pathogens, flocculating kaolin suspension, and emulsification at any of pH range. Lin et al. [107] found a cell wall EPS, termed DeinoPol, from radiation-tolerant *Deinococcus radiodurans* R1, that was highly effective to protect human keratinocytes in response to stress-induced apoptosis by scavenging reactive oxygen species. *Virgibacillus salarius* BM02 was isolated from Tunisian hypersaline environments as a producer of viscous EPS composed of Man, Ara, and Glc at a molar ratio of 1.0:0.26:0.08 with antioxidant and emulsification activities [62]. Interestingly, acid hydrolysate of the EPS promoted biomass production and nutritional enrichment in a cyanobacteria, *Arthrospira* (*Spirulina*) *platensis*, which is applicable as diet for human and livestock consumption. A psychrotrophic bacterium, *Pseudomonas* sp. BGI-2, has been isolated from ice samples collected from Batura glacier in Pakistan, as a producer of EPS, whose major monosaccharide constituents were found to be Glc, Gal, and GlcN [6]. As was expected, the EPS showed a significant cryoprotective effect on *Escherichia coli* k12, as being comparable to the effect of 20% glycerol.

Several attractive EPS-producing bacteria have been explored in various non-extreme environments, including soil, spring water, and ocean. *Bacillus mycoides* BS4, which was isolated from soil at gas station in Giza governorates, Egypt, produced EPS composed of Gal, Man, Glc and glucuronic acid (GlcA) with an inhibitory effect against HepG2 and Colorectal adenocarcinoma cells (Caco-2) [51]. Hu et al. [78] reported isolation of *Rhodococcus erythropolis* HX-2 also from an oil field in China. An EPS produced by the strain was composed by Glc, Gal, Fuc, Man, and GlcA, possessing anticancer activity against A549, Hela, and SMMC-7721 cancer cells. *Gracilibacillus* sp. SCU50 was isolated from a sample of saline soil [54]. The strain produced an EPS (named mhEPS), which consisted of Man, Gal, Glc, and Fuc in a molar ratio of 90.81:5.76:2.22:1.21, showed emulsifying activity. *Kosakonia* sp. CCTCC M2018092 was isolated from spring water, secreting a unique fucose-containing EPS, which was composed of L-Fuc, D-Glc, D-Gal, D-GlcA, and pyruvic acid in the molar ratio of 2.03:1.00:1.18:0.64:0.67 [103]. Partial acid hydrolysate of the EPS was conjugated with silver nanoparticles, resulting in a production of stable biodegradable antibacterial film. A cyanobacterium *Nostoc* sp. produces a complex EPS, in which (1 → 4)-linked α-L-arabinopyranose (Ara*p*), β-D-glucopyranose (Glc*p*), β-D-xylopyranose (Xyl*p*) and (1 → 3)-linked β-D-mannopyranose (Man*p*), two different uronic acids and a lactyl group, with (1 → 4,6)-linked β-D-Glc*p* as the only branch point was identified [208]. Interestingly, the EPS showed antitussive and bronchodilator activities dose-dependently when it orally administrated in guinea pigs. An endophytic actinobacterium *Glutamicibacter halophytocola* KLBMP 5180 was isolated from surface-sterilized root of a coastal halophyte, *Limonium sinense*, by Xiong et al. [227]. The strain produced two antioxidant EPSs with different size, namely 5180EPS-1 and 5180EPS-2, both comprising Rha, galacturonic acid (GalA), Glc, GlcA, Xyl, and Ara, as monosaccharide constituents. A marine bacteria *Pantoea* sp. YU16-S3 was investigated for the production of EPS (EPS-S3) in terms of the wound healing applications [173]. EPS-S3, composed of Glc, Gal, N-acetyl galactosamine (GalNAc) and GlcA, could induce re-epithelialization of injured tissue in rats. *Pediococcus pentosaceus* M41 was also isolated from marine source as a producer of viscous EPS (EPS-M41), which composed of Ara, Man, Glc, and Gal in a molar ratio of 1.2:1.8:15.1:1.0 [18]. It showed antioxidant and antitumor activities and inhibitory activities towards α-amylase and α-glucosidase.

2.2 Fungal EPSs

Fungi are the second major isolation source of EPS-producing microbes. Interestingly, an arachidonic acid-producing fungus, *Mortierella alpina* CBS 528.72 was found to be a producer of a chitin-like EPS, a linear polymer of β-(1 → 4)-linked N-acetyl-D-glucosamine (GlcNAc) residues [66]. The acetylation degree of the EPS was determined as being over 90%. It showed antioxidant and antitumor activities. Wang et al. [212] reported that Chinese edible fungus *Monascus purpureus* secreted two EPSs (EPS-1 and EPS-2) when the fungus was cultivated in a liquid medium. Monosaccharide composition of EPS-1 and EPS-2 were identified to be Man, Glc, and Gal in the molar ratio of 8:1:11 and Rha, Ara, Xyl, Man, Glc, and Gal in the molar ratio of 2.2:1.9:1:5.6:1.9:7.7, respectively. It was turned out that EPS-1 promoted cytokines secretion such as IL-6, IL-10, and TNF-α by improving the related mRNA expression levels in RAW 264.7 cell lines. A strain of Chinese medicinal and edible basidiomycete, *Floccularia luteovirens* Sacc QH was found to produce two EPSs (termed ALF1 and ALF2) in its liquid culture [109]. Among them, ALF1 was composed of Man, Fuc, Ara, Gal, and Glc, exhibiting antioxidant and antitumor activities. Zheng et al. [244] reported that an edible fungus *Tremella fuciformis* strain tyc63 conidium cells can produce three fractions of EPS, termed TFP-1, TFP-2 and TFP-3. Monosaccharide composition of TFP-1 was mainly Glc, Xyl, Man, and Fuc, while those of TFP-2 and TFP-3 are mainly Rha, Ara, Man, Gal, and Glc. All the three EPSs showed antioxidant activities.

2.3 Yeast EPSs

Most of the EPSs reported from yeasts were on homo-EPS such as mannan [97] and pullulan [186], whereas reports on several hetero-EPSs composed of monosaccharides including other than mannose or glucose have been reported recently. For example, Ragavan and Das [162] reported that a probiotic yeast *Lipomyces starkeyi* VIT-MN03 produced a viscous hetero-EPS, which is a mixture of glucan, mannan, and rhamnan. The EPS showed antioxidant, biosurfactant, cholesterol-removal, and mutagen-binding abilities. *Rhodotorula* sp. strain CAH2 was composed of Glc, Man, and Gal, although its bioactivities are still unclear [182]. Hamidi et al. [70] reported a cold-adapted yeast *Rhodotorula mucilaginosa* sp. GUMS16 was capable of producing an EPS composed of Glc and Man in a molar ratio of 85:15, showing antioxidant and antiproliferative activities. As pointed out in Sect. 2.1, cell wall mannan is highly possible to be contaminated in the purified EPS fraction when the EPSs are purified from yeast, and hence the purification steps of those yeast-derived EPSs should be validated by some appropriate means.

2.4 *Microalgal EPSs*

The number of reports on EPSs isolated from micro algae has been increasing recently. Capek et al. [33] reported that an EPS isolated from a micro algae or phytoplankton, *Chlorella vulgaris*, was highly branched α-L-arabino-α-L-rhamno-α,β-D-galactan. Namely, the backbone repeating unit is 2)-α-L-Rha (1 → 3)-α-L-Rha(1 → with a long 1,6-linked α-D-Gal*p* side chains. The side chains further branched at C2, C3 or C4 by α-D-Gal*f*, β-D-Gal*f* and α-L-Ara*f*, which formed longer 1,2-linked chains branched at C3, C4, or C5. The EPS showed immunomodulative activity in an experimental asthma model that used guinea pigs. On the other hand, presence of galactans which consisted of 1,3-, 1,6- and 1,3,6-linked Gal residues have been reported in *Chlorella vulgaris* by Ferreira et al. [53]. The EPS was extracted by treating the algal cells with KOH as well as recovered from the culture medium. It showed an immunostimulatory effect on B lymphocytes. An olive-green color micro alga, *Porphyridium sordidum*, was found to be a viscous EPS-producer [123]. Monosaccharides composed of the EPS were revealed as Xyl, Gal, Glc, and GlcA with modifications of sulfation and methylation. The largest advantage in microalgal EPS biosynthesis is utilizing carbon fixation by photosynthesis, however, development of efficient technique for harvesting microalgal cells from the culture medium is challenging for the application use [136].

2.5 *Archaeal EPSs*

Archaeal EPSs were reviewed by Poli et al. [158], but since that time, the number of newly reported archaeal EPSs are much scarce than EPSs isolated from other microbes. Squillaci et al. [195] reported that an extreme halophilic archaeon *Haloterrigena turkmenica* DSM-5511 produced a sulfated hetero-EPS composed of Glc, Gal, GlcN, GalN, and GlcA. The EPS showed emulsifying, water retention, and antioxidant properties. A haloarchaea *Haloferax mucosum* (DSM 27191) was found to produce a viscous EPS exhibiting emulsification activity against several organic solvents, but its monosaccharide composition was unrevealed yet [116]. No doubt to say that archaea is an attractive source of EPSs due to the potential for bioactivities and physicochemical properties, the major concern for the industrial application should be on finding appropriate culture conditions with low cost, if feasible.

3 Factors Affecting Cost-Benefit Ratio in Industrial Application of Microbial EPSs

3.1 Production Process

The major cost in the industrial use of microbial EPSs occurs in the production process. It includes costs for culture medium, reagents needed for purification procedures and packaging. In addition, costs on safety tests and transportation of the products should be considered. In case that microbes can grow and produce EPS in the required circumstances such as food matrices, soil, and water, cost concerning cultivation and purification could be negligible. To reduce the cultivation cost, utilization of agricultural byproducts and waste materials should be a prospective approach [210]. Use of recombinant microbes for EPS production is highly risky in food industry as yet, especially in such a society that introduction of genetically modified foods is controversial issue. Thorough purification of EPS is sometimes very difficult and costly, and hence it could be a big hurdle when it required for commercialization of EPSs. Furthermore, food-grade stuff should be used when the purified EPS is considered as a food additive. Scale-up problem is a common concern to be solved for mass production of EPSs, especially when the culture medium shows high viscosity [179]. It should be pointed out that degradation of EPSs during storage period and/or contamination of endotoxins in the final product could occur. Those should be problematic if they give adverse effect on the final product [46, 125].

3.2 Customer's Preference

It is apparent that customer's keen interests towards microbial EPSs are currently on their role as health promoting agents in food, cosmetic, pharmaceutical and livestock industries and as eco-friendly materials in agricultural and environmental industries. Among the health-promoting effects of EPSs, immunomodulative and antimicrobial activities are of great interests to confront issues involving aging societies and antimicrobial resistances. Biodegradable materials made by EPSs and antimicrobial EPSs can partially accommodate environmental issues to achieve SDGs. Commercialization of EPSs with such functions is most likely to bring about high economic profits to the company. As was pointed out in Sect. 3.1, usage of recombinant microbial EPSs will adverse effects on customer's preference especially in agricultural, cosmetic, and food industries.

4 Future Perspectives

The ultimate goal of industrial production of microbial EPSs should be establishment of rational process for tailor-made synthesis of favorable EPSs. To achieve this goal, deep insights and understanding in structure-function relationship of EPSs are essential. As a matter of fact, this approach is still too ambitious to date, but it is of course worthwhile continuing to try. To improve the current situation in realistic, it should be acceptable means to develop efficient protocols for screening ideal microbes which produce enough amount of EPS with beneficial functionalities. In this context, one may leave precise structural analysis open at the initial characterization of EPSs. However, to avoid misunderstanding due to possible contaminants with activities in "purified" EPS, adequate evaluation on the purity of EPS is highly recommended, in fact as much as possible, at earlier stage of R&D. Furthermore, several technical breakthroughs are needed for implementation of microbial EPS in several industries. Emerging and developing technologies aiming to suppress the production costs and to enhance the benefits of microbial EPSs in this last decade will be summarized in this section. Further, industries in which future application and implementation of microbial EPSs are coveted will be outlined.

4.1 Technical Breakthrough Expected to Confer Benefits on Industrial Application of Microbial EPSs

Enormous efforts have been performed and still continued to enhance EPS production by optimizing culture conditions, according to the fermentation engineering approach by using central composite design (CCD) integrated with suitable models such as Box-Behnken design [62], response surface design [9, 12], Plackett-Burman design [149], Taguchi design [124], etc. Intriguingly, an economic approach based on Pareto optimality theory was conducted to estimate the optimal productivity, which gave the lowest cost for culture medium with maintained bioactivity of an EPS [108]. Discouragingly, no common rule has been figured out yet to optimize culture conditions of microbes in relation to EPS production. By such conventional approaches, most of hetero-EPSs achieved several folds greater EPS production than those obtained under pre-optimized conditions. These were usually less amount required enough for industrial applications, in contrast to homo-EPSs that were successfully mass produced in the industrial levels. This is because biosynthesis of hetero-EPS cannot be independent from microbial cell growth, which has a substantial limitation to maximize the production yields. To overcome this bottleneck, diverse innovations should be needed. Hereby, several trials in such approaches are introduced in this subsection. Eradicating EPS production is contrary pursued in relation to reduce pathogenicity of EPS-producing bacteria, such as *Agrobacterium tumefaciens* [74], *Pseudomonas aeruginosa* [160] and *Vibrio cholerae* [220]. This issue is of great importance in agriculture and pharmaceutical industries, but not addressed here, because it is outside of the scope of this chapter.

4.1.1 Metabolic Engineering to Enhance EPS Yield

To date, two different pathways, Wzx/Wzy-dependent pathway and ABC-transporter-dependent pathway, have been illustrated for hetero-EPS biosynthetic pathways [175, 176]. Some of the related genes have been demonstrated to modify the sugar flux biased into EPS production. Modulation of Gal metabolism is likely to be a favorable target. For example, α-phosphoglucomutase is an enzyme capable of converting glucose 1-phosphate (Glc-1-P) to glucose 6-phosphate (Glc-6-P) vice versa. Overexpression of *pgm*, a gene encoding α-phosphoglucomutase, in combination with *galU*, encoding UDP glucose pyrophosphorylase, could increase EPS production several-folds when lactose (Lac) was used as a sole carbon source in *Streptococcus thermophilus* LY03 [98]. Gene expression levels of UDP-glucose pyrophosphorylase and UDP-galactose 4-epimerase were highly induced when EPS-producing *Lactobacillus casei* CRL 87 was cultivated at acidic pH of 5.0 and using Gal as a carbon source, giving better production of EPS, therefore these genes can be candidates for further metabolic engineering in the strain [133]. As is well known that EPS production is induced by abiotic stresses such as bile acid, freezing, peroxides, pH and salt, and hence the related metabolism can be a potential target for metabolic engineering. Jang et al. [84] showed that increased H_2O_2 level in alkyl hydroperoxide reductase encoding gene deletion mutant enhanced EPS production in *Acinetobacter oleivorans* DR1. Although the mechanism is still unclear, elevated expression level of NADH oxidase encoding gene increased EPS production, probably due to reduced lactate production flux [102]. Some nutrient uptake pathways and/or chemosensory systems in microbes are also potential targets for metabolic engineering in terms of enhancing EPS production. A gene, *SMc00722*, which was putatively involved with maintenance of intracellular magnesium concentration, was found in a Gram-negative α-proteobaceteria, *Ensifer meliloti* Rm1021 (Hawkins and Oresnik [73]. Mutation in the gene caused enhanced production of succinoglycan in the strain. A signal transducer protein, DifA, was revealed to regulate EPS production in a Gram-negative -proteobaceteria, *Myxococcus xanthus* [228]. Interestingly, Islam et al. [82] found that bacterial cellular density could alter the production of different type of EPSs in *Myxococcus xanthus*. Namely, an EPS produced via Wzx/Wzy-dependent polysaccharide-assembly pathway was preferentially produced by the bacterial cells located at the lower-density swarm periphery, whereas a novel biosurfactant polysaccharide, a type IV pilus-inhibited acidic polymer, was mainly produced in the higher-density swarm interior. This knowledge will provide novel options regarding metabolic engineering for targeting enhanced EPS production levels.

4.1.2 Nanoparticles

Highly attractive technique is development of EPS covalently conjugated with nanoparticles made of metals such as copper, gold, iron, platinum, selenium, silver, titanium, zinc, and so on, via biosynthesis in microbial cells, so-called "green synthesis" [184]. EPS is capable of reducing and stabilizing metal nanoparticles, exhibiting a wide spectrum of beneficial properties, e.g., antibacterial [24, 59, 103, 139, 163, 164, 168, 190, 209], anticancer [4, 31], antioxidant [164, 168, 187, 190, 226], bioremediation [34, 217], ferrous ion supply to truffles [155], food packaging [180], functionalization in yogurt [58], inhibition of biofilm formation [1, 28, 40, 59, 110, 164, 209], mosquito larvicidal [1] and waste-water treatment [65, 174]. A wide structural and functional varieties, high biocompatibility, and biodegradability of microbial EPS conjugated with nanoparticles seem to be the most promising technique for future applications in various sorts of industries.

4.1.3 Chemical and Enzymatic Modifications

Chemical and enzymatic modifications are relatively conventional techniques to modulate physicochemical and bioactive properties of EPSs. For example, EPSs have commonly been acetylated, carboxymethylated, phosphorylated, and sulfated to confer anticoagulant [30], antimicrobial [32], antitumor [215], antioxidant [32, 86, 215, 225, 240], hypoglycemic and hypolipidemic effects [218]. Curiously, N-phthaloyl derivative of polysaccharide derived from *Lachnum* YM262 could enhance its immunomodulating property in mice treated with cyclophosphamide [37]. Zhang et al. [240] found antitumor activity in selenized derivative of EPS which was purified from liquid culture of an edible mushroom, *Grifola frondosa*. Partial degradation of EPS by the action of certain enzymes is also potential to promote commercial values of microbial EPSs. A sort of EPS can be hydrolyzed by glucanases and/or depolymerized by polysaccharide lyases via β-elimination reaction [127], and even modified [222]. Recent findings in phage-encoded depolymerases acting on microbial EPSs can expand the horizons of enzymatic modification of EPS and its application [75, 126, 144, 145, 157].

4.1.4 Ultrasound and Microwave Treatments

Ultrasound treatment has been used for increasing aqueous solubility of EPS by means of partial degradation of EPS molecules [211]. Fortunately, some partially degraded EPSs showed higher bioactivities, including antioxidant [36], anti-inflammation [45], cryo- and bile-protective [192], prebiotic [120, 121, 191], compared to the intact ones. Interestingly, favorable effects, such as increased production of bioactive compounds including EPS, were found in ultrasound-treated milk with regard to biodegradation of milk components by starter bacteria during fermentation process [60, 159]. Lu et al. [113] indicated direct effects of ultrasound treatment on promotion of mycelial growth and EPS production in *Agaricus bitorquis* (Quél.) Sacc. Chaidam as a consequence of increased bacterial cell permeability and mass transfer rate. It has also been reported that EPS extraction efficiency could be enhanced by ultrasound treatment [72, 216]. Furthermore, Song et al. [194] showed that ultrasound treatment promoted nanoemulsification of EPS with medicinal substances which could be used for vulvovaginal candidiasis treatment. In the similar manner, microwave treatment was applied for enhancement of EPS production in a green microalga, *Scenedesmus* sp. [189].

4.2 Substantial Markets for Microbial EPSs

4.2.1 Food Industry

In the area of food industry, EPS-producing microbes are in general used as starters to manufacture EPS-containing products. Therefore, costs involved with purifying EPS can be excluded from considering the B/C. However, precise control on adequate amount of EPS production should be needed under the fermentation conditions. In addition, stability and insusceptibility of EPS against heat, pH, and salt should be favorable. It should be noted that microbial EPSs may be degraded by the action of endogenous and/or phage-derived EPS-degrading enzymes during fermentation and storage periods [94, 153], resulting in giving unfavorable characteristics to the products.

Viscosity, emulsifying, and flocculation abilities are expected for microbial EPSs mainly to improve the rheological and sensory characteristics of the food products, such as bread [25, 83, 115, 177]; Tinzl-Mal [9], cheese [7, 15, 140, 202, 219], ice cream [42, 242], juice [23, 48], sausage [43, 76, 77, 207] and yogurt [5, 22, 93, 118, 148, 229, 241, 245]. On the other hand, antimicrobial [2, 14, 49, 178], antioxidant [16, 49, 66, 90, 107] and hypocholesterolemic [26, 27, 106, 239] activities are expected for microbial EPSs to commit their health claims. Among them, use of microbial EPSs as antimicrobial agents matches the concept of SDGs owing to the less environmental load than conventional antibiotics, because of the high biodegradability. Morifuji et al. [131] showed an effect of administrating fermented milk to enhance adsorption of dietary carotenoid in human and rat. Interestingly, the authors found a significant positive correlation between the amount of EPS in the fermented milk and the serum β-carotene level in rats. Prebiotic property of microbial EPS is an emerging interest. Until recently, prebiotic effects observed for EPSs were very limited, presumably because researchers examine the effect towards individual probiotic strains. Whereas, recent studies revealed a significant role of EPSs towards microbial consortium such as intestinal microbiota by using the omics approach [99, 121, 152, 161]. Prebiotic property seems to be a promising new beneficial aspect of microbial EPSs when it is utilized in food industry, but it requires further evidence to convince the phenomenon. In addition, aflatoxin-binding capacity of galactan produced by *Weissella confusa* KR780676 may be useful for removal of aflatoxin from food products [92].

Besides pullulan film which has already been established for the purpose of food preservation, some microbial EPSs were developed recently to extend the shelf life by directly coating food products [21, 236] or being applied as packaging materials [61, 142]. For this purpose, EPSs should be highly pure and hence costs on their purification process and the yields influence significantly on the practical application. Chemical modification of EPSs will aid for development of EPS-based films and plastics.

4.2.2 Pharmaceutical Industry

Up to date, a great number of research has been performed to elucidate health beneficial aspects of microbial EPSs for human beings, such as antidiabete [16, 18, 79], anti-inflammatory [101, 237], antitumor [71, 81, 114, 167, 201, 231, 235], immunomodulative [38, 57, 147, 150, 166] and wound healing [173, 200] activities, in addition to antimicrobial, antioxidant, and hypocholesterolemic activities which have already described in the Sect. 4.2.1. Furthermore, even the molecular mechanisms how EPSs influence host immune system in a favorable way have been partially unraveled. Nevertheless, implementation of microbial EPSs seems to remain still a long way to the goal, because drug requires significantly high purity, safety, and medical efficacy unlike as the use in food manufacturing. In general, medical efficacies of microbial EPSs are not so high as like as natural and chemically modified/synthesized drugs. Therefore, it seems too ambitious yet to develop an EPS as a drug unless any definitive technical breakthrough, which can make the yields and/or the medical efficacies of EPSs leaped, is achieved. As being such a technique, conjugation of EPSs with metal nanoparticles is very likely as has been already mentioned in Sect. 4.1.2.

Another way to use microbial EPSs in pharmaceutical industry is as an agent for drug delivery. From this point of view, emulsification activity of EPS is attractive to solubilize nonpolar medical compounds. Recently, Song et al. [193] showed that an EPS produced by a marine mangrove derived bacterium, *Bacillus amyloliquefaciens* ZWJ, could emulsify calcipotriol (CPT), which is a synthetic vitamin D derivative that could relieve symptom of psoriasis vulgaris by enhancing the accumulation of CPT in psoriatic skin lesions and diminishing the expression levels of inflammatory cells and inflammatory factors. The same research group also reported emulsification nystatin, a drug for vulvovaginal candidiasis, by using an EPS derived from *Bacillus vallismortis* WF4 [194]. Further, natural and chemically modified EPSs capable of forming hydrogels are highly potential for construction of drug delivery system. Besides derivatives of well-established EPSs such as curdlan [196] and pullulan [205]. Zykwinska et al. [246] recently reported that a glycosaminoglycan-like EPS derived from a marine bacterium *Alteromonas infernus* could form microparticles in which transforming growth factor-β1 (TGF-β1) were successfully encapsulated.

Curiously, Guerreiro et al. [67] demonstrated that *Enterobacter* A47 produced a Fuc-containing polysaccharide, FucoPol, which showed ice growth inhibition at the water-ice interface, owing to the shear-thinning behavior and polyanionicity of the EPS. Therefore, the EPS is feasible as a cryoprotective agent for cell lines.

4.2.3 Agricultural Industry

Similar to EPSs intended to use in food industry, the purification process is generally unnecessary in agricultural industry, that is advantageous in terms of achieving better B/C. Hence, it becomes significant how efficiently EPS can be produced by microbes under certain circumstances assumed. It should be stressed that influence of microbial load to the environment must be carefully assessed, despite of the biodegradability, especially when exogenous microbes are introduced, because such EPS-producing microbiota will be distributed in open-air field.

Microbial EPSs are expected as crop fertilizer under both normal and harsh environments. For example, substances produced by EPS-producing strain *Kosakonia cowanii* LT-1 during solid-state fermentation promoted the rate of seed germination and growth of maize [56]. As some EPSs show good water retention capacity that promotes soil aggregation, they are potential to confer tolerance against drought to plants, such as rice [198] and Arabidopsis [112]. In fact, high salt concentration in saline soil, which elicits ionic charge, osmotic, oxidative, and water stress in plants, is a serious concern for the efficient land use in agriculture at arid area suffering by drought, coastal area with elevating seawater level and paddy field. Several observations have been reported on the role of EPSs derived mainly from halo-tolerant bacteria in terms of ameliorating salt stress on crops, including maize [3], quinoa [232], rice [134, 197], sunflower [204], tomato [47] and wheat [8, 188].

4.2.4 Environmental Industry

The same as agricultural industry, purification process is not necessarily required in the application of EPSs in environmental industry, and hence the production yield and the environmental load should be the major concern for B/C. There are several reports on EPS-producing bacteria which can assimilate oils as carbon sources [41, 80, 50] or EPS showing oil-holding capacity [183, 54, 55], whereas it is scarce to demonstrate EPS itself could show an effect of enhanced oil recovery in application level. Li et al. [104] reported potential capability of diutan gum, produced by *Sphingomonas* species, to enhance heavy oil recovery at high temperature and high salt concentration. EPS produced by *Pseudomonas stutzeri* XP1 originally found in Chinese oil reservoirs showed good oil recovery in situ [243]. On the other hand, evidences are accumulating on bioremediation by using microbial EPSs, which can recover and/or remove environmental contaminants such as toxic heavy metals [35,

80, 172, 181, 185, 213]. Waste water treatment seems also highly potential for the application of microbial EPSs [10, 65, 100, 170, 223].

4.2.5 Cosmetic Industry

Requirements for EPSs in cosmetic industry are water retentivity, antioxidant activity, and ultraviolet ray protection, mainly focusing on skin care. There are many microbial EPSs with such activities, but their production yields, costs for production, and safety assessment hamper the product commercialization with rare exception such as kefiran [156]. Recently, cosmetic application has been suggested on EPSs produced by halotolerant bacterium, including *Gracilibacillus* sp. SCU50 [54], *Halomonas elongata* S6 [90] and *Halomonas saliphila* LCB169^{T} [55], *Phyllobacterium* sp. 921F [39] and *Vibrio* sp. MO245 [122]. In addition, potential in cosmetic application of EPSs derived from *Polaribacter* sp. SM1127 found in Arctic area [199] and an extreme radiation-resistant bacterium *Deinococcus radiodurans* R1 [107] are fascinating in terms of their beneficial activities on skin care.

4.2.6 Livestock Industry

In contrast to food industry, improvement of texture and organoleptic characteristics of animal feed by using microbial EPS is not necessarily well considered. However, it is common for both industries to pursue favorable nutritional values and health beneficial functions such as antioxidant, antipathogenic, and immunomodulative activities in EPSs. Most studies have been focused on human health benefits in food and pharmaceutical areas until recently, whereas studies alike in livestock industry are behind. Actually, restriction on the purity and safety of EPSs concerned in livestock industry is not so high as EPSs intended for human use, that is, the practical application of EPSs in livestock industry should be easier than for human use. Hence, an accelerated growth of EPS utilization in this area can be expected, especially in relation to reduction of AMR. Not only porcine [63, 130, 206] and poultry [13, 141] industries, but also aquaculture industry is targeted for the purpose of promoting animal production by feeding microbial EPSs [9, 52, 119, 203].

5 Concluding Remarks

CBA plays a pivotal role when one considers industrial application of microbial EPSs. Precise evaluation on factors affecting cost and benefit gives meaningful output from CBA. To achieve a better B/C, mass production of EPS is prerequisite. It is rather advantageous unless purification of EPS is required. There are several problems which stem from the great diversity of microbial EPSs, especially in generalizing the preparation process and in evaluation of the benefits, still remained to be solved. Accumulation of basic knowledge on chemical structures, genetic information, bioactivities, and physicochemical properties of EPSs, systematic analysis on them, and compiling a database of them should be performed further in the future. Possible solutions for implementing microbial EPSs in industrial applications are (i) improvement of preparation process of microbial EPSs, (ii) innovation for modification techniques on their beneficial properties by chemical and/or enzymatic treatments, (iii) reconstruction of microbes as factories to achieve efficient EPS production by synthetic biological approach and (iv) accurate market survey and analysis on potential demands for microbial EPSs. One of immediate solutions that lead lowering costs and increasing benefits effectively could be a combination of conventional optimization analysis for EPS production, such as the CCD approach, and advanced technologies e.g., EPS conjugation with metal nanoparticles.

References

1. Abinaya M, Vaseeharan B, Divya M et al (2018) Bacterial exopolysaccharide (EPS)-coated ZnO nanoparticles showed high antibiofilm activity and larvicidal toxicity against malaria and Zika virus vectors. J Trace Elem Med Biol 45:93–103
2. Adebayo-Tayo B, Fashogbon R (2020) *In vitro* antioxidant, antibacterial, *in vivo* immunomodulatory, antitumor and hematological potential of exopolysaccharide produced by wild type and mutant *Lactobacillus delbureckii* subsp. *bulgaricus*. Heliyon 6:e03268
3. Akhtar SS, Andersen MN, Naveed M et al (2015) Interactive effect of biochar and plant growth-promoting bacterial endophytes on ameliorating salinity stress in maize. Funct Plant Biol 42:770–781
4. Akturk O (2020) Colloidal stability and biological activity evaluation of microbial exopolysaccharide levan-capped gold nanoparticles. Colloids Surf B 192:111061
5. Ale EC, Perezlindo MJ, Pavón Y et al (2016) Technological, rheological and sensory characterizations of a yogurt containing an exopolysaccharide extract from *Lactobacillus fermentum* Lf2, a new food additive. Food Res Int 90:259–267
6. Ali P, Shah AA, Hasan F et al (2020) A glacier bacterium produces high yield of cryoprotective exopolysaccharide. Front Microbiol 10:3096
7. Allam MGM, Darwish AMG, Ayad EHE et al (2017) *Lactococcus* species for conventional Karish cheese conservation. LWT-Food Sci Technol 79:625–631
8. Amna, Ud Din B, Sarfraz S et al (2019) Mechanistic elucidation of germination potential and growth of wheat inoculated with exopolysaccharide and ACC-deaminase producing *Bacillus* strains under induced salinity stress. Ecotoxicol Environ Saf 183:109466
9. Ang CY, Sano M, Dan S et al (2020) Postbiotics applications as infectious disease control agent in aquaculture. Biocontrol Sci 25:1–7
10. Araújo D, Concórdio-Reis P, Marques AC et al (2020) Demonstration of the ability of the bacterial polysaccharide FucoPol to flocculate kaolin suspensions. Environ Technol 41:287–295
11. Arena A, Maugeri TL, Pavone B et al (2006) Antiviral and immunoregulatory effect of a novel exopolysaccharide from a marine thermotolerant *Bacillus licheniformis*. Int Immunopharm 6:8–13

12. Asgher M, Urooj Y, Qamar SA et al (2020) Improved exopolysaccharide production from *Bacillus licheniformis* MS3: optimization and structural/functional characterization. Int J Biol Macromol 151:984–992
13. Ashfaq I, Amjad H, Ahmad W et al (2020) Growth Inhibition of common enteric pathogens in the intestine of broilers by microbially produced dextran and levan exopolysaccharides. Curr Microbiol 77:2128–2136
14. Aullybux AA, Puchooa D, Bahorun T et al (2019) Phylogenetics and antibacterial properties of exopolysaccharides from marine bacteria isolated from Mauritius seawater. Annal Microbiol 69:957–972
15. Ayyash M, Abu-Jdayil B, Hamed F et al (2018) Rheological, textural, microstructural and sensory impact of exopolysaccharide-producing *Lactobacillus plantarum* isolated from camel milk on low-fat akawi cheese. LWT-Food Sci Technol 87:423–431
16. Ayyash M, Abu-Jdayil B, Itsaranuwat P et al (2020) Exopolysaccharide produced by the potential probiotic *Lactococcus garvieae* C47: Structural characteristics, rheological properties, bioactivities and impact on fermented camel milk. Food Chem 333:
17. Ayyash M, Abu-Jdayil B, Itsaranuwat P et al (2020) Characterization, bioactivities, and rheological properties of exopolysaccharide produced by novel probiotic *Lactobacillus plantarum* C70 isolated from camel milk. Int J Biol Macromol 144:938–946
18. Ayyash M, Abu-Jdayil B, Olaimat A et al (2020) Physicochemical, bioactive and rheological properties of an exopolysaccharide produced by a probiotic *Pediococcus pentosaceus* M41. Carbohydr Polym 229:
19. Ayyash M, Stathopoulos C, Abu-Jdayil B et al (2020) Exopolysaccharide produced by potential probiotic *Enterococcus faecium* MS79: characterization, bioactivities and rheological properties influenced by salt and pH. LWT 131:109741
20. Bacelar AH, Silva-Correia J, Oliveira JM et al (2016) Recent progress in gellan gum hydrogels provided by functionalization strategies. J Mater Chem B 4:6164–6174
21. Balti R, Ben Mansour M, Zayoud N et al (2020) Active exopolysaccharides based edible coatings enriched with red seaweed (*Gracilaria gracilis*) extract to improve shrimp preservation during refrigerated storage. Food Biosci 34:100522
22. Bancalari E, Alinovi M, Bottari B et al (2020) Ability of a wild *Weissella* strain to modify viscosity of fermented milk. Front Microbiol 10:3086
23. Bancalari E, Castellone V, Bottari B et al (2020) Wild *Lactobacillus casei* group strains: potentiality to ferment plant derived juices. Foods 9:9030314
24. Banerjee A, Das D, Rudra SG et al (2020) Characterization of exopolysaccharide produced by *Pseudomonas* sp. PFAB4 for synthesis of EPS-coated AgNPs with antimicrobial properties. J Polym Environ 28:242–256
25. Belz MCE, Axel C, Arendt EK et al (2019) Improvement of taste and shelf life of yeasted low-salt bread containing functional sourdoughs using *Lactobacillus amylovorus* DSM 19280 and *Weisella cibaria* MG1. Int J Food Microbiol 302:69–79
26. Bhat B, Bajaj BK (2018) Hypocholesterolemic and bioactive potential of exopolysaccharide from a probiotic *Enterococcus faecium* K1 isolated from kalarei. Biores Technol 254:264–267
27. Bhat B, Bajaj BK (2019) Hypocholesterolemic potential and bioactivity spectrum of an exopolysaccharide from a probiotic isolate *Lactobacillus paracasei* M7. Bioact Carbohydr Diet Fibre 19:100191
28. Bhattacharyya P, Agarwal B, Goswami M et al (2018) Zinc oxide nanoparticle inhibits the biofilm formation of *Streptococcus pneumoniae*. Antonie Van Leeuwenhoek 111:89–99
29. Boardman NE (2006) Cost-benefit analysis: concepts and practice, 3rd edn. Prentice Hall, Upper Saddle River, NJ
30. Brandi J, Oliveira EC, Monteiro NK et al (2011) Chemical modification of botryosphaeran: structural characterization and anticoagulant activity of a water-soluble sulfonated (1 → 3)(1 → 6)-β-d-glucan. J Microbiol Biotechnol 21:1036–1042
31. Buttacavoli M, Albanese NN, Di Cara G et al (2018) Anticancer activity of biogenerated silver nanoparticles: an integrated proteomic investigation. Oncotarget 9:9685–9705

32. Calegari GC, Santos VAQ, Barbosa-Dekker AM et al (2019) Sulfonated(1 → 6;)-β-d-glucan (lasiodiplodan): preparation, characterization and bioactive properties. Food Technol Biotechnol 57:490–502
33. Capek P, Matulová M, Šutovská M et al (2020) Chlorella vulgaris α-L-arabino-α-L-rhamno-α, β-D-galactan structure and mechanisms of its anti-inflammatory and anti-remodelling effects. Int J Biol Macromol 162:188–198
34. Casentini B, Gallo M, Baldi F (2019) Arsenate and arsenite removal from contaminated water by iron oxides nanoparticles formed inside a bacterial exopolysaccharide. J Environ Chem Eng 7:102908
35. Cavallero GJ, Ferreira ML, Casabuono AC et al (2020) Structural characterization and metal biosorptive activity of the major polysaccharide produced by *Pseudomonas veronii* 2E. Carbohydr Polym 245:116458
36. Chen X, Siu KC, Cheung YC et al (2014) Structure and properties of a (1 → 3)-β-d-glucan from ultrasound-degraded exopolysaccharides of a medicinal fungus. Carbohydr Polym 106:270–275
37. Chen T, Wang Y, Li J et al (2016) Phthaloyl modification of a polysaccharide from *Lachnum* YM262 and immunomodulatory activity. Process Biochem 51:1599–1609
38. Chen N, Zhao X, Wang F et al (2020) Proteomic study of sulfated polysaccharide from *Enterobacter cloacae* Z0206 against H_2O_2-induced oxidative damage in murine macrophages. Carbohydr Polym 237:116147
39. Chi Y, Ye H, Li H et al (2019) Structure and molecular morphology of a novel moisturizing exopolysaccharide produced by *Phyllobacterium* sp. 921F. Int J Biol Macromol 135:998–1005
40. Cusimano MG, Ardizzone F, Nasillo G et al (2020) Biogenic iron-silver nanoparticles inhibit bacterial biofilm formation due to Ag^+ release as determined by a novel phycoerythrin-based assay. Appl Microbiol Biotechnol 104:6325–6336
41. Darwish AA, Al-Bar OA, Yousef RH et al (2019) Production of antioxidant exopolysaccharide from *Pseudomonas aeruginosa* utilizing heavy oil as a solo carbon source. Pharmacogn Res 11:378–383
42. Dertli E, Toker OS, Durak MZ et al (2016) Development of a fermented ice-cream as influenced by *in situ* exopolysaccharide production: rheological, molecular, microstructural and sensory characterization. Carbohydr Polym 136:427–440
43. Dertli E, Yilmaz MT, Tatlisu NB et al (2016) Effects of *in situ* exopolysaccharide production and fermentation conditions on physicochemical, microbiological, textural and microstructural properties of Turkish-type fermented sausage (sucuk). Meat Sci 121:156–165
44. Dilna SV, Surya H, Aswathy RG et al (2015) Characterization of an exopolysaccharide with potential health-benefit properties from a probiotic *Lactobacillus plantarum* RJF4. LWT 64:1179–1186
45. Du B, Zeng H, Yang Y et al (2016) Anti-inflammatory activity of polysaccharide from *Schizophyllum commune* as affected by ultrasonication. Int J Biol Macromol 91:100–105
46. Du AGL, Zykwinska A, Sinquin C et al (2017) Purification of the exopolysaccharide produced by *Alteromonas infernus*: identification of endotoxins and effective process to remove them. Appl Microbiol Biotechnol 101:6597–6606
47. EL Arroussi H, Benhima R, Elbaouchi A et al (2018) *Dunaliella salina* exopolysaccharides: a promising biostimulant for salt stress tolerance in tomato (*Solanum lycopersicum*). J Appl Phycol 30:2929–2941
48. Elizaquível P, Sánchez G, Salvador A et al (2011) Evaluation of yogurt and various beverages as carriers of lactic acid bacteria producing 2-branched (1,3)-β-d-glucan. J Dairy Sci 94:3271–3278
49. Elsehemy IA, Noor El Deen AM, Awad HM et al (2020) Structural, physical characteristics and biological activities assessment of scleroglucan from a local strain *Athelia rolfsii* TEMG. Int J Biol Macromol 163:1196–1207

50. Fan Y, Wang J, Gao C et al (2020) A novel exopolysaccharide-producing and long-chain n-alkane degrading bacterium Bacillus licheniformis strain DM-1 withpotential application for in-situ enhanced oil recovery. Sci Rep 10:8519
51. Farag MMS, Moghannem SAM, Shehabeldine AM et al (2020) Antitumor effect of exopolysaccharide produced by *Bacillus mycoides*. Microb Pathog 140:
52. Feng J, Cai Z, Chen Y et al (2020) Effects of an exopolysaccharide from *Lactococcus lactis* Z-2 on innate immune response, antioxidant activity, and disease resistance against *Aeromonas hydrophila* in *Cyprinus carpio* L. Fish Shellfish Immunol 98:324–333
53. Ferreira AS, Ferreira SS, Correia A et al (2020) Reserve, structural and extracellular polysaccharides of *Chlorella vulgaris*: a holistic approach. Algal Res 45:101757
54. Gan L, Li X, Wang H et al (2020) Structural characterization and functional evaluation of a novel exopolysaccharide from the moderate halophile *Gracilibacillus* sp. SCU50. Int J Biol Macromol 154:1140–1148
55. Gan L, Li X, Zhang H et al (2020) Preparation, characterization and functional properties of a novel exopolysaccharide produced by the halophilic strain *Halomonas saliphila* LCB169^{T}. Int J Biol Macromol 156:372–380
56. Gao H, Lu C, Wang H et al (2020) Production exopolysaccharide from *Kosakonia cowanii* LT-1 through solid-state fermentation and its application as a plant growth promoter. Int J Biol Macromol 150:955–964
57. Garcia-Castillo V, Marcial G, Albarracín L et al (2020) The exopolysaccharide of *lactobacillus fermentum* UCO-979C is partially involved in its immunomodulatory effect and its ability to improve the resistance against *Helicobacter pylori* infection. Microorganisms 8:479
58. Garcés V, González A, Sabio L et al (2020) Magnetic and golden yogurts. Food as a potential nanomedicine carrier. Materials 13:481
59. Garza-Cervantes JA, Escárcega-González CE, Díaz Barriga Castro E et al (2019) Antimicrobial and antibiofilm activity of biopolymer-Ni, Zn nanoparticle biocomposites synthesized using *R. mucilaginosa* UANL-001L exopolysaccharide as a capping agent. Int J Nanomed 14:2557–2571
60. Gholamhosseinpour A, Hashemi SMB, Raoufi Jahromi L et al (2020) Conventional heating, ultrasound and microwave treatments of milk: fermentation efficiency and biological activities. Int Dairy J 110:104809
61. Giro TM, Beloglazova KE, Rysmukhambetova GE et al (2020) Xanthan-based biodegradable packaging for fish and meat products. Foods Raw Mater 8:75
62. Gomaa M, Yousef N (2020) Optimization of production and intrinsic viscosity of an exopolysaccharide from a high yielding *Virgibacillus salarius* BM02: study of its potential antioxidant, emulsifying properties and application in the mixotrophic cultivation of *Spirulina platensis*. Int J Biol Macromol 149:552–561
63. González-Ortiz G, Pérez JF, Hermes RG et al (2014) Screening the ability of natural feed ingredients to interfere with the adherence of enterotoxigenic *Escherichia coli* (ETEC) K88 to the porcine intestinal mucus. Br J Nutr 111:633–642
64. Goto K, Inoue S, Watanabe O (2009) Kinousei shokuhin sesshu to sentaku ni kansuru Kokusaihikaku (in Japanese). J Food Sys Res 16:27–31
65. Govarthanan M, Jeon C-H, Jeon Y-H et al (2020) Non-toxic nano approach for wastewater treatment using *Chlorella vulgaris* exopolysaccharides immobilized in iron-magnetic nanoparticles. Int J Biol Macromol 162:1241–1249
66. Goyzueta MLD, Noseda MD, Bonatto SJR et al (2020) Production, characterization, and biological activity of a chitin-like EPS produced by *Mortierella alpina* under submerged fermentation. Carbohydr Polym 247:116716
67. Guerreiro BM, Freitas F, Lima JC et al (2020) Demonstration of the cryoprotective properties of the fucose-containing polysaccharide FucoPol. Carbohydr Polym 245:116500
68. Gupta P, Diwan B (2017) Bacterial exopolysaccharide mediated heavy metal removal: a review on biosynthesis, mechanism and remediation strategies. Biotechnol Rep 13:58–71

69. Halder U, Banerjee A, Bandopadhyay R (2017) Structural and functional properties, biosynthesis, and patenting trends of bacterial succinoglycan: a review. Indian J Microbiol 57:278–284
70. Hamidi M, Gholipour AR, Delattre C et al (2020) Production, characterization and biological activities of exopolysaccharides from a new cold-adapted yeast: *Rhodotorula mucilaginosa* sp. GUMS16. Int J Biol Macromol 151:268–277
71. Hao Y, Huang Y, Chen J et al (2020) Exopolysaccharide from *Cryptococcus heimaeyensis* S20 induces autophagic cell death in non-small cell lung cancer cells via ROS/p38 and ROS/ERK signalling. Cell Prolif 00:e12869
72. Hasheminya SM, Dehghannya J (2020) Novel ultrasound-assisted extraction of kefiran biomaterial, a prebiotic exopolysaccharide, and investigation of its physicochemical, antioxidant and antimicrobial properties. Mater Chem Phys 243:122645
73. Hawkins JP, Oresnik IJ (2017) Characterisation of a gene encoding a membrane protein that affects exopolysaccharide production and intracellular Mg^{2+} concentrations in *Ensifer meliloti*. FEMS Microbiol Lett 364:fnx061
74. Heckel B, Tomlinson AD, Morton ER et al (2014) *Agrobacterium tumefaciens* ExoR controls acid response genes and impacts exopolysaccharide synthesis, horizontal gene transfer, and virulence gene expression. J Bacteriol 196:3221–3233
75. Hernandez-Morales AC, Lessor LL, Wood TL et al (2018) Genomic and biochemical characterization of *Acinetobacter* podophage Petty reveals a novel lysis mechanism and tail-associated depolymerase activity. J Virol 92:e01064-17
76. Hilbig J, Gisder J, Prechtl RM et al (2019) Influence of exopolysaccharide-producing lactic acid bacteria on the spreadability of fat-reduced raw fermented sausages (Teewurst). Food Hydrocoll 93:422–431
77. Hilbig J, Hildebrandt L, Herrmann K et al (2020) Influence of homopolysaccharide-producing lactic acid bacteria on the spreadability of raw fermented sausages (onion mettwurst). J Food Sci 85:289–297
78. Hu X, Li D, Qiao Y et al (2020) Purification, characterization and anticancer activities of exopolysaccharide produced by *Rhodococcus erythropolis* HX-2. Int J Biol Macromol 145:646–654
79. Huang Z, Lin F, Zhu X et al (2020) An exopolysaccharide from *Lactobacillus plantarum* H31 in pickled cabbage inhibits pancreas α-amylase and regulating metabolic markers in HepG2 cells by AMPK/PI3K/Akt pathway. Int J Biol Macromol 143:775–784
80. Ibrahim IM, Konnova SA, Sigida EN et al (2020) Bioremediation potential of a halophilic *Halobacillus* sp. strain, EG1HP4QL: exopolysaccharide production, crude oil degradation, and heavy metal tolerance. Extremophiles 24:157–166
81. Ibrahim AY, Youness ER, Mahmoud MG et al (2020) Acidic exopolysaccharide produced from marine *Bacillus amyloliquefaciens* 3MS 2017 for the protection and treatment of breast cancer. Breast Cancer (Auckl) 14:1178223420902075
82. Islam ST, Alvarez IV, Saïdi F et al (2020) Modulation of bacterial multicellularity *via* spatio-specific polysaccharide secretion. PLoS Biol 18:e3000728
83. İspirli H, Özmen D, Yılmaz M et al (2020) Impact of glucan type exopolysaccharide (EPS) production on technological characteristics of sourdough bread. Food Control 107:
84. Jang IA, Kim J, Park W (2016) Endogenous hydrogen peroxide increases biofilm formation by inducing exopolysaccharide production in *Acinetobacter oleivorans* DR1. Sci Rep 6:21121
85. Jansson PE, Lindberg B, Sandford PA (1983) Structural studies of gellan gum, an extracellular polysaccharide elaborated by *Pseudomonas elodea*. Carbohydr Res 124:135–139
86. Jia X, Wang C, Bai Y et al (2017) Sulfation of the extracellular polysaccharide produced by the king oyster culinary-medicinal mushroom, *Pleurotus eryngii* (Agaricomycetes), and its antioxidant properties *in vitro*. Int J Med Mushrooms 19:355–362
87. Jiang J, Guo S, Ping W et al (2020) Optimization production of exopolysaccharide from *Leuconostoc lactis* L2 and its partial characterization. Int J Biol Macromol 159:630–639

88. Jiang B, Tian L, Huang X et al (2020) Characterization and antitumor activity of novel exopolysaccharide APS of *Lactobacillus plantarum* WLPL09 from human breast milk. Int J Biol Macromol 163:985–995
89. Jouault SC, Chevolot L, Helley D et al (2001) Characterization, chemical modifications and *in vitro* anticoagulant properties of an exopolysaccharide produced by *Alteromonas infernus*. Biochim Biophys Acta Gen Subj 1528:141–151
90. Joulak I, Azabou S, Finore I et al (2020) Structural characterization and functional properties of novel exopolysaccharide from the extremely halotolerant *Halomonas elongata* S6. Int J Biol Macromol 164:95–104
91. Kavitake D, Balyan S, Devi PB et al (2020) Evaluation of oil-in-water (O/W) emulsifying properties of galactan exopolysaccharide from *Weissella confusa* KR780676. J Food Sci Technol 57:1579–1585
92. Kavitake D, Singh SP, Kandasamy S, et al (2020b) Report on aflatoxin-binding activity of galactan exopolysaccharide produced by *Weissella confusa* KR780676. 3 Biotech 10:181
93. Khanal SN, Lucey JA (2017) Evaluation of the yield, molar mass of exopolysaccharides, and rheological properties of gels formed during fermentation of milk by *Streptococcus thermophilus* strains St-143 and ST-10255y. J Dairy Sci 100:6906–6917
94. Knecht LE, Veljkovic M, Fieseler L (2020) Diversity and function of phage encoded depolymerases. Front Microbiol 10:2949
95. Kogan G, Šoltés L, Stern R et al (2007) Hyaluronic acid: a natural biopolymer with a broad range of biomedical and industrial applications. Biotechnol Lett 29:17–25
96. Leonardi P, Lugli F, Iotti M et al (2020) Effects of biogenerated ferric hydroxides nanoparticles on truffle mycorrhized plants. Mycorrhiza 30:211–219
97. Lesage G, Bussey H (2006) Cell wall assembly in *Saccharomyces cerevisiae*. Microbiol Mol Biol Rev 70:317–343
98. Levander F, Svensson M, Rådström P (2002) Enhanced exopolysaccharide production by metabolic engineering of *Streptococcus thermophilus*. Appl Environ Microbiol 68:784–790
99. Li B, Chen H, Cao L (2020) Effects of an *Escherichia coli* exopolysaccharide on human and mouse gut microbiota *in vitro*. Int J Biol Macromol 150:991–999
100. Li C, Chen D, Ding J et al (2020) A novel hetero-exopolysaccharide for the adsorption of methylene blue from aqueous solutions: isotherm, kinetic, and mechanism studies. J Clean Prod 265:121800
101. Li LQ, Song AX, Yin JY et al (2020) Anti-inflammation activity of exopolysaccharides produced by a medicinal fungus *Cordyceps sinensis* Cs-HK1 in cell and animal models. Int J Biol Macromol 149:1042–1050
102. Li N, Wang Y, Zhu P et al (2015) Improvement of exopolysaccharide production in *Lactobacillus casei* LC2W by overexpression of NADH oxidase gene. Microbiol Res 171:73–77
103. Li S, Xia H, Xie A et al (2020) Structure of a fucose-rich polysaccharide derived from EPS produced by *Kosakonia* sp. CCTCC M2018092 and its application in antibacterial film. Int J Biol Macromol 159:295–303
104. Li Y, Xu L, Gong H (2017) A microbial exopolysaccharide produced by *Sphingomonas* species for enhanced heavy oil recovery at high temperature and high salinity. Energy Fuels 31:3960–3969
105. Li S, Xia H, Xie A, et al (2020) Structure of a fucose-rich polysaccharide derived from EPS produced by Kosakonia sp. CCTCC M2018092 and its application inantibacterial film. Int J Biol Macromol 159:295–303
106. Lim J, Kale M, Kim DH et al (2017) Antiobesity effect of exopolysaccharides isolated from Kefir grains. J Agric Food Chem 65:10011–10019
107. Lin SM, Baek CY, Jung J-H et al (2020) Antioxidant activities of an exopolysaccharide (DeinoPol) produced by the extreme radiation-resistant bacterium *Deinococcus radiodurans*. Sci Rep 10:55
108. Lin T, Chen C, Chen B et al (2019) Optimal economic productivity of exopolysaccharides from lactic acid bacteria with production possibility curves. Food Sci Nutr 7:2336–2344

109. Liu Z, Jiao Y, Lu H et al (2020) Chemical characterization, antioxidant properties and anticancer activity of exopolysaccharides from *Floccularia luteovirens*. Carbohydr Polym 229:115432
110. Liu L, Li JH, Zi SF et al (2019) AgNP combined with quorum sensing inhibitor increased the antibiofilm effect on *Pseudomonas aeruginosa*. Appl Microbiol Biotechnol 103:6195–6204
111. Liu L, Liu Y, Li J et al (2011) Microbial production of hyaluronic acid: current state, challenges, and perspectives. Microb Cell Fact 10:99
112. Lu X, Liu SF, Yue L et al (2018) *Epsc* involved in the encoding of exopolysaccharides produced by *Bacillus amyloliquefaciens* FZB42 act to boost the drought tolerance of *Arabidopsis thaliana*. Int J Mol Sci 19:3795
113. Lu H, Lou H, Wei T et al (2020) Ultrasound enhanced production of mycelia and exopolysaccharide by *Agaricus bitorquis* (Quél.) Sacc. Chaidam. Ultrason Sonochem 64:105040
114. Lu Y, Zhang X, Wang J, et al (2020) Exopolysaccharides isolated from *Rhizopus nigricans* induced colon cancer cell apoptosis *in vitro* and *in vivo via* activating the AMPK pathway. Biosci Rep 40:BSR20192774
115. Lynch KM, Coffey A, Arendt EK (2018) Exopolysaccharide producing lactic acid bacteria: Their techno-functional role and potential application in gluten-free bread products. Food Res Int 110:52–61
116. López-Ortega MA, Rodríguez-Hernández AI, Camacho-Ruíz RM et al (2020) Physicochemical characterization and emulsifying properties of a novel exopolysaccharide produced by haloarchaeon *Haloferax mucosum*. Int J Biol Macromol 142:152–162
117. López-Ortega MA, Rodríguez-Hernández AI, Camacho-Ruíz RM, et al (2020) Physicochemical characterization and emulsifying properties of a novelexopolysaccharide produced by haloarchaeon Haloferax mucosum. Int J Biol Macromol 142:152–162
118. Madhubasani GBL, Prasanna PHP, Chandrasekara A et al (2020) Exopolysaccharide producing starter cultures positively influence on microbiological, physicochemical, and sensory properties of probiotic goats' milk set-yoghurt. J Food Process Preserv 44:e14361
119. Mahdhi A, Chakroun I, Espinosa-Ruiz C et al (2020) Dietary administration effects of exopolysaccharide from potential probiotic strains on immune and antioxidant status and nutritional value of European sea bass (*Dicentrarchus labrax* L.). Res Vet Sci 131:51–58
120. Mao YH, Song AX, Li LQ et al (2020) A high-molecular weight exopolysaccharide from the Cs-HK1 fungus: ultrasonic degradation, characterization and *in vitro* fecal fermentation. Carbohydr Polym 246:116636
121. Mao YH, Song AX, Li LQ et al (2020) Effects of exopolysaccharide fractions with different molecular weights and compositions on fecal microflora during *in vitro* fermentation. Int J Biol Macromol 144:76–84
122. Martin-Pastor M, Ferreira AS, Moppert X et al (2019) Structure, rheology, and copper-complexation of a hyaluronan-like exopolysaccharide from *Vibrio*. Carbohydr Polym 222:114999
123. Medina-Cabrera EV, Rühmann B, Schmid J et al (2020) Characterization and comparison of *Porphyridium sordidum* and *Porphyridium purpureum* concerning growth characteristics and polysaccharide production. Algal Res 49:101931
124. Medina-Cabrera EV, Rühmann B, Schmid J et al (2020) Optimization of growth and EPS production in two *Porphyridum* strains. Biores Technol Rep 11:100486
125. Mengi B, Ikeda S, Murayama D et al (2020) Factors affecting decreasing viscosity of the culture medium during the stationary growth phase of exopolysaccharide-producing *Lactobacillus fermentum* MTCC 25067. Biosci Microbiota Food Health 39:160–168
126. Mi L, Liu Y, Wang C et al (2019) Identification of a lytic *Pseudomonas aeruginosa* phage depolymerase and its anti-biofilm effect and bactericidal contribution to serum. Virus Genes 55(3):394–405
127. Michaud P, Da Costa A, Courtois B et al (2003) Polysaccharide lyases: recent developments as biotechnological tools. Crit Rev Biotechnol 23:233–266

128. Min WH, Fang XB, Wu T et al (2019) Characterization and antioxidant activity of an acidic exopolysaccharide from *Lactobacillus plantarum* JLAU103. J Biosci Bioeng 127:758–766
129. Min WH, Fang XB, Wu T, et al (2019) Characterization and antioxidant activity of an acidic exopolysaccharide from Lactobacillus plantarum JLAU103. J BiosciBioeng. 127:758–766
130. Mizuno H, Tomotsune K, Islam MA et al (2020) Exopolysaccharides from *Streptococcus thermophilus* ST538 modulate the antiviral innate immune response in porcine intestinal epitheliocytes. Front Microbiol 11:894
131. Morifuji M, Ichikawa S, Kitade M et al (2020) Exopolysaccharides from milk fermented by lactic acid bacteria enhance dietary carotenoid bioavailability in humans in a randomized crossover trial and in rats. Am J Clin Nutr 111:903–914
132. Moscovici M (2015) Present and future medical applications of microbial exopolysaccharides. Front Microbiol 6:01012
133. Mozzi F, Savoy de Giori G, Font de Valdez G (2003) UDP-galactose 4-epimerase: a key enzyme in exopolysaccharide formation by *Lactobacillus casei* CRL 87 in controlled pH batch cultures. J Appl Microbiol 94:175–183
134. Mukherjee P, Mitra A, Roy M (2020) *Halomonas* rhizobacteria of *Avicennia marina* of indian sundarbans promote rice growth under saline and heavy metal stresses through exopolysaccharide production. Front Microbiol 10:1207
135. Murphy K, Curley D, O'Callaghan TF et al (2017) The composition of human milk and infant faecal microbiota over the first three months of life: a pilot study. Sci Rep 7:40597
136. Muylaert K, Bastiaens L, Vandamme D et al (2017) "Harvesting of microalgae: overview of process options and their strengths and drawbacks" in Microalgae-Based Biofuels and Bioproducts From Feedstock Cultivation to End-products, Woodhead Publishing Series in Energy:113–132
137. Nachtigall C, Surber G, Herbi F et al (2020) Production and molecular structure of heteropolysaccharides from two lactic acid bacteria. Carbohydr Polym 236:116019
138. Naessens M, Cerdobbel A, Soetaert W et al (2005) *Leuconostoc* dextransucrase and dextran: production, properties and applications. J Chem Technol Biotechnol 80:845–860
139. Navarro Gallón SM, Alpaslan E, Wang M et al (2019) Characterization and study of the antibacterial mechanisms of silver nanoparticles prepared with microalgal exopolysaccharides. Mater Sci Eng, C 99:685–695
140. Nepomuceno RSC, Costa Junior LCG, Costa RGB (2016) Exopolysaccharide-producing culture in the manufacture of Prato cheese. LWT-Food Sci Technol 72:383–389
141. Neveling DP, Ahire JJ, Laubscher W et al (2020) Genetic and phenotypic characteristics of a multi-strain probiotic for broilers. Curr Microbiol 77:369–387
142. Nešić A, Cabrera-Barjas G, Dimitrijević-Branković S et al (2020) Prospect of polysaccharide-based materials as advanced food packaging. Molecules 25:135
143. Nowak B, Ciszek-Lenda M, Śróttek M et al (2012) *Lactobacillus rhamnosus* exopolysaccharide ameliorates arthritis induced by the systemic injection of collagen and lipopolysaccharide in DBA/1 mice. Arch Immunol Ther Exp 60:211–220
144. Oliveira H, Costa AR, Ferreira A et al (2019) Functional analysis and antivirulence properties of a new depolymerase from a myovirus that infects *Acinetobacter baumannii* capsule K45. J Virol 93:e0116318
145. Oliveira H, Pinto G, Mendes B et al (2020) A tailspike with exopolysaccharide depolymerase activity from a new *Providencia stuartii* phage makes multidrug-resistant bacteria susceptible to serum-mediated killing. Appl Environ Microbiol 86:e00073-20
146. Öner ET, Hernández L, Combie J (2016) Review of Levan polysaccharide: From a century of past experiences to future prospects. Biotechnol Adv 34:827–844
147. Ou Y, Zhu L, Xu S et al (2020) Activation of RAW264.7 macrophage by exopolysaccharide from *Aphanothece halaphytica* (EPSAH) and the underlying mechanisms. Fund Clin Pharmacol fcp.12550
148. Pachekrepapol U, Lucey JA, Gong Y et al (2017) Characterization of the chemical structures and physical properties of exopolysaccharides produced by various *Streptococcus thermophilus* strains. J Dairy Sci 100:3424–3435

149. Padmanaban S, Balaji N, Muthukumaran C et al (2015) Statistical optimization of process parameters for exopolysaccharide production by *Aureobasidium pullulans* using sweet potato based medium. 3 Biotech 5:1067–1073
150. Paik W, Alonzo F III, Knight KL (2020) Suppression of *Staphylococcus aureus* superantigen-independent interferon gamma response by a probiotic polysaccharide. Infect Immun 88:e00661-19
151. Palaniraj A, Jayaraman V (2011) Production, recovery and applications of xanthan gum by *Xanthomonas campestris*. J Food Eng 106:1–12
152. Pan L, Han Y, Zhou Z (2020) *In vitro* prebiotic activities of exopolysaccharide from *Leuconostoc pseudomesenteroides* XG5 and its effect on the gut microbiota of mice. J Funct Foods 67:103853
153. Pham PL, Dupont I, Roy D et al (2000) Production of exopolysaccharide by *Lactobacillus rhamnosus* R and analysis of its enzymatic degradation during prolonged fermentation. Appl Environ Microbiol 66:2302–2310
154. De Philippis R, Colica G, Micheletti E (2011) Exopolysaccharide-producing cyanobacteria in heavy metal removal from water: molecular basis and practical applicability of the biosorption process. Appl Microbiol Biotechnol 92:697–708
155. Picceri GG, Leonardi P, Iotti M et al (2018) Bacteria-produced ferric exopolysaccharide nanoparticles as iron delivery system for truffles (*Tuber borchii*). Appl Microbiol Biotechnol 102:1429–1441
156. Piermaría J, Bengoechea C, Abraham AG et al (2016) Shear and extensional properties of kefiran. Carbohyd Polym 152:97–104
157. Pires DP, Oliveira H, Melo LDR et al (2016) Bacteriophage-encoded depolymerases: their diversity and biotechnological applications. Appl Microbiol Biotechnol 100:2141–2151
158. Poli A, Di Donato P, Abbamondi GR et al (2011) Synthesis, production, and biotechnological applications of exopolysaccharides and polyhydroxyalkanoates by archaea. Archaea 2011:693253
159. Potoroko I, Kalinina I, Botvinnikova V et al (2018) Ultrasound effects based on simulation of milk processing properties. Ultrason Sonochem 48:463–472
160. Powell LC, Pritchard MF, Ferguson EL et al (2018) Targeted disruption of the extracellular polymeric network of *Pseudomonas aeruginosa* biofilms by alginate oligosaccharides. NPJ Biofilms Microbiomes 4:13
161. Püngel D, Treveil A, Dalby MJ et al (2020) *Bifidobacterium breve* UCC2003 exopolysaccharide modulates the early life microbiota by acting as a potential dietary substrate. Nutrients 12:948
162. Ragavan ML, Das N (2020) *In vitro* studies on therapeutic potential of probiotic yeasts isolated from various sources. Curr Microbiol 77(10):2821–2830
163. Rajivgandhi GN, Maruthupandy M, Li JL et al (2020) Photocatalytic reduction and anti-bacterial activity of biosynthesized silver nanoparticles against multi drug resistant *Staphylococcus saprophyticus* BDUMS 5 (MN310601). Mater Sci Eng, C 114:111024
164. Rajivgandhi GN, Ramachandran G, Maruthupandy M et al (2020) Anti-oxidant, anti-bacterial and anti-biofilm activity of biosynthesized silver nanoparticles using *Gracilaria corticata* against biofilm producing *K. pneumoniae*. Colloids Surf A Physicochem Eng Asp 600:124830
165. Ravyts F, Vuyst LD, Leroy F (2012) Bacterial diversity and functionalities in food fermentations. Eng Life Sci 12:356–367
166. Ren Q, Tang Y, Zhang L et al (2020) Exopolysaccharide produced by *Lactobacillus casei* promotes the differentiation of CD^{4+} T cells into Th17 cells in BALB/c mouse Peyer's patches *in vivo* and *in vitro*. J Agric Food Chem 68:2664–2672
167. Riaz Rajoka MS, Mehwish HM, Fang H et al (2020) Characterization and anti-tumor activity of exopolysaccharide produced by *Lactobacillus kefiri* isolated from Chinese kefir grains. J Func Foods 63:103588

168. Riaz Rajoka MS, Mehwish HM, Zhang H et al (2020) Antibacterial and antioxidant activity of exopolysaccharide mediated silver nanoparticle synthesized by *Lactobacillus brevis* isolated from Chinese koumiss. Colloids Surf B 186:110734
169. Rodríguez C, Medici M, Rodríguez AV et al (2009) Prevention of chronic gastritis by fermented milks made with exopolysaccharide-producing *Streptococcus thermophilus* strains. J Dairy Sci 92:2423–2434
170. Rollemberg SLS, de Oliveira QL, de Barros NA et al (2020) Pilot-scale aerobic granular sludge in the treatment of municipal wastewater: optimizations in the start-up, methodology of sludge discharge, and evaluation of resource recovery. Bioresour Technol 311:123467
171. Rop O, Mlcek J, Jurikova T (2009) Beta-glucans in higher fungi and their health effects. Nutr Rev 67:624–631
172. Rusinova-Videva S, Nachkova S, Adamov A et al (2020) Antarctic yeast *Cryptococcus laurentii* (AL65): biomass and exopolysaccharide production and biosorption of metals. J Chem Technol Biotechnol 95:1372–1379
173. Sahana TG, Rekha PD (2020) A novel exopolysaccharide from marine bacterium *Pantoea* sp. YU16-S3 accelerates cutaneous wound healing through Wnt/β-catenin pathway. Carbohydr Polym 238:116191
174. Saravanan C, Rajesh R, Kaviarasan T et al (2017) Synthesis of silver nanoparticles using bacterial exopolysaccharide and its application for degradation of azo-dyes. Biotechnol Rep 15:33–40
175. Schilling C, Badri A, Sieber V et al (2020) Metabolic engineering for production of functional polysaccharides. Curr Opin Biotechnol 66:44–51
176. Schmid J, Sieber V, Rehm B (2015) Bacterial exopolysaccharides: biosynthesis pathways and engineering strategies. Front Microbiol 6:496
177. Seitter M, Fleig M, Schmidt H et al (2020) Effect of exopolysaccharides produced by *Lactobacillus sanfranciscensis* on the processing properties of wheat doughs. Eur Food Res Technol 246:461–469
178. Sengupta D, Datta S, Biswas D (2020) Surfactant exopolysaccharide of *Ochrobactrum pseudintermedium* C1 has antibacterial potential: Its bio-medical applications *in vitro*. Microbiol Res 236:126466
179. Seviour RJ, McNeil B, Fazenda ML et al (2011) Operating bioreactors for microbial exopolysaccharide production. Crit Rev Biotechnol 31:170–185
180. Shahabi-Ghahfarrokhi I, Babaei-Ghazvini A (2019) Using photo-modification to compatibilize nano-ZnO in development of starch-kefiran-ZnO green nanocomposite as food packaging material. Int J Biol Macromol 124:922–930
181. Shukla A, Parmar P, Goswami D et al (2020) Characterization of novel thorium tolerant *Ochrobactrum intermedium* AM7 in consort with assessing its EPS-Thorium binding. J Hazard Mater 388:122047
182. Silambarasan S, Logeswari P, Cornejo P et al (2019) Evaluation of the production of exopolysaccharide by plant growth promoting yeast *Rhodotorula* sp. strain CAH2 under abiotic stress conditions. Int J Biol Macromol 121:55–62
183. da Silva Fonseca M, Marchioro MLK, Guimarães DKS et al (2020) *Neodeightonia phoenicum* CMIB-151: isolation, molecular identification, and production and characterization of an exopolysaccharide. J Polym Environ 28:1954–1966
184. Singh J, Dutta T, Kim K et al (2018) 'Green' synthesis of metals and their oxide nanoparticles: applications for environmental remediation. J Nanobiotechnol 16:84
185. Singh S, Kumar V (2020) Mercury detoxification by absorption, mercuric ion reductase, and exopolysaccharides: a comprehensive study. Environ Sci Pollut Res 27:27181–27201
186. Singh RS, Saini GK, Kennedy JF (2008) Pullulan: microbial sources, production and applications. Carbohydr Polym 73:515–531
187. Singh S, Sran KS, Pinnaka AK et al (2019) Purification, characterization and functional properties of exopolysaccharide from a novel halophilic *Natronotalea sambharensis* sp. nov. Int J Biol Macromol 136:547–558

188. Singh RP, Jha PN (2016) A halotolerant bacterium *Bacillus licheniformis* HSW-16 augments induced systemic tolerance to salt stress in wheat plant (*Triticum aestivum*) Front Plant Sci 7:1890
189. Sivaramakrishnan R, Suresh S, Pugazhendhi A et al (2020) Response of *Scenedesmus* sp. to microwave treatment: enhancement of lipid, exopolysaccharide and biomass production. Bioresour Technol 312:123562
190. Sivasankar P, Seedevi P, Poongodi S et al (2018) Characterization, antimicrobial and antioxidant property of exopolysaccharide mediated silver nanoparticles synthesized by *Streptomyces violaceus* MM72. Carbohydr Polym 181:752–759
191. Song AX, Mao YH, Siu KC et al (2018) Bifidogenic effects of *Cordyceps sinensis* fungal exopolysaccharide and konjac glucomannan after ultrasound and acid degradation. Int J Biol Macromol 111:587–594
192. Song AX, Mao YH, Siu KC et al (2019) Protective effects of exopolysaccharide of a medicinal fungus on probiotic bacteria during cold storage and simulated gastrointestinal conditions. Int J Biol Macromol 133:957–963
193. Song B, Song R, Cheng M et al (2020) Preparation of calcipotriol emulsion using bacterial exopolysaccharides as emulsifier for percutaneous treatment of psoriasis vulgaris. Int J Mol Sci 21:77
194. Song R, Yan F, Cheng M et al (2020) Ultrasound-assisted preparation of exopolysaccharide/nystatin nanoemulsion for treatment of vulvovaginal candidiasis. Int J Nanomed 15:2027–2044
195. Squillaci G, Finamore R, Diana P et al (2016) Production and properties of an exopolysaccharide synthesized by the extreme halophilic archaeon *Haloterrigena turkmenica*. Appl Microbiol Biotechnol 100:613–623
196. Suflet DM, Popescu I, Prisacaru AI et al (2020) Synthesis and characterization of curdlan–phosphorylated curdlan based hydrogels for drug release. Int J Polym Mater Polym Biomater:1–10
197. Sun L, Lei P, Wang Q et al (2020) The endophyte *Pantoea alhagi* NX-11 alleviates salt stress damage to rice seedlings by secreting exopolysaccharides. Front Microbiol 10:3112
198. Sun L, Yang Y, Wang R et al (2020) Effects of exopolysaccharide derived from *Pantoea alhagi* NX-11 on drought resistance of rice and its efficient fermentation preparation. Int J Biol Macromol 162:946–955
199. Sun ML, Zhao F, Zhang XK et al (2020) Improvement of the production of an Arctic bacterial exopolysaccharide with protective effect on human skin cells against UV-induced oxidative stress. Appl Microbiol Biotechnol 104:4863–4875
200. Sun ML, Zhao F, Chen XL et al (2020) Promotion of wound healing and prevention of frostbite injury in rat skin by exopolysaccharide from the Arctic marine bacterium *Polaribacter* sp. SM1127. Mar Drugs 18:48
201. Tahmourespour A, Ahmadi A, Fesharaki M (2020) The anti-tumor activity of exopolysaccharides from *Pseudomonas* strains against HT-29 colorectal cancer cell line. Int J Biol Macromol 149:1072–1076
202. Tarazanova M, Huppertz T, Kok J et al (2018) Influence of lactococcal surface properties on cell retention and distribution in cheese curd. Int Dairy J 85:73–78
203. Taufek NM, Harith HH, Abd Rahim, MH et al (2020) Performance of mycelial biomass and exopolysaccharide from Malaysian *Ganoderma lucidum* for the fungivore red hybrid Tilapia (*Oreochromis* sp.) in Zebrafish embryo. Aquac Rep 17:100322
204. Tewari S, Arora NK (2014) Multifunctional exopolysaccharides from *Pseudomonas aeruginosa* PF23 involved in plant growth stimulation, biocontrol and stress amelioration in sunflower under saline conditions. Curr Microbiol 69:484–494
205. Tiwari S, Patil R, Dubey SK et al (2020) Derivatization approaches and applications of pullulan. Adv Colloid Interface Sci 269:296–308
206. Tkáčiková Ľ, Mochnáčová E, Tyagi P et al (2020) Comprehensive mapping of the cell response to *E. coli* infection in porcine intestinal epithelial cells pretreated with exopolysaccharide derived from *Lactobacillus reuteri*. Vet Res 51:49

207. Trabelsi I, Ktari N, Triki M et al (2018) Physicochemical, techno-functional, and antioxidant properties of a novel bacterial exopolysaccharide in cooked beef sausage. Int J Biol Macromol 111:11–18
208. Uhliariková I, Šutovská M, Barboríková J et al (2020) Structural characteristics and biological effects of exopolysaccharide produced by cyanobacterium *Nostoc* sp. Int J Biol Macromol 160:364–371
209. Vazquez-Rodriguez A, Vasto-Anzaldo XG, Leon-Buitimea A et al (2020) Antibacterial and antibiofilm activity of biosynthesized silver nanoparticles coated with exopolysaccharides obtained from *Rhodotorula mucilaginosa*. IEEE Trans Nanobiosci 19:498–503
210. Ventorino V, Nicolaus B, Di Donato P et al (2019) Bioprospecting of exopolysaccharide-producing bacteria from different natural ecosystems for biopolymer synthesis from vinasse. Chem Biol Technol Agric 6:18
211. Wang ZM, Cheung YC, Leung PH et al (2010) Ultrasonic treatment for improved solution properties of a high-molecular weight exopolysaccharide produced by a medicinal fungus. Bioresour Technol 101:5517–5522
212. Wang N, Jia G, Wang C et al (2020) Structural characterisation and immunomodulatory activity of exopolysaccharides from liquid fermentation of *Monascus purpureus* (Hong Qu). Food Hydrocoll 103:105636
213. Wang Q, Li Q, Lin Y et al (2020) Biochemical and genetic basis of cadmium biosorption by *Enterobacter ludwigii* LY6, isolated from industrial contaminated soil. Environ Pollut 264:114637
214. Wang K, Li W, Rui X (2014) Characterization of a novel exopolysaccharide with antitumor activity from *Lactobacillus plantarum* 70810. Int J Biol Macromol 63:133–139
215. Wang K, Li W, Rui X et al (2015) Chemical modification, characterization and bioactivity of a released exopolysaccharide (r-EPS1) from *Lactobacillus plantarum* 70810. Glycoconj J 32:17–27
216. Wang YY, Ma H, Ding ZC et al (2019) Three-phase partitioning for the direct extraction and separation of bioactive exopolysaccharides from the cultured broth of *Phellinus baumii*. Int J Biol Macromol 123:201–209
217. Wang Y, Shu X, Hou J et al (2018) Selenium nanoparticle synthesized by *Proteus mirabilis* YC801: an efficacious pathway for selenite biotransformation and detoxification. Int J Mol Sci 19:3809
218. Wang Y, Su N, Hou G et al (2017) Hypoglycemic and hypolipidemic effects of a polysaccharide from *Lachnum* YM240 and its derivatives in mice, induced by a high fat diet and low dose STZ. MedChemComm 8:964–974
219. Wang J, Wu T, Fang X et al (2019) Manufacture of low-fat Cheddar cheese by exopolysaccharide-producing *Lactobacillus plantarum* JLK0142 and its functional properties. J Dairy Sci 102:3825–3838
220. Watnick PI, Lauriano CM, Klose KE et al (2001) The absence of a flagellum leads to altered colony morphology, biofilm development and virulence in *Vibrio cholerae* O139. Mol Microbiol 39:223–235
221. Wei Y, Li F, Li L et al (2019) Genetic and biochemical characterization of an exopolysaccharide with *in vitro* antitumoral activity produced by *Lactobacillus fermentum* YL-11. Front Microbiol 10:2898
222. Whitfield GB, Marmont LS, Howell PL (2015) Enzymatic modifications of exopolysaccharides enhance bacterial persistence. Front Microbiol 6:596
223. Wu X, Wu X, Zhou X et al (2020) The roles of extracellular polymeric substances of *Pandoraea* sp. XY-2 in the removal of tetracycline. Bioproc Biosys Eng 43(11):1951–1960
224. Xiao R, Grinstaff MW (2017) Chemical synthesis of polysaccharides and polysaccharide mimetics. Prog Polym Sci 74:78–116
225. Xiao L, Han S, Zhou J et al (2020) Preparation, characterization and antioxidant activities of derivatives of exopolysaccharide from *Lactobacillus helveticus* MB2-1. Int J Biol Macromol 145:1008–1017

226. Xiao Y, Huang Q, Zheng Z et al (2017) Construction of a *Cordyceps sinensis* exopolysaccharide-conjugated selenium nanoparticles and enhancement of their antioxidant activities. Int J Biol Macromol 99:483–491
227. Xiong YW, Ju XY, Li XW et al (2020) Fermentation conditions optimization, purification, and antioxidant activity of exopolysaccharides obtained from the plant growth-promoting endophytic actinobacterium *Glutamicibacter halophytocola* KLBMP 5180. Int J Biol Macromol 153:1176–1185
228. Xu Q, Black WP, Nascimi HM et al (2011) DifA, a methyl-accepting chemoreceptor protein-like sensory protein, uses a novel signaling mechanism to regulate exopolysaccharide production in *Myxococcus xanthus*. J Bacteriol 193:759–767
229. Xu Z, Guo Q, Zhang H et al (2018) Exopolysaccharide produced by *Streptococcus thermophiles* S-3: molecular, partial structural and rheological properties. Carbohydr Polym 194:132–138
230. Xu X, Peng Q, Zhang Y et al (2020) A novel exopolysaccharide produced by *Lactobacillus coryniformis* NA-3 exhibits antioxidant and biofilm-inhibiting properties *in vitro*. Food Nutr Res 64:3744
231. Yahya SMM, Abdelnasser SM, Hamed AR et al (2020) Newly isolated marine bacterial exopolysaccharides enhance antitumor activity in HepG2 cells *via* affecting key apoptotic factors and activating toll like receptors. Mol Biol Rep 46:6231–6241
232. Yang A, Akhtar SS, Iqbal S et al (2016) Enhancing salt tolerance in quinoa by halotolerant bacterial inoculation. Funct Plant Biol 43:632–642
233. You X, Li Z, Ma K et al (2020) Structural characterization and immunomodulatory activity of an exopolysaccharide produced by *Lactobacillus helveticus* LZ-R-5. Carbohydr Polym 235:115977
234. You X, Yang L, Zhao X et al (2020) Isolation, purification, characterization and immunostimulatory activity of an exopolysaccharide produced by *Lactobacillus pentosus* LZ-R-17 isolated from Tibetan kefir. Int J Biol Macromol 158:408–419
235. Yousef RH, Baothman OA, Abdulaal WH et al (2020) Potential antitumor activity of exopolysaccharide produced from date seed powder as a carbon source for *Bacillus subtilis*. J Microbiol Methods 170:105853
236. Yuan Z, Shi Y, Cai F et al (2020) Isolation and identification of polysaccharides from *Pythium arrhenomanes* and application to strawberry fruit (*Fragaria ananassa* Duch.) preservation. Food Chem 309:125604
237. Zampieri RM, Adessi A, Caldara F et al (2020) Anti-inflammatory activity of exopolysaccharides from *Phormidium* sp. ETS05, the most abundant cyanobacterium of the therapeutic euganean thermal muds, using the zebrafish model. Biomolecules 10:582
238. Zehir Şentürk D, Dertli E, Erten H et al (2019) Structural and technological characterization of ropy exopolysaccharides produced by *Lactobacillus plantarum* strains isolated from Tarhana. Food Sci Biotechnol 29:121–129
239. Zhang C, Li J, Wang J et al (2017) Antihyperlipidaemic and hepatoprotective activities of acidic and enzymatic hydrolysis exopolysaccharides from *Pleurotus eryngii* SI-04. BMC Complement Altern Med 17:403
240. Zhang W, Lu Y, Zhang Y et al (2016) Antioxidant and antitumour activities of exopolysaccharide from liquid-cultured *Grifola frondosa* by chemical modification. Int J Food Sci Technol 51:1055–1061
241. Zhang H, Ren W, Guo Q et al (2018) Characterization of a yogurt-quality improving exopolysaccharide from *Streptococcus thermophilus* AR333. Food Hydrocoll 81:220–228
242. Zhang J, Zhao W, Guo X et al (2017) Survival and effect of exopolysaccharide-producing *Lactobacillus plantarum* YW11 on the physicochemical properties of ice cream. Pol J Food Nutr Sci 67:191–200
243. Zhao F, Guo C, Cui Q et al (2018) Exopolysaccharide production by an indigenous isolate *Pseudomonas stutzeri* XP1 and its application potential in enhanced oil recovery. Carbohydr Polym 199:375–381

244. Zheng Q, He BL, Wang JY et al (2020) Structural analysis and antioxidant activity of extracellular polysaccharides extracted from culinary-medicinal white jelly mushroom *Tremella fuciformis* (Tremellomycetes) conidium cells. Int J Med Mushrooms 22:489–500
245. Zhu Y, Wang X, Pan W et al (2019) Exopolysaccharides produced by yogurt-texture improving *Lactobacillus plantarum* RS20D and the immunoregulatory activity. Int J Biol Macromol 121:342–349
246. Zykwinska A, Marquis M, Godin M et al (2020) Microcarriers based on glycosaminoglycan-like marine exopolysaccharide for TGF-β1 long-term protection. Mar Drugs 17:65